Fatigue and Fracture Mechanics of Marine Structures

Fatigue and Fracture Mechanics of Marine Structures

Editors

Sören Ehlers
Moritz Braun

MDPI • Basel • Beijing • Wuhan • Barcelona • Belgrade • Manchester • Tokyo • Cluj • Tianjin

Editors
Sören Ehlers
Hamburg University of Technology
Germany

Moritz Braun
Hamburg University of Technology
Germany

Editorial Office
MDPI
St. Alban-Anlage 66
4052 Basel, Switzerland

This is a reprint of articles from the Special Issue published online in the open access journal *Journal of Marine Science and Engineering* (ISSN 2077-1312) (available at: https://www.mdpi.com/journal/jmse/special_issues/fatigue_fracture_mechanics_marine_structures).

For citation purposes, cite each article independently as indicated on the article page online and as indicated below:

LastName, A.A.; LastName, B.B.; LastName, C.C. Article Title. *Journal Name* **Year**, *Volume Number*, Page Range.

ISBN 978-3-0365-7596-4 (Hbk)
ISBN 978-3-0365-7597-1 (PDF)

Contents

About the Editors

Sören Ehlers

Sören Ehlers, D.Sc., is a professor of design and analysis of ships and offshore structures at the Hamburg University of Technology (TUHH) and the director of the DLR institute of maritime energy systems. He is an expert in consequence assessment for accidental events and in the field of material modelling for non-linear finite element simulations and small and large-scale experimental research. He has about 300 publications in the corresponding fields. He is a member of the ISSC V.2 on Experimental Methods, a symposium coordinator for the Arctic Technology Symposium at OMAE, the deputy chairman of the board of the German Association for Marine Technology, the ICSOS conference co-chair and co-founder, an editor for the Ship Technology Research Journal, an associate editor for Marine Structures and for Ships and Offshore Structures, and a reviewer for several international journals.

Moritz Braun

Moritz Braun, Dr., is the head of the department for ship reliability at the German Aerospace Center of Maritime Energy Systems in Geesthacht, Germany. He holds a B.Sc. in mechanical engineering (2012), an M.Sc. in Marine Technology (2014) and a Doctor of Engineering (2021). For his doctoral thesis, he received the Curt Bartsch Award of the German Society for Maritime Technology. His research focuses on the structural integrity assessment of large engineering structures and in particular on fatigue and fracture mechanics. He has about 70 publications in the corresponding fields. He is the delegate of the German Association for the German Welding Association (DVS) at the International Institute of Welding (IIW) in 'Commission XV: Design, Analysis, and Fabrication of Welded Structures', the German representative in the International Ship and Offshore Structures Congress (ISSC) technical committee 'V.7 Structural Assessment During Operations', and the deputy chairman of the working group 'Damages of Ship Structures' of DVS.

Journal of
Marine Science and Engineering

MDPI

Editorial

Special Issue on Fatigue and Fracture Mechanics of Marine Structures

Moritz Braun [1,2,*] and Sören Ehlers [1,2]

1 German Aerospace Centre (DLR), Institute of Maritime Energy Systems, D-21502 Geesthacht, Germany
2 Institute for Ship Structural Design and Analysis, Hamburg University of Technology, D-21073 Hamburg, Germany
* Correspondence: moritz.braun@dlr.de

Fatigue and fracture are important design criteria for marine structures subjected to static and cyclic loading. In particular, the ability to ensure structural integrity under all design conditions is of utmost importance for the safety of humans and the environment. Much effort has been allocated to the development of fatigue and fracture mechanic methods over the last decades; however, discrepancies between predicted and actual strength remain due to various reasons. Hence, this Special Issue aims at describing the fatigue and fracture mechanic behaviours of marine structures. In this connection, studies aiming at novel methods (either analytical, empirical, or numerical), as well as presentations of case studies or innovative test studies, are presented. This includes considerations on parameters which have a large influence on the fatigue and fracture behaviour of materials, welded joints or structural components, such as loading conditions, scale effects, etc.

Xu et al. [1] propose an improved fatigue reliability analysis method for deepwater risers using data-driven models based on response surface methods to replace physical-based models, improving computational efficiency. An annual crack growth model based on fracture mechanics considers crack inspection data, while a crack growth dynamic Bayesian network evaluates and updates the fatigue reliability of the riser. Results indicate that the proposed method accurately and efficiently analyzes riser fatigue, with crack inspection results updating random parameter distributions and the fatigue reliability of deepwater risers through Bayesian inference.

Liu and Chang [2] propose a new part-cutting constraint rule based on partial cutting for the optimization of the hull parts CNC laser cutting path, overcoming the drawbacks of traditional algorithms. A novel toolpath model for hull parts called HPCPO is proposed, and a segmented genetic algorithm based on reinforcement learning (RLSGA) is used to solve the HPCPO problem, with the population viewed as an intelligent agent. The results showed that RLSGA outperformed other algorithms and effectively solved the HPCPO problem, offering a new approach for optimizing the CNC laser cutting path of hull parts.

The study by Ahola et al. [3] focuses on improving the weld quality and lifespan of ship structures made of high-strength and ultra-high-strength steels by investigating the fatigue strength of a longitudinal stiffener detail. The research involved the use of two types of high-strength steel and gas metal arc welding with rutile-cored wires, followed by fatigue testing on small and large-scale specimens under axial and four-point bending loading conditions. The high-frequency mechanical impact (HFMI) treatment was also considered a post-weld treatment technique. The study found that using solid filler wires could potentially improve the welding quality and that repaired and post-weld-treated welds performed better in re-tests, with fatigue strength being almost twice that of the as-welded condition.

The study by Beiler et al. [4] discusses the importance of fatigue life extension methods for cruise ships or yachts that operate in freezing temperatures. The thermal cut edges of the outer shell are particularly vulnerable to crack growth, and there are currently no

Citation: Braun, M.; Ehlers, S. Special Issue on Fatigue and Fracture Mechanics of Marine Structures. *J. Mar. Sci. Eng.* **2023**, *11*, 974. https://doi.org/10.3390/jmse11050974

Received: 25 April 2023
Accepted: 28 April 2023
Published: 4 May 2023

design curves for the fatigue assessment of thermal cut steel edges at sub-zero temperatures. To address this, fatigue tests were conducted at −20 °C and −50 °C, as well as at room temperature. The results showed that the fatigue strength at sub-zero temperatures was significantly higher compared to room temperature, suggesting that sub-zero temperatures down to −50 °C do not reduce the fatigue life of thermal cut steel edges.

The study by Gericke et al. [5] investigates the effect of thermally sprayed aluminum coatings (Al99%) on the fatigue strength and corrosion protection of gas metal arc welded (GMAW) non-alloyed structural steel specimens used for offshore wind turbines. Wöhler tests were conducted on specimens with different weld details in the as-welded condition and after thermal spray coating, along with different types of surface preparation. The study found that thermal spraying significantly increased fatigue strength, especially for specimens with welded transverse stiffeners. The corrosion studies demonstrated that thermally sprayed Al99% coatings have high resistance to corrosion in seawater environments and are suitable as sacrificial anodes for bare steel. The combination of fatigue strength improvement and corrosion protection makes thermally sprayed Al coatings a promising choice for offshore structures design and operation.

Yu et al. [6] propose a fatigue spectrum analysis and reliability analysis method for multilayer composite flexible risers. The study describes a complete fatigue and reliability analysis process for multilayer structures, which are complex structures due to the influence between layers and the range of uncertain factors. Finite element software is used to analyze the stress of the nine-layer flexible riser under the influence of upper platform structure movement and ocean current. The study also analyzes the uncertainty in the process of fatigue damage calculation and provides future research directions.

Morita et al. [7] examine the cyclic deformation behavior of mid-carbon steel under fatigue loading. Uniaxial tensile and fatigue tests are conducted to investigate the influence of loading history. The material exhibits different uniaxial ratcheting behavior depending on loading history, with a smooth and gradual increase in cyclic softening under smaller stress/strain conditions. Numerical simulations are carried out to reproduce the cyclic stress-strain behavior using the Fatigue SS Model, which can describe cyclic softening behavior within a macroscopically elastic stress state. The model is validated by the good agreement between experimental and numerical results, and a method is introduced for predicting fatigue crack initiation life under variable loading conditions based on cumulative plastic work.

Putranto et al. [8] present the application of the equivalent single layer (ESL) approach for the ultimate strength assessment of ship hull girder in numerical finite element (FE) simulations. The ESL approach replaces the stiffened panel with a single plate having equivalent stiffness, reducing the time in pre-processing and FE analysis. Two case studies, one compartment model and a full-sized double-hull tanker model, are used to demonstrate the applicability of the ESL approach. The ultimate strength predictions obtained from the ESL approach show good agreement with full three-dimensional finite element method (3D FEM) results, providing up to 3 times computational efficiency and ease of modeling.

Cho et al. [9] discuss the use of tandem submerged arc welding (SAW) to increase the welding productivity process in shipbuilding; however, the weld bead profiles produced by this process often exceed acceptance criteria specified in international regulations such as AWS D1.1, ISO 5817 and NORSOK M-101. These regulations limit weld bead profiles, particularly weld bead height, due to concerns about weldment fatigue properties. To investigate the effect of weld profiles on fatigue properties, the authors performed experiments and statistical analyses, proposing new criteria for weld bead profiles that satisfy the E curve as the design S-N curve for Tandem SAW welding.

Welding is widely used in different industries, and structural integrity assessment is vital to ensure safety due to the existence of defects in welded structures. Fracture toughness is a crucial part of structural integrity assessment, but it requires a lot of time and effort. Kim et al. [10] show that the Charpy impact test can estimate fracture behavior well in the ductile to brittle transition region for API 2W Gr.50 steel with different welding conditions; however, differences in prediction accuracy were obtained, which require additional consideration to guarantee the safety of welded structures.

The study by Petry et al. [11] aims to optimize the Voce–Chaboche (V-C) material model parameter for high-strength steel welded joints that undergo cyclic loading. The researchers used an optimization algorithm based on the Newton trust region method and an accumulated true strain parameter to determine the model parameters of each material zone in an S690 steel butt-welded joint. The results indicate that when the elastic modulus is optimized as a V-C parameter, the model slightly underestimates the strain range, leading to conservative fatigue life estimates; however, using an elastic modulus obtained experimentally can result in a non-conservative but more accurate fatigue life estimation. The study concludes that the NTR-based accumulated true strain approach successfully determined the V-C model parameters for different material zones in the welded joint, closely estimating the strain range and the fatigue life for a variable amplitude load history.

Johannesson et al. [12] focus on the mechanical life of flexible marine power cables, which are designed for mechanical loads due to movements caused by connecting devices to hubs. The Variational Mode and Effect Analysis (VMEA) reliability design method is used to identify and quantify uncertainties in fatigue life, including scatter, model, and statistical uncertainties. The VMEA method implements a load-strength approach combining numerical simulations to assess cable loads and experimental tests to assess cable strength. The study evaluates uncertainties in fatigue life for WEC system cables during the design phase using the VMEA method, which identifies weak spots in the reliability assessment and provides a foundation for evaluating safety against fatigue. Results show that the fatigue life model is a major contributor to overall uncertainty, both in terms of scatter and model uncertainty.

As Guest Editors for this Special Issue, we would like to express our sincere appreciation to the authors who have contributed their work to this collection. It has been an honor and a privilege to work with such a talented group of researchers, and we are truly grateful for the effort, time, and dedication they have put into their contributions. Each article offers unique insights and perspectives on the topic, reflecting a wide range of approaches and methodologies on fatigue and fracture mechanics of marine structures.

We have been impressed by the quality of the submissions, as well as the rigor and attention to detail demonstrated by each author. Their contributions have made this Special Issue a truly exceptional collection, one that we believe will be of great interest and value to readers in the field. Thank you again for your contributions to this Special Issue, and for your ongoing commitment to advancing scientific knowledge and understanding.

Funding: This research received no external funding.

Conflicts of Interest: The authors declare no conflict of interest.

References

1. Xu, L.; Hu, P.; Li, Y.; Qiu, N.; Chen, G.; Liu, X. Improved Fatigue Reliability Analysis of Deepwater Risers Based on RSM and DBN. *J. Mar. Sci. Eng.* **2023**, *11*, 688. [CrossRef]
2. Liu, X.; Chang, D. An Improved Method for Optimizing CNC Laser Cutting Paths for Ship Hull Components with Thicknesses up to 24 mm. *J. Mar. Sci. Eng.* **2023**, *11*, 652. [CrossRef]
3. Ahola, A.; Lipiäinen, K.; Lindroos, J.; Koskimäki, M.; Laukia, K.; Björk, T. On the Fatigue Strength of Welded High-Strength Steel Joints in the As-Welded, Post-Weld-Treated and Repaired Conditions in a Typical Ship Structural Detail. *J. Mar. Sci. Eng.* **2023**, *11*, 644. [CrossRef]
4. Beiler, M.; Grimm, J.-H.; Bui, T.-N.; von Bock und Polach, F.; Braun, M. Fatigue Strength of AH36 Thermal Cut Steel Edges at Sub-Zero Temperatures. *J. Mar. Sci. Eng.* **2023**, *11*, 346. [CrossRef]
5. Gericke, A.; Hauer, M.; Ripsch, B.; Irmer, M.; Nehlsen, J.; Henkel, K.-M. Fatigue Strength of Structural Steel-Welded Connections with Arc-Sprayed Aluminum Coatings and Corrosion Behavior of the Corresponding Coatings in Sea Water. *J. Mar. Sci. Eng.* **2022**, *10*, 1731. [CrossRef]
6. Yu, J.; Wang, F.; Yu, Y.; Li, H.; Liu, X.; Sun, R. Study on Fatigue Spectrum Analysis and Reliability Analysis of Multilayer Flexible Riser. *J. Mar. Sci. Eng.* **2022**, *10*, 1561. [CrossRef]

7. Morita, K.; Mouri, M.; Fincato, R.; Tsutsumi, S. Experimental and Numerical Study of Cyclic Stress–Strain Response and Fatigue Crack Initiation Life of Mid-Carbon Steel under Constant and Multi-Step Amplitude Loading. *J. Mar. Sci. Eng.* **2022**, *10*, 1535. [CrossRef]
8. Putranto, T.; Kõrgesaar, M.; Tabri, K. Application of Equivalent Single Layer Approach for Ultimate Strength Analyses of Ship Hull Girder. *J. Mar. Sci. Eng.* **2022**, *10*, 1530. [CrossRef]
9. Cho, D.; Nam, S.; Cho, C.; Lee, D.; Jeong, E.; Jeong, Y.; Lee, J.; Cridland, M.; Kim, B.; Kim, K.; et al. A New Study on the Fatigue Properties of SA Weld Joints by Considering the Effects of Welded Bead Shape. *J. Mar. Sci. Eng.* **2022**, *10*, 1527. [CrossRef]
10. Kim, J.; Park, J.; Kim, M. Estimation of Fracture Toughness of API 2W Gr.50 Steel in Ductile to Brittle Transition Behavior Using Master Curve Approach. *J. Mar. Sci. Eng.* **2022**, *10*, 1066. [CrossRef]
11. Petry, A.; Gallo, P.; Remes, H.; Niemelä, A. Optimizing the Voce–Chaboche Model Parameters for Fatigue Life Estimation of Welded Joints in High-Strength Marine Structures. *J. Mar. Sci. Eng.* **2022**, *10*, 818. [CrossRef]
12. Johannesson, P.; Lang, X.; Johnson, E.; Ringsberg, J.W. Mechanical Reliability Analysis of Flexible Power Cables for Marine Energy. *J. Mar. Sci. Eng.* **2022**, *10*, 716. [CrossRef]

Journal of
Marine Science and Engineering

MDPI

Article

Improved Fatigue Reliability Analysis of Deepwater Risers Based on RSM and DBN

Liangbin Xu [1], Pengji Hu [2], Yanwei Li [2], Na Qiu [2], Guoming Chen [2] and Xiuquan Liu [2,*]

1 Offshore Engineering and Technology, Sun Yat-Sen University, Zhuhai 528478, China
2 Centre for Offshore Engineering and Safety Technology, China University of Petroleum, Qingdao 266580, China
* Correspondence: lxqmcae@163.com

Abstract: The fatigue reliability assessment of deepwater risers plays an important role in the safety of oil and gas development. Physical-based models are widely used in riser fatigue reliability analyses. However, these models present some disadvantages in riser fatigue reliability analyses, such as low computational efficiency and the inability to introduce inspection data. An improved fatigue reliability analysis method was proposed to conduct the fatigue reliability assessment of deepwater risers. The data-driven models were established based on response surface methods to replace the original physical-based models. They are more efficient than the physics-based model, because a large number of complex numerical and iterative solutions are avoided in fatigue reliability analysis. The annual crack growth model of the riser based on fracture mechanics was established by considering the crack inspection data as a factor, and the crack growth dynamic Bayesian network was established to evaluate and update the fatigue reliability of the riser. The performance of the proposed method was demonstrated by applying the method to a case. Results showed that the data-driven models could be used to analyze riser fatigue accurately, and the crack growth model could be performed to analyze riser fatigue reliability efficiently. The crack inspection results update the random parameters distribution and the fatigue reliability of deepwater risers by Bayesian inference. The accuracy and efficiency of fatigue analysis of deepwater risers can be improved using the proposed method.

Keywords: deepwater riser; response surface method; fatigue damage; dynamic bayesian network; fatigue reliability analysis; fatigue reliability updating

Citation: Xu, L.; Hu, P.; Li, Y.; Qiu, N.; Chen, G.; Liu, X. Improved Fatigue Reliability Analysis of Deepwater Risers Based on RSM and DBN. *J. Mar. Sci. Eng.* **2023**, *11*, 688. https://doi.org/10.3390/jmse11040688

Academic Editors: Sören Ehlers and Moritz Braun

Received: 3 February 2023
Revised: 13 March 2023
Accepted: 13 March 2023
Published: 24 March 2023

1. Introduction

Deepwater risers are important equipment connecting the floating platform with subsea wellhead during offshore oil and gas development. Their main function is to provide fluid passage between the surface platform and subsea wellhead in drilling and production processes [1]. Deepwater risers are sensitive to dynamic loads, and fatigue damage due to ocean waves and movements of the floating platform is one of their main failure modes. The long term accumulation of fatigue damage leads to the fatigue fracture of risers and even leads to major accidents [2]. Several accidents, including oil leaks and subsea wellhead breakage, have occurred due to riser fatigue [3,4]. Consequently, studying the fatigue reliability of risers during drilling and production processes is essential.

Fatigue analyses are the basis of fatigue reliability analysis for deepwater risers. Physical-based models are widely used in riser fatigue analyses at present. The fatigue analysis schemes used by these models can be divided into frequency and time domain methods. The frequency domain method is rarely used because of possible linearization errors in the calculation. By contrast, the time domain method is often employed in riser fatigue analyses due to its high computational accuracy. In the time domain method, the dynamic response of the riser system is first analyzed using the finite element method, the finite difference method, and other approaches [5,6]. Second, the dynamic response is processed through the rain flow counting method. Lastly, fatigue damage is calculated based on the S–N curve or fracture mechanics [7,8]. The time domain fatigue of deepwater risers, including drilling, steel catenary, and flexible risers, has been widely investigated in previous studies.

Research on riser fatigue reliability has mainly focused on catenary and top-tensioned risers. The process of fatigue reliability analysis is relatively complex. Generally, the assessment of the fatigue capacity and fatigue failure of risers is conducted by applying two different methodologies, namely, crack growth rate curves using the fracture mechanics model and S–N curves using the Palmgren–Miner damage model [9,10]. During the fatigue reliability analysis of risers, the randomness of general dynamic model parameters, such as wall thickness, the additional mass coefficient, and the drag force coefficient, is often considered [11–14]. The uncertainty of the fatigue performance parameters of the S–N curve method and that of the Paris parameters of the fracture mechanics method are also considered [15–18]. In particular, the uncertainty of geotechnical model parameters and the nonlinear behavior of the pipe–seabed interaction are also considered when evaluating the fatigue reliability of catenary risers in the contact area [19–21]. The uncertainty of these parameters results in the uncertainty of riser fatigue. Fatigue assessment using certain parameters only cannot meet the needs of safety design and riser assessment. The fatigue reliability of risers must be assessed reasonably based on statistical theory [22]. Several practical reliability analysis approaches, including numerical simulation and analytical algorithms, have been developed for riser fatigue reliability assessment. The Monte Carlo method belongs to the former type of reliability analysis, whereas the first order and second order reliability methods belong to the latter type. The three methods are widely used, and each has its own advantages and disadvantages. The Monte Carlo method has the highest solving accuracy but the lowest solving efficiency, and the first order reliability method has a solving performance that is opposite to that of the Monte Carlo method; the second order reliability method is relatively mediocre when compared with the two other methods [23,24].

In previous research, physical-based models were widely used for fatigue and reliability analyses of risers. However, the calculation for single fatigue analysis using physical-based models is time consuming [25]. The calculation cost of fatigue reliability analysis using physical-based models increases significantly with the increase in the number of random parameters and the degree of random distribution of the parameters. These reasons account for the low efficiency of the fatigue reliability analysis of risers. Therefore, some scholars have attempted to improve the efficiency of fatigue reliability analysis by reducing the time of single fatigue analysis or the scale of fatigue analysis [26,27]. Data-driven models have been introduced as substitutes for time consuming physical-based models [25,28–30]. Notably, risers need to be inspected in a fixed time to guarantee their safety, and using the inspection results to update the fatigue reliability is essential to the safety of risers in the process of drilling and production. A problem at present is the difficulty of introducing inspection results into riser fatigue reliability analysis. The dynamic Bayesian network (DBN) is an optional method for solving this difficult problem. The DBN has increasingly been applied to probability analysis because of its functions of forward and backward analyses [31]. Fatigue reliability can be calculated based on the prior probabilities of root nodes in forward analysis, and the posterior reliability can be updated based on a few observations. Thus, the DBN was selected for riser fatigue reliability analysis and updating based on the inspection results in this study.

In this study, data-driven models based on the RSM were established to substitute for physical-based models in the fatigue analysis of risers. The direct mapping between riser dynamic analysis parameters and fatigue damage was established by the data-driven models, which can greatly improve the efficiency of fatigue analysis. The annual fatigue crack growth model was then established based on fracture mechanics theory and data-driven models that consider the crack inspection data as a factor. The crack growth DBN model was then established to obtain the fatigue reliability of the risers. The crack inspection data of the risers were introduced into the DBN model to update the fatigue reliability of the risers.

The rest of the paper is organized in the following manner. The riser fatigue reliability analysis methodology is introduced in Section 2. The physical-based models are introduced in Section 3. The data-driven models established by RSM are introduced in Section 4, and the fatigue reliability analysis method based on DBN is introduced in Section 5. A case is introduced for fatigue reliability analysis and updating in Section 6. The conclusions are summarized in Section 7.

2. Methodology

The methodology for the fatigue reliability analysis of risers is illustrated in Figure 1. Physical-based models were used to establish fatigue datasets of risers based on the S–N curve method. Data-driven models were established based on the fatigue datasets as a replacement for physical-based models to improve fatigue analysis efficiency. However, the fatigue response surface model based on S–N curve method was unsuitable for fatigue reliability analysis and updating of a cracked riser. The riser annual fatigue crack growth model based on fracture mechanics theory was established. Then, fatigue reliability analysis was performed by DBN method based on the fatigue crack growth model and inspection results. Bayesian forward inference was adopted for riser fatigue reliability analysis, and Bayesian backward inference was applied for fatigue reliability updating based on the inspection results. The physical-based models, data-driven models, and crack growth DBN model are introduced in detail in Sections 3–5, respectively.

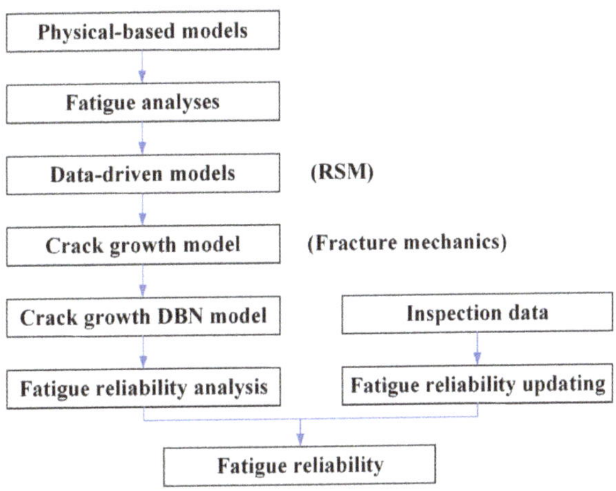

Figure 1. Analysis methodology.

3. Physical-Based Models

Figure 2 shows a deepwater drilling platform/riser coupling system that comprises a platform, a drilling riser system, and tensioners. The top end of the riser system is hung on the drilling platform via the tensioners and upper flex joint, and the bottom end of the riser system is connected to the Blowout preventer (BOP), the subsea wellhead, and the conductor via the lower marine riser package (LMRP). The riser system vibrates transversely and axially under environment loads and platform motions.

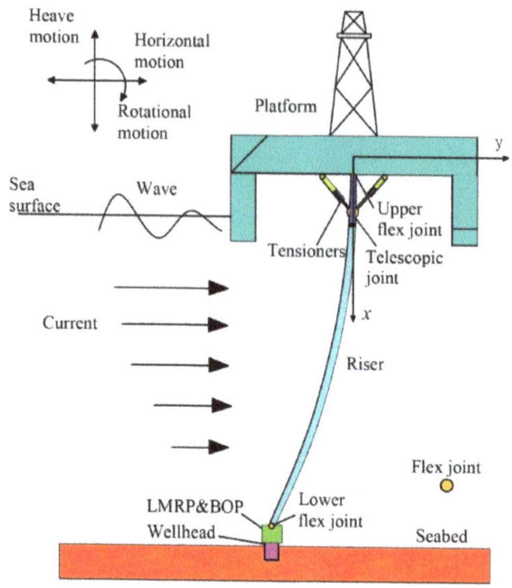

Figure 2. Deepwater drilling platform/riser coupling system.

The transverse vibration equation of the riser system can be written as [32,33]

$$m(x)\frac{\partial^2 y}{\partial t^2} + c\frac{\partial y}{\partial t} + \frac{\partial^2}{\partial x^2}\left(EI(x)\frac{\partial^2 y}{\partial x^2}\right) - \frac{\partial}{\partial x}\left(T(x)\frac{\partial y}{\partial x}\right) = F_{\text{sea}}(x,t) \tag{1}$$

where m is the mass per unit length of the riser, c is the damping coefficient, E is the elastic modulus of the riser, $I(x)$ is the second moment of area, $T(x)$ is the effective axial tension, and $F_{\text{sea}}(x,t)$ is the marine load applied per unit length of the riser.

The axial vibration equation of the riser system can be written as [32–34]

$$\frac{\partial}{\partial x}\left(EA(x)\frac{\partial u(x,t)}{\partial x}\right)dx - c\frac{\partial u(x,t)}{\partial t} - m\frac{\partial^2 u(x,t)}{\partial t^2} = F + mg \tag{2}$$

where $u(x,t)$ is the axial vibration, F is the axial hydrodynamic load on the riser system, and $A(x)$ is the area of the riser cross-section.

The effective axial tension, $T(x)$, can be expressed as

$$T(x) = T_{\text{top}} - \int_0^x (F + mg)dx \tag{3}$$

where T_{top} is the tension at the top of the riser system, and L is the length of the riser system.

The marine load applied per unit length of the riser, $F_{\text{sea}}(x,t)$, can be written as

$$F_{\text{sea}}(x,t) = \frac{\pi}{4}\rho C_M D_h^2 \dot{u}_w - \frac{\pi}{4}\rho C_a D_h^2 \ddot{y} + \frac{1}{2}\rho D_h C_d (u_w + u_c - \dot{y})|u_w + u_c - \dot{y}| \tag{4}$$

where ρ is the density of seawater, C_M is the coefficient of inertia force, C_a is the coefficient of additional mass ($C_a = C_M$-1), D_h is the hydrodynamic outer diameter, C_d is the coefficient of drag force, u_w is the water point velocity, $\dot{u}_w$ is the water point acceleration, u_c is the current velocity, $\dot{y}$ is the velocity of the riser, and $\ddot{y}$ is the acceleration of the riser.

The top end of the risers is connected to the platform by the upper flex joint, and the bottom end is connected to the LMRP via the lower flex joint, as is shown in Figure 2. Therefore, the top and bottom boundary conditions of the riser system are expressed as [35]

$$\begin{cases} u(x,t)\big|_{x=0} = S_{\mathrm{px}}(t) \\ y(x,t)\big|_{x=0} = S_{\mathrm{py}}(t) \\ EI(x)\frac{\partial^2 y(x,t)}{\partial x^2}\big|_{x=0} = K_{\mathrm{u}}\frac{\partial y(x,t)}{\partial x}\big|_{x=0} \end{cases} \tag{5}$$

$$\begin{cases} u(x,t)\big|_{x=L} = 0 \\ y(x,t)\big|_{x=L} = 0 \\ EI(x)\frac{\partial^2 y(x,t)}{\partial x^2}\big|_{x=L} = K_{\mathrm{b}}\frac{\partial y(x,t)}{\partial x}\big|_{x=L} \end{cases} \tag{6}$$

where K_{b} is the rotation stiffness of the lower flexible joint, K_{u} is the rotation stiffness of the upper flexible joint, L is the length of the risers, $S_{\mathrm{px}}(t)$ is the floating platform's heave displacement, and $S_{\mathrm{py}}(t)$ is the floating platform's drift distance. $S_{\mathrm{px}}(t)$ and $S_{\mathrm{py}}(t)$ can be calculated by Equation (7) [36].

$$S_{\mathrm{p}}(t) = \sum_{n=0}^{Nw} S_{\mathrm{n}} \cos(k_{\mathrm{n}} S_{\mathrm{p}}(t) - \omega_{\mathrm{n}} t + \phi_{\mathrm{n}} + \alpha_{\mathrm{n}}) \tag{7}$$

where S_{n} is a single wave's amplitude, N_{w} is the number of random waves, k_{n} is the number of constituent waves, ω_{n} is the wave frequency, φ_{n} is the initial phase angle of the wave, and α_{n} is the phase angle between the platform movement and wave.

On the basis of the motion equation of the riser system, the finite element method was used to establish the finite element analysis model of the riser system [37,38]. The whole dynamical equation of the riser system is written as

$$[\mathbf{M}]\{\ddot{\mathbf{u}}\} + [\mathbf{C}]\{\dot{\mathbf{u}}\} + [\mathbf{K}]\{\mathbf{u}\} = \{\mathbf{F}\} \tag{8}$$

where $\{u\}$ is the displacement vector of the riser system; $[\mathbf{M}]$, $[\mathbf{C}]$, and $[\mathbf{K}]$ are the mass, damping, and stiffness matrices of the riser system, respectively; and $\{F\}$ is the load vector of the riser system.

The dynamic response of the risers was obtained by solving the whole motion finite element model of the riser system. The rain flow counting method was then used to count the fatigue stress time history and obtain the stress cycle and amplitude [7,8].

$$D_{\mathrm{L}} = \frac{1}{n_{\mathrm{f}}} = \frac{\sum(\Delta\sigma_{\mathrm{i}})^{m_{\mathrm{f}}}}{C} \tag{9}$$

where D_{L} is the long term fatigue damage, n_{f} is the fatigue life, and C and m_{f} are the S–N curve method parameters.

4. Data-Driven Models

4.1. RSM

The numerical simulation of the nonlinear dynamic equation of the risers in Equation (8) is complex, which is the main reason a fatigue analysis based on physical-based models is time consuming. The application of data-driven models based on the RSM in the fatigue damage calculation of risers can solve the low efficiency problem of physical-based models. In practical application, the response surface is often constructed by polynomials of different orders, which is the most common application of the RSM. Inaccurate calculations will occur if lower order polynomial functions are used to deal with complex nonlinear problems. However, the calculation efficiency will be reduced due to the need to determine multiple parameters and conduct multiple tests if a higher order response surface equation is established. Considering the above two factors, the response function of the quadratic polynomial was used to establish the riser fatigue data-driven models in this paper, because it can solve nonlinear problems and reduce the number of experiments [28]. The second order response surface model is written as

$$D_{\mathrm{L}} = \beta_0 + \sum_{\mathrm{i}=1}^{k} \beta_{\mathrm{i}} x_{\mathrm{i}} + \sum_{\mathrm{i}=1}^{k} \beta_{\mathrm{ii}} {x_{\mathrm{ii}}}^2 \tag{10}$$

where β is the coefficient of the response surface, x refers to random parameters, and k is the number of parameters for response surface fitting.

4.2. Fatigue Crack Growth Modeling

The fatigue damage of risers is often calculated based on the S–N curve. However, riser inspection results are generally represented as fatigue crack size. Using the inspection results to update riser fatigue reliability is difficult when the fatigue damage is calculated based on the S–N curve. Riser fatigue damage based on the S–N curve needs to be transformed into fatigue damage based on fracture mechanics. Fracture mechanics theory is a mathematical method to study the functional relationship among crack shape, load magnitude, and number of cycles. The growth of fatigue cracks for cyclic loading is commonly described by the Paris law. The crack growth rate is defined as follows [39]

$$\frac{da}{dN} = C_0(\Delta K)^m \tag{11}$$

where a is the crack size, N is the cycle number, C_0 and m are fracture mechanics method parameters, and ΔK is the amplitude of the stress intensity factor.

The amplitude of the stress intensity factor describes the stress condition around crack tips, and it can be written as

$$\Delta K = Y(\zeta)\Delta\sigma\sqrt{\pi a} \tag{12}$$

where $Y(\zeta)$ is a geometric function related to crack size, and $\Delta\sigma$ is the applied stress range.

Risers are generally subjected to a combination of axial tensile and bending loads. A riser crack is considered a type I crack, because the load direction is perpendicular to the crack surface, and the upper and lower sides of the crack open relative to each other under the combined load. The geometric function $Y(\zeta)$ is shown as [40,41]

$$Y(\zeta) = 1.122 - 1.4(\zeta) + 7.33(\zeta)^2 - 13.08(\zeta)^3 + 14(\zeta)^4 \tag{13}$$

where ζ is the ratio of crack depth a of the riser to wall thickness t.

The crack growth due to a single cycle load is described as

$$\frac{da}{dN} = C_0\left(Y(\zeta)\Delta\sigma\sqrt{\pi a}\right)^m \tag{14}$$

The amount of crack growth in a year is written as

$$a_{\mathrm{yn}} = \sum_{\mathrm{i}=1}^{k} C_0\left(Y(\xi)\Delta\sigma_{\mathrm{i}}\sqrt{\pi a_{\mathrm{y}-1}}\right)^m \tag{15}$$

where k is the cycle load number in a year.

The geometric function is assumed to be constant in one year. The crack growth in a year is then simplified as

$$a_{\mathrm{yn}} = C_0\left(Y(\xi)\sqrt{\pi a_{\mathrm{y}-1}}\right)^m \sum_{\mathrm{i}=1}^{k} (\Delta\sigma_{\mathrm{i}})^m \tag{16}$$

On the basis of the S–N curve and the data-driven model in Equation (10), the relationship between Δσ and the established fatigue data-driven models can be obtained by substituting Equation (10) into Equation (9) as follows:

$$D_L = \frac{1}{n_f} = \frac{1}{C}\sum_{i=1}^{k}(\Delta\sigma_i)^{m_f} = \beta_0 + \sum_{i=1}^{k}\beta_i x_i + \sum_{i=1}^{k}\beta_{ii}{x_{ii}}^2 \quad (17)$$

S–N curve parameter m_f in Equation (9) serves as an intermediate calculation variable and has the same value as fracture mechanics parameter m. Therefore, the sum of the stress amplitude is shown as

$$\sum_{i=1}^{k}(\Delta\sigma_i)^m = C\left(\beta_0 + \sum_{i=1}^{k}\beta_i x_i + \sum_{i=1}^{k}\beta_{ii}{x_{ii}}^2\right) \quad (18)$$

The annual crack growth can be obtained by substituting Equation (18) into Equation (16) as follows:

$$a_{yn} = C_0\left(Y(\xi)\sqrt{\pi a_{y-1}}\right)^m C\left(\beta_0 + \sum_{i=1}^{k}\beta_i x_i + \sum_{i=1}^{k}\beta_{ii}{x_{ii}}^2\right) \quad (19)$$

Crack depth is the sum of all crack extension depths, and it is expressed as

$$a_y = a_{y-1} + a_{yn} \quad (20)$$

where a_y is the crack depth in the current year, and $a_{y\text{-}1}$ is the crack depth in the previous year.

The crack size increases with time in the process of crack growth. When crack size reaches its failure limit, riser failure occurs. The failure criterion is described as fatigue crack depth being greater than the calculated critical crack size of the risers. The corresponding fatigue crack limit state equation is shown as

$$Z = a_c - a_y \quad (21)$$

where Z is the fatigue failure function, and a_c is the critical size of riser fatigue crack failure.

Generally, the Level 2A critical assessment criterion in BS 7910 [42] is used to obtain the critical size a_c when material information is complete, and it is expressed as

$$\begin{cases} K_r = (1 - 0.14L_r^2)\left\{0.3 + 0.7e^{(-0.65L_r^6)}\right\} & L_r \le L_{rmax} \\ K_r = 0 & L_r > L_{rmax} \end{cases} \quad (22)$$

where K_r is the fracture ratio, L_r is the ratio of applied load to yield load, and L_{rmax} is the permitted limit of L_r.

These parameters are defined in detail as

$$L_{rmax} = \frac{\sigma_Y + \sigma_u}{2\sigma_Y} \quad (23)$$

where σ_Y is the yield stress, and σ_u is the ultimate stress of the material.

$$L_r = \frac{\sigma_{ref}}{\sigma_Y} \quad (24)$$

where σ_{ref} is the reference stress.

$$K_r = \frac{K_I}{K_{mat}} \quad (25)$$

where K_I is the stress intensity factor, and K_{mat} is the material fracture toughness (if a valid K_{Ic} is available, K_{mat} should be taken as K_{Ic}).

The reference stress, σ_{ref}, is defined as

$$\sigma_{\mathrm{ref}} = \frac{P_{\mathrm{m}}\left\{\pi(1-\frac{a}{t}) + 2(\frac{a}{t})\sin(\frac{c_{\mathrm{L}}}{r_{\mathrm{m}}})\right\}}{(1-\frac{a}{t})\left\{\pi - (\frac{c_{\mathrm{L}}}{r_{\mathrm{m}}})(\frac{a}{t})\right\}} + \frac{2P_{\mathrm{b}}}{3(1-a'')^2} \tag{26}$$

where P_{m} is the primary membrane stress, P_{b} is the primary banding stress, r_{m} is the riser mean radius, a is the crack depth, c_{L} is the half of crack length, and a'' is written as

$$\begin{cases} a'' = \frac{\frac{a}{t}}{\left\{1+(\frac{t}{c_{\mathrm{L}}})\right\}} & \pi r_{\mathrm{m}} \geq c_{\mathrm{L}} + t \\ a'' = (\frac{a}{t})(\frac{c_{\mathrm{L}}}{\pi r_{\mathrm{m}}}) & \pi r_{\mathrm{m}} < c_{\mathrm{L}} + t \end{cases} \tag{27}$$

The stress intensity factor in the Level 2 critical assessment criterion is written as

$$K_{\mathrm{I}} = \left\{(Y\sigma)_{\mathrm{p}} + (Y\sigma)_{\mathrm{s}}\right\}\sqrt{\pi a} \tag{28}$$

where $(Y\sigma)_{\mathrm{p}}$ and $(Y\sigma)_{\mathrm{s}}$ represent contributions from primary and secondary stresses, respectively.

$$\begin{aligned} (Y\sigma)_{\mathrm{p}} &= M_1 f_{\mathrm{w}}[k_{\mathrm{tm}} M_{\mathrm{km}} M_{\mathrm{m}} P_{\mathrm{m}} + k_{\mathrm{tb}} M_{\mathrm{kb}} M_{\mathrm{b}}\{P_{\mathrm{b}} + (k_{\mathrm{m}}-1)P_{\mathrm{m}}\}] \\ (Y\sigma)_{\mathrm{s}} &= M_{\mathrm{m}} Q_{\mathrm{m}} + M_{\mathrm{b}} Q_{\mathrm{b}} \end{aligned} \tag{29}$$

where Q_{m} is the secondary membrane stress, and Q_{b} is the secondary bending stress. For the value of parameters M_1, f_{w}, M_{m}, M_{b}, M_{km}, M_{kb}, k_{tm}, k_{tb}, and k_{m}, readers may refer to BS7910.

5. Dynamic Bayesian Network

The Bayesian network (BN) is utilized to analyze the fatigue reliability of risers due to its forward and backward inference capability. The BN is a combination of probability and graph theories and is widely used in data analysis [43]. The structure of a BN model is clearly represented by graph theory, and the problems to be solved are analyzed by probability theory [44–46]. A basic BN is shown in Figure 3.

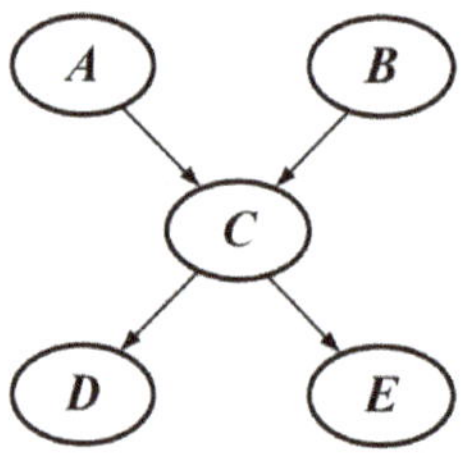

Figure 3. Basic BN.

Each node is attached to a probability distribution. Root nodes A and B are attached to their marginal probability distributions $P(A)$ and $P(B)$, respectively. The non-root node X (C, D, E) is attached to the conditional probability distribution $P(X \mid pa(X))$, and $pa(X)$ is the father node of node X [47]. A logical analysis is performed by analyzing the relationship between nodes in the Bayesian model.

Prior and posterior probabilities are relative to a certain set of evidence in Bayesian theory. In the presence of two different random variables A and D, $A = a$ is an event, and $D = d$ is evidence. The probability estimation $P(A = a)$ of event $A = a$ before considering evidence is called prior probability. Conversely, the probability estimation $P(A = a \mid D = d)$

of event $D = d$ after considering evidence is called posterior probability. The relation between prior and posterior probabilities is described as

$$P(A = a|D = d) = \frac{P(A = a)P(D = d|A = a)}{D = d} \tag{30}$$

where $P\ (D = d\ A = a)$ is the likelihood of event $A = a$ and can be called $L\ (A = a \mid D = d)$. In reality, the likelihood of the event is easier to obtain than the posterior probability.

The DBN is an extension of static Bayesian network modeling in the time dimension. The state of the DBN is updated over time to analyze dynamic stochastic processes. Some assumptions need to be made before establishing a DBN model:

1. The conditional probability is uniformly stable for all variable t over a finite period of time.
2. The DBN is a Markov process. Future probabilities are determined by present probabilities and not by past probabilities, namely, $P(Y[t+1] \mid Y[1],Y[2], \ldots ,Y[t]) = P(Y[t+1] \mid Y[t])$.
3. The conditional probability of adjacent time slices is stable. $P(Y[t+1] \mid Y[t])$ is independent of time.

The DBN is composed of a prior network and a transfer network. The prior network is a joint probability distribution defined at the initial state. The transfer network is the conditional transition probability $P(Y[t+1] \mid Y[t])$ defined by the variables $Y[t]$ and $Y[t+1]$. The DBN is shown in Figure 4.

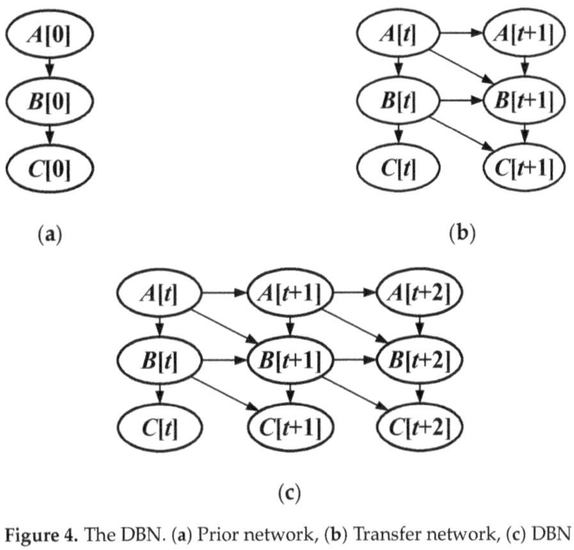

Figure 4. The DBN. (**a**) Prior network, (**b**) Transfer network, (**c**) DBN inference.

A DBN model of riser fatigue crack was established in this study to analyze the fatigue reliability of risers, as is shown in Figure 5. The crack depth in Equation (21) is the parent node, and the other variables are the child nodes. The distribution of fatigue crack size can be calculated through forward Bayesian inference. The inspection results can also be introduced to the DBN model, and the influence of the inspection results on riser failure probability is examined using backward Bayesian inference.

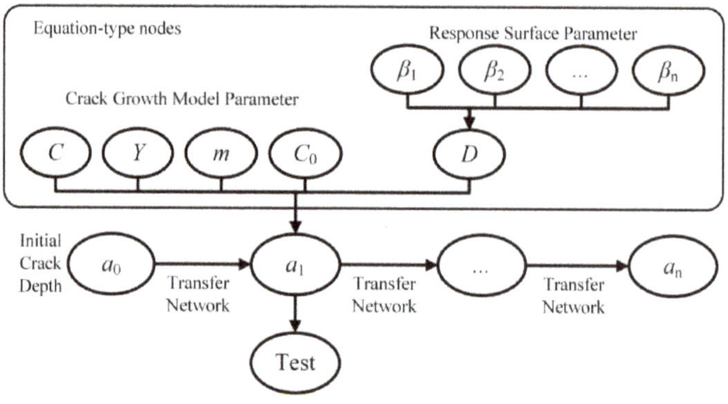

Figure 5. Fatigue crack DBN model of riser.

6. Case Analysis

6.1. Response Surface Analysis

A riser system in the South China Sea was employed as an example for case analysis. The dynamic response analysis parameters are listed in Table 1. The wave conditions for the fatigue analysis are listed in Table 2. The response amplitude operator (RAO) was used for platform motion simulation and is shown in Figure 6.

Table 1. Dynamic response analysis parameters.

Parameters	Value
Length of each deepwater riser	22.86 m
Length of deepwater riser system	1531.62 m
Riser outer diameter	0.5334 m
Wall thickness	22.225 mm
Top tension	6 MN
Element number of single riser	20
Time step size	0.02

Table 2. Wave conditions.

Sea condition number	1	2	3	4	5	6
Wave height (m)	1.23	2.58	3.91	4.95	5.95	6.53
Wave period (s)	3.5	4.5	5.5	6.5	7	7.5
Probability of occurrence (%)	3.99	13.48	38.32	32.56	10.64	1.01

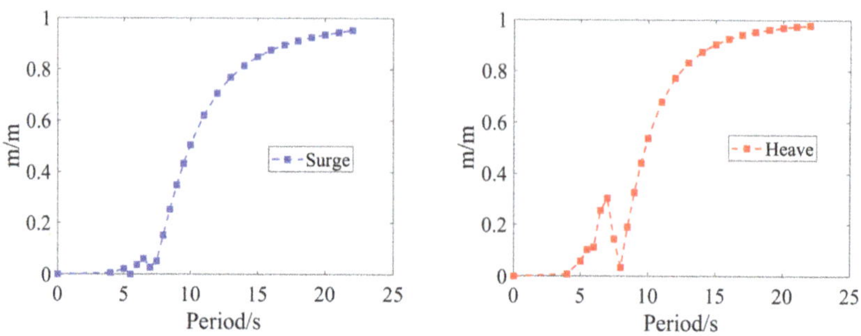

Figure 6. RAO.

The physical-based equations shown in Equation (8) are often solved with the generalized α method, the Newmark β method, or other numerical integral methods. Some software programs, such as Orcaflex and Deepriser, have also been developed especially for riser dynamic analysis. Orcaflex was used for the current analysis due to its ability to analyze the dynamic response and calculate the fatigue damage of risers. The fatigue damage of the deepwater riser system is shown in Figure 7.

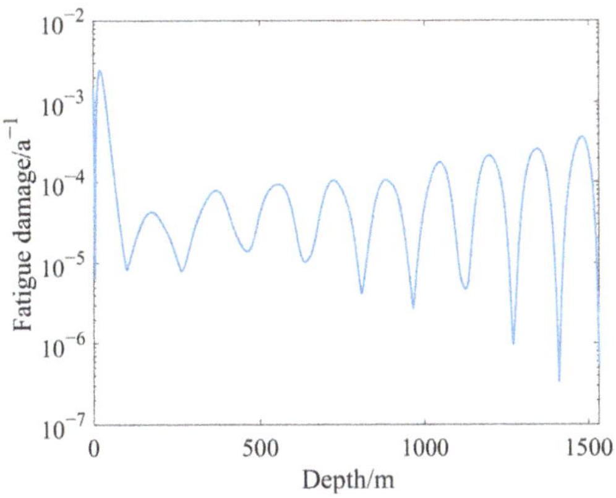

Figure 7. Riser fatigue damage.

It can be seen from Figure 7 that a large portion of the fatigue damage was concentrated at the top of the riser system. The main reason is that the top of the riser system is directly subjected to dynamic wave loads. The following fatigue sensitivity and reliability analysis was conducted at the largest fatigue point. Riser wall thickness t, additional mass coefficient C_a, S–N curve coefficient C, and drag coefficient C_d were adopted as random parameters. The distributions, mean values, and coefficients of variation (COVs) of these random parameters are listed in Table 3 [15,48].

Table 3. Random parameters distributions.

Parameters	Distribution	Mean (μ)	COV
Wall thickness t	Normal	22.225 mm	0.05
Additional mass coefficient C_a	Lognormal	1	0.05
S–N curve coefficient C	Lognormal	1.023×10^{12}	0.05
Drag force coefficient C_d	Normal	1	0.05

μ_{Xi}-1.5σ_{Xi} and μ_{Xi}+1.5σ_{Xi} were adopted as critical values of each parameter X_i. One of the parameters was set to $\mu_{Xi}\pm1.5\sigma_{Xi}$, and the others were set to be the mean value in one calculation round. The fatigue damage of the risers under two critical values was calculated. The other parameters were calculated in the same manner. The standard deviation of riser fatigue under each parameter is shown as follows

$$\sigma_{X_i} = \sqrt{\sum\left(\frac{\partial D_L}{\partial x_i}\right)^2 {\sigma_{x_i}}^2} \tag{31}$$

The sensitivity of riser fatigue to each parameter is expressed as

$$I_i = \frac{\left(\frac{\partial D_L}{\partial x_i}\right)^2 \sigma_{x_i}^2}{D_L^2} \tag{32}$$

The sensitivity of riser fatigue considering t, C_a, C, and C_d as random parameters and the others as the mean values, respectively, is shown in Figures 8–12.

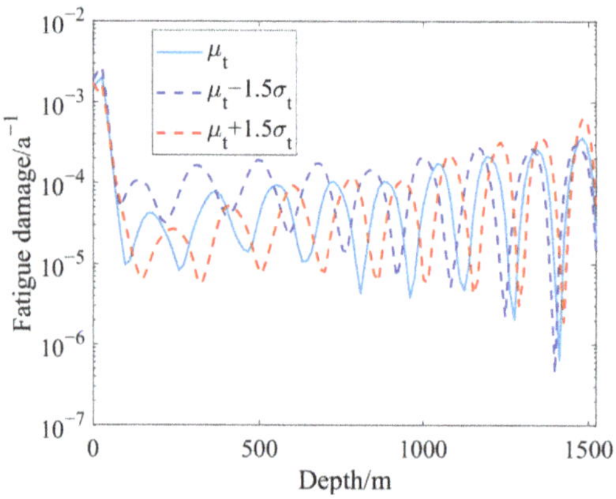

Figure 8. Effect of t on fatigue damage.

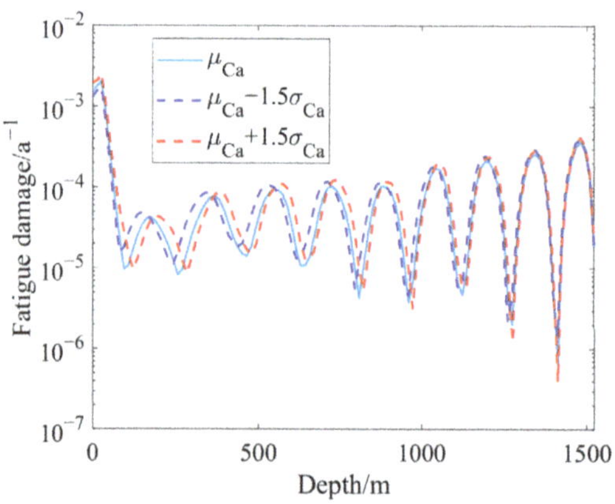

Figure 9. Effect of C_a on fatigue damage.

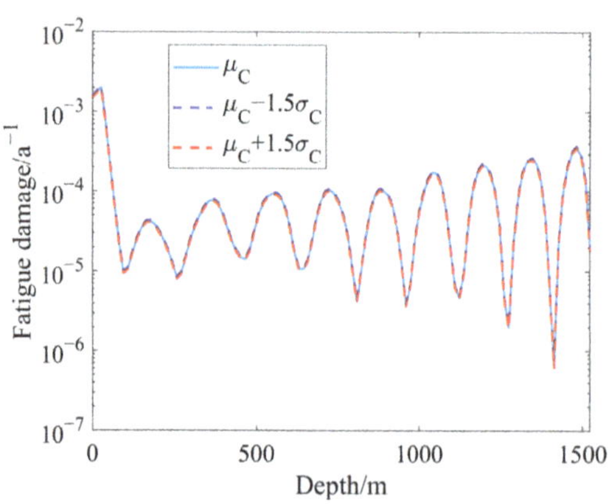

Figure 10. Effect of C on fatigue damage.

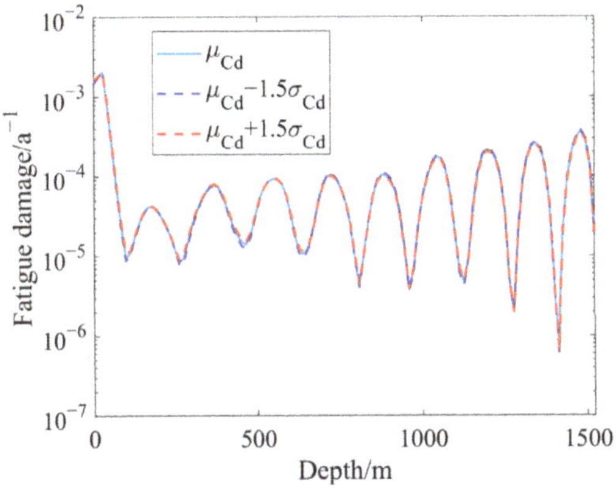

Figure 11. Effect of C_d on fatigue damage.

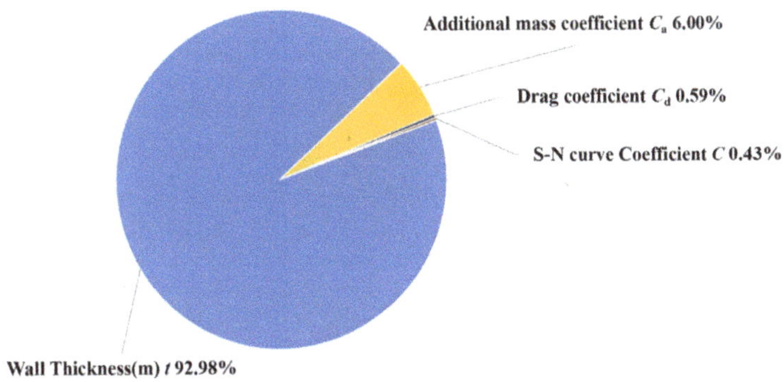

Figure 12. Parameter sensitivity analysis of fatigue damage.

The parameter sensitivity analysis results showed that wall thickness t had the greatest influence on fatigue damage, followed by additional mass coefficient C_a, S–N curve parameter C, and drag coefficient C_d. Drag coefficient C_d and S–N curve parameter C had an almost negligible influence on the fatigue damage of risers relative to the two other parameters. Therefore, wall thickness t and additional mass coefficient C_a were employed as the parameters for establishing the response surface model of fatigue damage. The random variable $\mu \pm h\sigma$ was also chosen, with h being 1.5. The parameters are shown in Table 4 [49].

Table 4. Data for the response surface model of fatigue.

Variable number	1	2	3	4	5
Wall thickness t (mm)	22.225	20.558	20.558	23.892	23.892
Additional mass coefficient C_a	1	0.925	1.075	0.925	1.075
Fatigue damage	0.002476	0.003056	0.003839	0.001683	0.002266

A fatigue damage data-driven model was established using the polynomial method shown in Equation (33) to determine the relationship between riser fatigue damage and random variables. Then, the riser fatigue crack growth model shown in Equation (34) was obtained with this data-driven model.

$$D_L = 0.0495 - 2.0641t + 36.4980t^2 - 0.0429C_a + 0.0237{C_a}^2 \quad (33)$$

$$a_y = a_{y-1} + C_0 Y^{m_0}(a)\left(\sqrt{\pi a_{y-1}}\right)^{m_0} CD_L \tag{34}$$

Five groups of fatigue damage based on data-driven and physical-based models are listed in Table 5 and were compared with one another. The results of the two methods are essentially the same. The fatigue response surface method had high computational speed, because the response surface model had no integration or differentiation process. The established response surface model could be used for the subsequent fatigue reliability analysis due to its high calculation accuracy and efficiency.

Table 5. Verification of the data-driven model of fatigue.

Variable number	1	2	3	4	5
Wall thickness t (mm)	22.225	21.114	21.114	23.336	23.336
Additional mass coefficient C_a	1	0.95	1.05	0.95	1.05
Data-driven models	0.002454	0.002824	0.003274	0.001842	0.002292
Physical-based models	0.002476	0.002735	0.003298	0.001909	0.002363
Relative error	0.1222%	4.391%	1.588%	2.714%	1.003%

6.2. Fatigue Reliability Analysis

In the process of critical crack assessment, initial crack depth a_0 is assumed to be 1 mm, and initial crack length c_0 is assumed to be 2 mm. The yield stress σ_Y of the riser is 555 MPa, the ultimate stress of the material σ_u is 625 MPa, and the outer diameter is 263.5 mm. According to Equation (22) to Equation (29), the riser reaches the critical crack depth of 12.5792 mm in the 40th year. The key parameters of the data-driven models were adopted to calculate the fatigue failure probability of deepwater risers on the basis of the established response surface model. The key parameters are listed in Table 6.

Table 6. Basic parameters for fatigue reliability analysis.

Variable	**Distribution**	**Mean**	**COV**
S–N curve parameter C	Constant	1.023×10^{12}	-
Fracture mechanics parameter m	Constant	3	-
Fracture mechanics parameter C_0	Constant	2.3×10^{-12}	-
Crack depth a_0	Normal	1 mm	0.1
Additional mass coefficient C_a	Log-normal	1	0.2
Wall thickness t	Normal	22.225 mm	0.05

Each variable was discretized, because continuously distributed variables cannot be calculated with the DBN. Variables C_a and t were normally distributed. $[\mu\text{-}1.5\sigma, \mu\text{+}1.5\sigma]$ was used as the truncated boundary, and the outer part was ignored in discrete processing. The variables were divided into 10 segments. Variable a was in half of the normal distribution, and a was divided into 20 segments. The sum of all segments is equal to critical crack size a_c. The segments that exceeded the critical crack size indicated riser fatigue failure. The riser fatigue crack limit state equation in Equation (21) was then transformed into the BN for fatigue reliability analysis, as indicated in Figure 13. The label of each root node represented the variables in the response surface model. The distribution of fatigue cracks could be calculated with the BN.

The BN was transformed into the DBN, because fatigue crack increases with time, as is shown in Figure 14. The total analysis time of the established DBN was set to 40 years, and the analysis time intervals were set to 1 year. The conditional probability of a was determined with Equation (34). A probability table was also established based on the corresponding distribution of discrete variables. The probability distribution of crack depth at each detection time could be calculated via Bayesian forward inference.

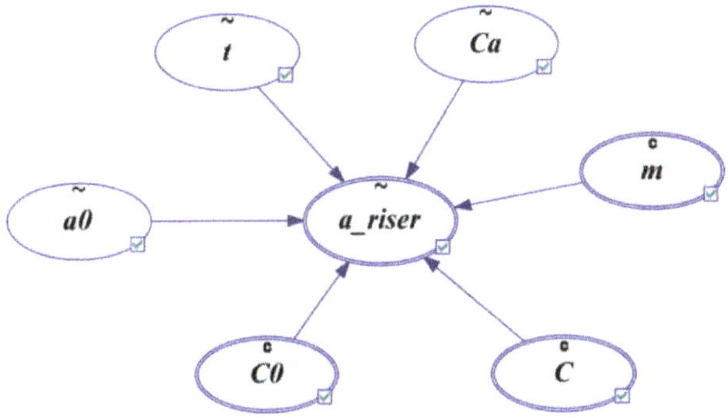

Figure 13. The BN model.

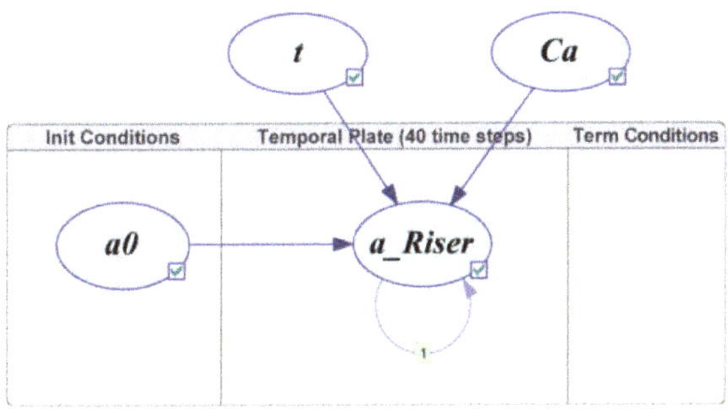

Figure 14. The DBN of fatigue failure.

The fatigue failure probability was calculated efficiently by using the established DBN, as is shown in Figure 15. The riser fatigue failure probability was low and changed slightly in the initial time for the small crack size. Then, the crack size and corresponding failure probability increased gradually with service time, and, finally, the crack reached the critical crack size, and the fatigue failure of the riser occurred. The riser system should be inspected before the riser failure probability reaches the probability limit and maintained afterward to avoid major accidents.

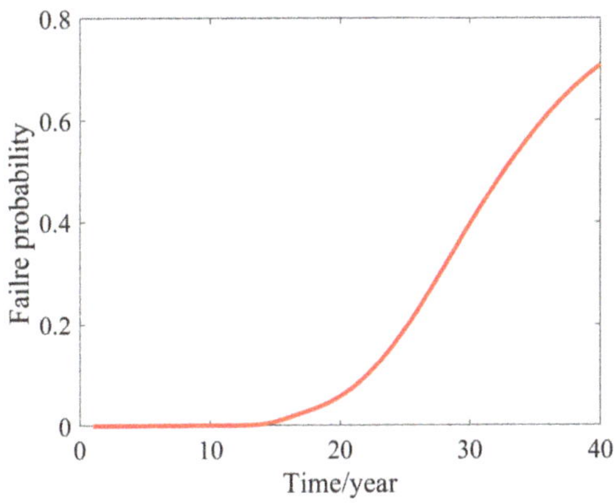

Figure 15. Riser fatigue failure probability.

6.3. Fatigue Reliability Updating

Risers are inspected periodically in actual riser engineering to guarantee their safety. The probability of detection (POD) is often used to describe the probability of detecting the corresponding crack size with a certain detection method under a given condition. It can be regarded as a function of defect crack size and shown as

$$POD(a) = 1 - 1.2\exp(-0.3a) \tag{35}$$

Two groups of inspection results (*Test1*: detection and *Test2*: no detection) of the riser fatigue crack were studied. *Test1* means a fatigue crack is detected, and *Test2* means no fatigue crack is detected. The inspection results were introduced into the fatigue crack DBN model, as is shown in Figure 16.

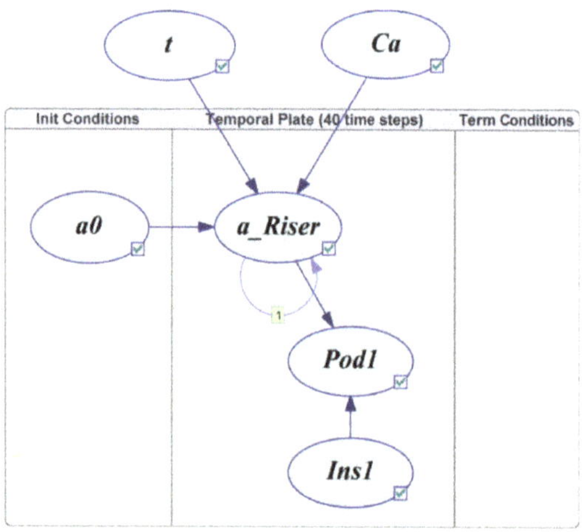

Figure 16. DBN model with inspection results.

Parameters t and C_a and riser failure probability can be updated based on the inspection results in the backward inference of the DBN model. The inspection results were introduced for parameter updating in the 20th year, as is shown in Figure 17.

Figure 17a,b show that the introduction of inspection results exerted a great influence on parameter t. The updated parameter t could not maintain a normal distribution after the inspection results were introduced. The inspection results had little influence on parameter C_a, and updated parameter C_a almost maintained a lognormal distribution after the inspection results were introduced. Figure 17c shows that the inspection results had a great influence on the fatigue failure probability of the deepwater risers. For *Test1* (detection), the updated fatigue reliability of the riser at this time was lower than the value before the inspection, and the failure probability would be higher than before the inspection, because the fatigue crack was detected. On the contrary, the updated fatigue failure probability of the risers in *Test2* (no detection) was lower than the value before the update, because the fatigue crack was not detected in the defect inspection. Riser fatigue reliability can be evaluated realistically by integrating simulation models and inspection results.

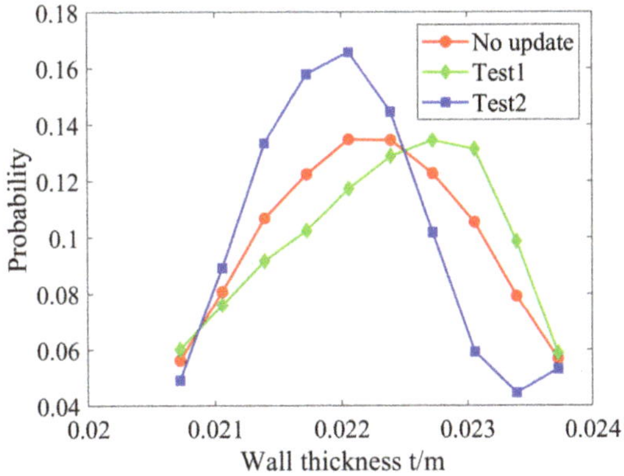

(**a**)

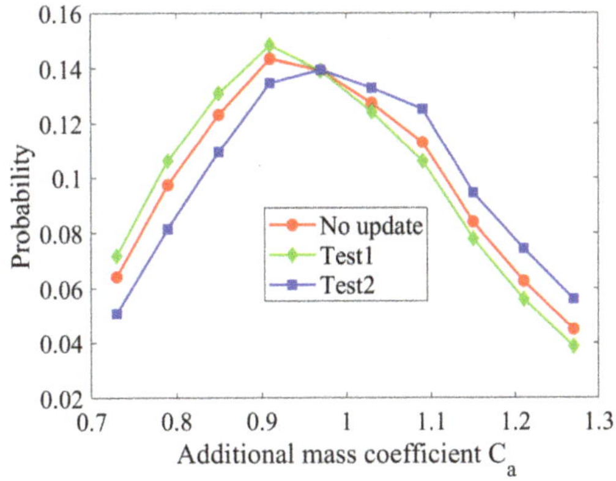

(**b**)

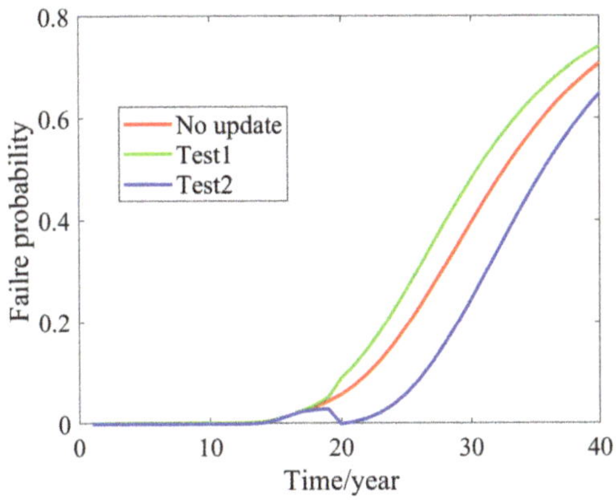

(**c**)

Figure 17. Parameter updating in the 20th year. (**a**) Updated t, (**c**) Updated failure probability, (**b**) Updated C_a.

7. Conclusions

This study attempts to solve a challenging problem: the fatigue reliability analysis of deepwater risers. An improved fatigue reliability analysis method based on RSM and the DBN for deepwater risers was proposed. A more efficient data-driven model was established based on RSM compared with physical-based models. A crack growth model of the riser was established by considering the crack inspection data as a factor. The crack DBN model was employed to conduct a fatigue reliability analysis and updating. The combination of data-driven models and the crack growth model could solve the problems of traditional methods in the fatigue reliability analysis of risers, such as low computational efficiency and the inability to introduce inspection data.

The whole fatigue reliability analysis process includes fatigue damage analysis, fatigue damage modeling, and fatigue reliability analysis. First, the fatigue damage of the riser system was analyzed based on physical-based models. Second, a data-driven model for riser fatigue damage was established based on the fatigue damage analysis results, and, further, a riser fatigue crack growth model was established. Lastly, a fatigue crack DBN model was established for fatigue reliability analysis based on the fatigue crack growth model, and the fatigue reliability of the riser was updated by taking the crack inspection data as evidence.

The proposed method was applied to the fatigue reliability analysis and updating of deepwater risers. The results showed that fatigue reliability could be calculated efficiently through dynamic Bayesian forward inference. Moreover, the distribution of random variables and fatigue reliability could be updated based on the inspection results via the backward inference of the DBN model. The proposed method could analyze and update the fatigue reliability of deepwater risers accurately and efficiently.

Author Contributions: Conceptualization, L.X.; methodology, X.L.; software, P.H.; validation, P.H., and Y.L.; formal analysis, N.Q.; investigation, L.X.; resources, L.X.; data curation, N.Q.; writing—original draft preparation, L.X.; writing—review and editing, P.H.; visualization, P.H.; supervision, G.C.; project administration, X.L.; funding acquisition, L.X. All authors have read and agreed to the published version of the manuscript.

Funding: This research was funded by the National Natural Science Foundation of China (Grant No: 52271300, 52071337, 51809279) and the Program for Changjiang Scholars and Innovative Research Teams at the University (Grant No. IRT14R58).

Institutional Review Board Statement: Not applicable.

Informed Consent Statement: Not applicable.

Data Availability Statement: Data associated with this research are available and can be obtained by contacting the corresponding author upon reasonable request.

Conflicts of Interest: The authors declare that they have no known competing financial interests or personal relationships that could have appeared to influence the work reported in this paper.

References

1. *ISO 13624-1*; Petroleum and Natural Gas Industries—Drilling and Production Equipment—Part 1: Design and Operation of Marine Drilling Riser Equipment. International Organization for Standardization: Geneva, Switzerland, 2009.
2. Hejazi, R.; Grime, A.; Randolph, M.; Efthymiou, M. An efficient probabilistic framework for the long-term fatigue assessment of large diameter steel risers. *Appl. Ocean Res.* **2021**, *118*, 102941. [CrossRef]
3. Chang, Y.; Wu, X.; Zhang, C.; Chen, G.; Liu, X.; Li, J.; Cai, B.; Xu, L. Dynamic Bayesian networks based approach for risk analysis of subsea wellhead fatigue failure during service life. *Reliab. Eng. Syst. Saf.* **2019**, *188*, 454–462. [CrossRef]
4. U.S. Department of the Interior, Minerals Management Service. *Deepwater Riser Design, Fatigue Life and Standards Study Report*; Doc. No. 86330-20-R-RP-005; U.S. Department of the Interior, Minerals Management Service: Washington, DC, USA, 2007.
5. Low, Y.M.; Narakorn, S. VIV fatigue reliability analysis of marine risers with uncertainties in the wake oscillator model. *Eng. Struct.* **2016**, *106*, 96–108. [CrossRef]
6. Chen, R.; Low, Y.M. Efficient long-term fatigue analysis of deepwater risers in the time domain including wave directionality. *Mar. Struct.* **2021**, *78*, 103002. [CrossRef]

7. Cortina, J.P.R.; Fernando, J.M.; Sagrilo, V.S. Neural networks applied to the wave-Induced fatigue analysis of steel risers. *Math. Probl. Eng.* **2018**, *2018*, 2719682. [CrossRef]
8. Zhang, S.; Liu, X.; Chang, Y. Wave-loading fatigue analysis on deep-water drilling platform/riser system. *J. Ship Mech.* **2019**, *23*, 843–850.
9. Khan, R.A.; Ahmad, S. Dynamic response and fatigue reliability analysis of marine riser under random loads. In Proceedings of the ASME 2007 26th International Conference on Offshore Mechanics and Arctic Engineering, San Diego, CA, USA, 10–15 June 2007.
10. Dong, Y.; Kong, X.; An, G.; Kang, J. Fatigue reliability of single-sided girth welds in offshore pipelines and risers accounting for non-destructive inspection. *Mar. Struct.* **2022**, *86*, 103268. [CrossRef]
11. Sen, T.K. Probability of fatigue failure in steel catenary risers in deep water. *J. Eng. Mech.* **2006**, *32*, 1001–1006. [CrossRef]
12. Akpan, U.O.; Koko, T.S.; Rushton, P.A. Probabilistic fatigue reliability of large diameter steel catenary risers (SCR) for ultra-deepwater operations. In Proceedings of the ASME 2007 26th International Conference on Offshore Mechanics and Arctic Engineering, San Diego, CA, USA, 10–15 June 2007.
13. Morooka, C.; Stefane, W.; Matt, C.G.C. An assessment for structural reliability of offshore risers due to probability of fatigue failure. In Proceedings of the 8th International Pipeline Conference, Calgary, AB, Canada, 27 September–1 October 2010.
14. Khan, R.A.; Ahmad, S. Probabilistic fatigue safety analysis of oil and gas risers under random loads. In Proceedings of the ASME 2010 29th International Conference on Ocean, Offshore and Arctic Engineering, Shanghai, China, 6–11 June 2010.
15. Yang, H.; Li, H. Fatigue reliability analysis of marine risers by approximation model. In Proceedings of the Twentieth 2010 International Offshore and Polar Engineering Conference, Beijing, China, 20–25 June 2010.
16. Singh, M.; Ahmad, S. Probabilistic analysis and risk assessment of deepwater composite production riser against fatigue limit state. In Proceedings of the ASME 2015 34th International Conference on Offshore Mechanics and Arctic Engineering, St. John's, NL, Canada, 31 May–5 June 2015.
17. Khan, R.A.; Ahmad, S. Nonlinear dynamic and bilinear fatigue reliability analyses of marine risers in deep offshore fields. *Ship Offshore Struct.* **2018**, *13*, 10–19. [CrossRef]
18. Chagas, B.L.B.; Morooka, C.K. A study on the main parameters that affects the reliability of fatigue failure in marine drilling riser. In Proceedings of the ASME 2021 40th International Conference on Offshore Mechanics and Arctic Engineering, Virtual, 21–30 June 2021.
19. Li, F.; Low, Y.M. Sensitivity study of critical parameters influencing the uncertainty of fatigue damage in steel catenary risers. In Proceedings of the ASME 2010 29th International Conference on Ocean, Offshore and Arctic Engineering, Shanghai, China, 6–11 June 2010.
20. Li, F.; Low, Y.M. Fatigue reliability analysis of a steel catenary riser at the touchdown point incorporating soil model uncertainties. *Appl. Ocean Res.* **2012**, *38*, 100–110. [CrossRef]
21. Elosta, H.; Huang, S.; Incecik, A. Wave loading fatigue reliability and uncertainty analyses for geotechnical pipeline models. *Ship Offshore Struct.* **2014**, *9*, 450–463. [CrossRef]
22. Yang, P.; Li, J.; Zhang, W.; Wu, D.; Gu, X.; Ma, Q. Analysis on statistical uncertainties of wave loads and structural fatigue reliability for a semi-submersible platform. *Ocean Eng.* **2021**, *237*, 109609. [CrossRef]
23. Xia, J.; Das, P.K. Probabilistic fatigue reliability analysis of deepwater steel catenary risers. In Proceedings of the ASME 2008 27th International Conference on Offshore Mechanics and Arctic Engineering, Estoril, Portugal, 15–20 June 2008.
24. Lim, H.; Manuel, L.; Low, Y.M.; Srinil, N. A surrogate model for estimating uncertainty in marine riser fatigue damage resulting from vortex-induced vibration. *Eng. Struct.* **2022**, *254*, 113796. [CrossRef]
25. Hejazi, R.; Grime, A.; Randolph, M.; Efthymiou, M. A Bayesian machine learning approach to rapidly quantifying the fatigue probability of failure for steel catenary risers. *Ocean Eng.* **2021**, *235*, 109353. [CrossRef]
26. Nazir, M.; Khan, F.; Amyotte, P. Fatigue reliability analysis of deep water rigid marine risers associated with morison-type wave loading. *Stoch. Environ. Res. Risk Assess.* **2008**, *22*, 379–390. [CrossRef]
27. Monsalve-Giraldo, J.S.; Videiro, P.M.; Mendes de Sousa, F.J.; dos Santos, C.M.P.M.; Sagrilo, L.V.S. Parametric interpolation method for probabilistic fatigue analysis of steel risers. *Appl. Ocean Res.* **2019**, *90*, 101838. [CrossRef]
28. Dong, Y.; Teixeira, A.P.; Guedes Soares, C. Time-variant fatigue reliability assessment of welded joints based on the PHI2 and response surface methods. *Reliab. Eng. Syst. Saf.* **2018**, *177*, 120–130. [CrossRef]
29. Su, Y.; Fu, G.; Wan, B.; Yu, T.; Zhou, W.; Wang, X. Fatigue reliability design for metal dual inline packages under random vibration based on response surface method. *Microelectron. Reliab.* **2019**, *100–101*, 113404. [CrossRef]
30. Zhang, Z.; Qiu, Z. Fatigue reliability analysis for structures with hybrid uncertainties combining quadratic response surface and polynomial chaos expansion. *Int. J. Fatigue* **2021**, *144*, 106071. [CrossRef]
31. Cai, B.; Liu, Y.; Zhang, Y.; Fan, Q.; Liu, Z.; Tian, X. A dynamic Bayesian networks modeling of human factors on offshore. *J. Loss Prev. Process Ind.* **2013**, *26*, 639–649. [CrossRef]
32. Liu, X.; Sun, H.; Yu, M. Mechanical analysis of deepwater drilling riser system based on multibody system dynamics. *Pet. Sci.* **2021**, *18*, 603–617. [CrossRef]
33. Xu, L.; Wang, J.; Li, Y. Development and sea trial investigation for deepwater drilling riser specialized soft hang-off system during transit. *Ocean Eng.* **2022**, *243*, 110310. [CrossRef]
34. Liu, X.; Li, Y.; Zhang, N. Improved axial dynamic analysis of risers based on finite element method and data-driven models. *Ocean Eng.* **2020**, *214*, 107782. [CrossRef]

35. Tian, D.; Fan, H.; Leira, B.J.; Sævik, S. Study on the static behavior of installing a deep-water drilling riser on a production platform. *J. Pet. Sci. Eng.* **2020**, *185*, 106652. [CrossRef]
36. Sexton, R.M.; Agbezuge, L.K. Random wave and vessel motion effects on drilling riser dynamics. In Proceedings of the Offshore Technology Conference, Houston, TX, USA, 2–5 May 1976.
37. Zhang, W.; Gao, D. Research on the dynamic behavior of riser in deepwater drilling under the condition of disconnection mode. *Pet. Drill. Tech.* **2010**, *38*, 7–9.
38. Wang, Y.; Gao, D.; Fang, J. Axial dynamic analysis of marine riser in installation. *J. Nat. Gas Sci. Eng.* **2014**, *21*, 112–117. [CrossRef]
39. Paris, P.; Erdogan, F. A critical analysis of crack propagation laws. *J. Basic Eng.* **1963**, *85*, 528–533. [CrossRef]
40. Liu, W.; Chen, G. Coupling analysis for vibration and fatigue of a cracked cantilever beam. *J. Vib. Shock* **2011**, *30*, 140–144.
41. Tada, H.; Paris, P.; Irwin, G.R. *The Stress Analysis of Cracks Handbook*, 3rd ed.; ASME Press: New York, NY, USA, 2000.
42. *BS 7910: 2005;* Guide to Methods for Assessing the Acceptability of Flaws in Metallic Structures. British Standards Institution: Britain, UK, 2005.
43. Chang, Y.; Chen, G.; Wu, X.; Ye, J.; Chen, B.; Xu, L. Failure probability analysis for emergency disconnect of deepwater drilling riser using Bayesian network. *J. Loss Prev. Process Ind.* **2018**, *51*, 42–53. [CrossRef]
44. Cooper, G.F. The computational complexity of probabilistic inference using bayesian belief networks. *Artif. Intell.* **1990**, *42*, 393–405. [CrossRef]
45. Xiao, Q.; Gao, S.; Gao, X. *Learning Theory and Application of Dynamic Bayesian Network Reasoning*, 1st ed.; National Defense Industry Press: Beijing, China, 2007.
46. Wang, Y.; Liu, S.; Chen, Z.; Wang, M.; Yang, J.; Chen, X. Dynamic Bayesian networks for reliability evaluation of subsea wellhead connector during service life based on Monte Carlo method. *J. Loss Prev. Process Ind.* **2021**, *71*, 104487. [CrossRef]
47. Liu, Z.; Liu, Y. A Bayesian network based method for reliability analysis of subsea blowout preventer control system. *J. Loss Prev. Process Ind.* **2019**, *59*, 44–53. [CrossRef]
48. Leira, B.J.; Lgland, R.T.; Ye, N. Fatigue safety factors for the tensile and pressure armours of flexible risers. In Proceedings of the International Conference on Offshore Mechanics and Arctic Engineering, San Diego, CA, USA, 10–15 June 2007.
49. Gui, J.; Kang, H. Response surface method for structural reliability analysis and its implementation with MATLAB. *Chin. J. Comput. Mech.* **2003**, *21*, 683–687.

Journal of
Marine Science and Engineering

Article

An Improved Method for Optimizing CNC Laser Cutting Paths for Ship Hull Components with Thicknesses up to 24 mm

Xuan Liu [1] and Daofang Chang [2,*]

[1] Institute of Logistics Science and Engineering, Shanghai Maritime University, Shanghai 201306, China
[2] Logistics Engineering College, Shanghai Maritime University, Shanghai 201306, China
* Correspondence: dfchang@shmtu.edu.cn

Abstract: In this paper, the essence and optimization objectives of the hull parts path optimization problem of CNC laser cutting are described, and the shortcomings of the existing optimization methods are pointed out. Based on the optimization problem of the hull parts CNC laser cutting path, a new part-cutting constraint rule based on partial cutting is proposed, which aims to overcome the drawbacks of the traditional algorithms with serial cutting constraint rules. This paper addresses the problem of optimizing the path for CNC laser cutting of hull parts, including an empty path and the order and directions used for the provided cut contours. Based on the discretization of the part contour segments, a novel toolpath model for hull parts called hull parts cutting path optimization problems based on partial cutting rules (HPCPO) is proposed in this paper. To solve the HPCPO problem, a segmented genetic algorithm based on reinforcement learning (RLSGA) is proposed. In RLSGA, the population is viewed as an intelligent agent, and the agent's state is the population's diversity coefficient. Three different segmented crossover operators are considered as the agent's actions, and the agent's reward is related to the changes in the population's fitness and diversity coefficients. Two benchmark problems for HPCPO were constructed to evaluate the performance of RLSGA and compared with four other algorithms. The results showed that RLSGA outperformed the other algorithms and effectively solved the HPCPO problem.

Keywords: laser cutting; path optimization; cutting constraint rules; genetic algorithm; reinforcement learning

Citation: Liu, X.; Chang, D. An Improved Method for Optimizing CNC Laser Cutting Paths for Ship Hull Components with Thicknesses up to 24 mm. *J. Mar. Sci. Eng.* **2023**, *11*, 652. https://doi.org/10.3390/jmse11030652

Academic Editor: Joško Parunov

Received: 20 February 2023
Revised: 12 March 2023
Accepted: 18 March 2023
Published: 20 March 2023

1. Introduction

Shipbuilding is an intricate and specialized industry where the manufacturing process's efficiency and accuracy significantly impact the final product's quality and cost-effectiveness [1]. In recent years, with the continuous advancement of manufacturing technology, computer numerical control (CNC) laser cutting has become a commonly used method for metal material processing in modern manufacturing. The CNC laser cutting computer-aided manufacturing system provides significant benefits such as shortened product development cycles, improved production efficiency, enhanced product quality, and reduced energy consumption [2]. According to the "2021 China Laser Industry Development Report," laser cutting has the most significant proportion in the laser industry's actual application, followed by laser welding and laser marking, as illustrated in Figure 1. CNC laser cutting technology has become a vital tool in the shipbuilding industry, capable of accurately and efficiently cutting various hull components, including the hull plate, bulkheads, and decks [3]. As a critical technology in laser cutting systems, automatic nesting and cutting path planning have received increasing attention from research units and scholars [4].

For flat laser cutting, the travel path of the laser head consists of the processing trajectory and the auxiliary processing path. The processing trajectory refers to the laser head's travel path as it cuts the graphic outline. In contrast, the auxiliary processing path, the non-cutting path, represents the laser head's movement path between different graphic outlines. The auxiliary processing path serves the critical function of quickly and accurately positioning the laser head, determining the cutting order of various outlines, and improving processing efficiency. Figure 2 illustrates the schematic diagram of the laser head non-cutting path.

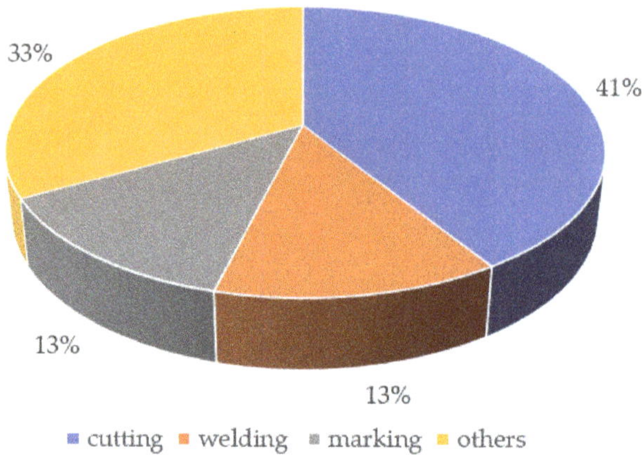

Figure 1. Industrial laser processing application distribution in China in 2021.

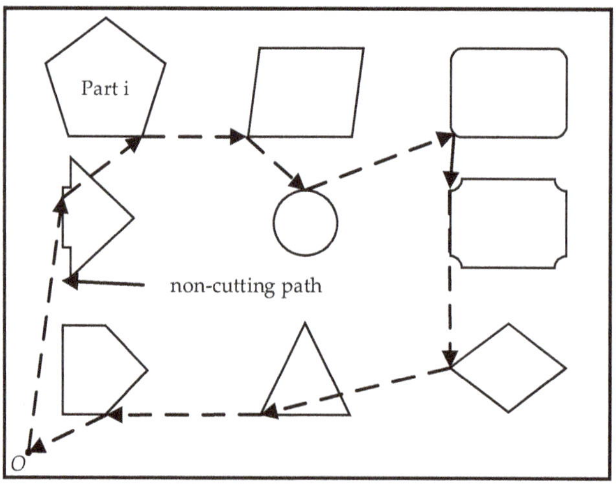

Figure 2. Schematic diagram of laser head non-cutting path.

This paper focuses on optimizing the non-cutting path for laser cutting, which involves finding the optimal path for the laser head movement based on the requirements of the cutting process in a prenested layout [5]. The size of the non-cutting path is related to the cutting order and the starting processing point. Reducing the non-cutting path and avoiding unnecessary movement of the laser head can significantly improve the efficiency of parts with complex trajectories, especially those with batch production requirements.

Laser cutting has higher production requirements and costs than wire, waterjet, and plasma. Therefore, reducing the size of the non-cutting path is essential in decreasing operating costs, conserving resources, and improving production efficiency. For instance, the cost analysis of CO_2 laser and fiber laser cutting processes for cutting a 5 mm stainless steel plate is shown in Table 1, highlighting the high cost of laser cutting. Thus, planning the cutting path is crucial to reduce production costs [6].

Table 1. CO_2 laser and fiber laser cutting process cost analysis.

Classification		Consumption	
		CO_2 Laser	Fiber Laser
laser gas consumption	He(99 .999%)/$L \cdot h^{-1}$	13	-
	N_2(99 .999%)/$L \cdot h^{-1}$	6	-
	CO_2(99 .999%)/$L \cdot h^{-1}$	1	-
	Compressed air/$Nm^3 \cdot h^{-1}$	- [2]	35
cutting gas consumption	Cutting speed/$m \cdot min^{-1}$	2.5	8.2
	N_2(99 .999%)/$Nm^3 \cdot h^{-1}$	20.27	39.25
Electricity consumption	Total power of equipment/kW	47.56	28.56
equipment depreciation	Total purchase and installation cost/¥·Y	650,000	700,000
equipment maintenance	Equipment repair and maintenance costs/¥·Y [1]	100,000	70,000
Vulnerable and consumable parts consumption	Focusing lens/piece·Y	belike 4	-
	Protective lens/day·Y	-	2.5
	Nozzle/h·piece^{-1}	40	40
	Other consumption/¥·Y	belike 4000	belike 4000

[1] Y stands for yearly. [2] "-" indicates that this consumption is not required.

The remaining parts of this paper are organized as follows: Section 2 reviews the relevant literature. Section 3 provides a detailed description of the cutting path optimization problem and cutting path optimization model based on partial cutting rules. Section 4 introduces the designed RLSGA algorithm. Section 5 verifies the effectiveness and efficiency of the proposed model and algorithm through experiments. Section 6 provides further conclusions and discussions.

2. Literature Review

Madić et al. [7] proposed a genetic programming (GP) approach to develop a mathematical model that describes the CO_2 laser cutting process for the aluminum alloy AlMg3. This study used GP to investigate the relationship between cutting speed, laser power, assist gas pressure, and kerf taper angle. The authors conducted a complete factorial design experiment to obtain the GP model evolution process database. The results showed that the fit between the experimental and GP model prediction values of the kerf taper angle was appropriate. Furthermore, 3D surface plots were generated using the derived GP mathematical model to analyze the effects of input parameters on the change in kerf taper angle values. This study demonstrated the potential of GP in developing empirical mathematical models for laser cutting process optimization. Sherif et al. [8] proposed a two-stage sequential optimization approach in laser cutting for the nesting and cutting sequence. The paper focuses on developing a solution technique for any layout's optimal cutting sequence. A simulated annealing algorithm (SAA) was considered to evolve the optimal cutting sequence. The proposed SAA was tested with five typical problems and was shown to provide near-optimal solutions. Comparing the two literature problems reveals that the proposed SAA can give improved results compared to GA and ACO algorithms. This study aims to maximize material utilization and minimize the ideal travel distance of the laser cut tool.

In 2016, Dewil et al. [9] classified the problem of laser cutting path planning into six categories: the continuous cutting problem (CCP), the endpoint cutting problem (ECP), the intermittent cutting problem (ICP), the touring polygons problem (TPP), the traveling salesman problem (TSP), and the generalized traveling salesman problem (GTSP). Each problem differs in selecting start and end points for the machining trajectory and in whether the graphic contour must be entirely cut. Various methods have been proposed to address each problem. Figure 3 shows the classification of the problems and their relationships. Currently, most scholars simplify the cutting path planning problem into two categories: the traveling salesman problem model and the generalized traveling salesman problem model. They have proposed a range of optimization algorithms to solve these two categories of problems.

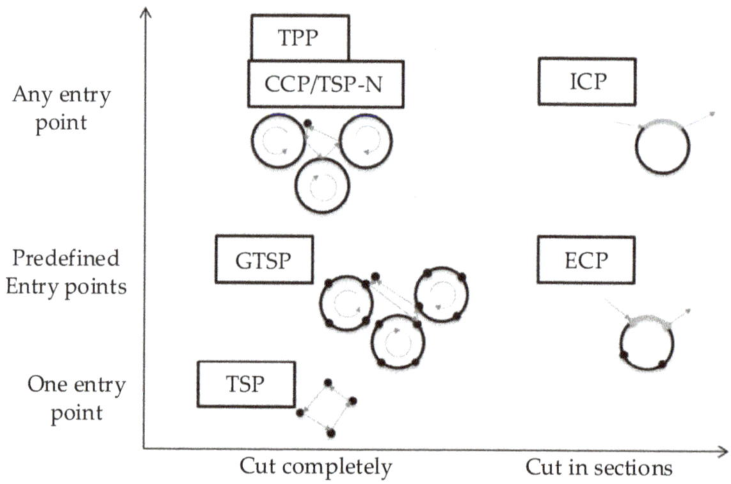

Figure 3. Classification of laser cutting path planning problems.

Establishing a mathematical model for the traveling salesman problem (TSP) and utilizing intelligent optimization algorithms to solve the laser cutting path planning problem is a common approach. Shen Lu [10] proposed two heuristic search algorithms, namely the "n"-shaped and "s"-shaped algorithms. These algorithms simplify the shape center of each city point and use a centroid genetic algorithm to solve the traveling salesman problem to obtain the sequence of the laser cutting air path. Based on the nearest principle, the characteristic points of each part are searched to determine the entry point of the laser head. To avoid the laser head passing through the already-cut area, an improved genetic algorithm is proposed to satisfy the laser cutting process requirements and minimize the air travel path. In contrast to the selection of contour control points mentioned earlier, Li Nini et al. [11] extracted a node from each closed contour to represent the entire contour, treating the laser cutting path planning as a traveling salesman problem and using an improved genetic algorithm to solve it. For non-closed contour graphics, Chen Ting et al. [12] combined the background of the laser cutting die industry and used the taboo search greedy heuristic algorithm to plan the path. To improve the global search ability of the greedy algorithm, the algorithm was optimized locally. Hu Shenghong [13] determined the processing order of the image group and the processing start point of each image group based on greedy and taboo search algorithms. Zhou Rui et al. [14] used a genetic algorithm as the framework and introduced local operators to solve the laser cutting collaborative operation path planning problem.

Compared to the traveling salesman problem, the generalized traveling salesman problem is more complex due to multiple control points in each graph. Chentsov [15] highlighted the limitations of using dynamic programming to solve such problems in the early days, as they were unable to handle large-scale data. He proposed a quasi-optimal greedy algorithm to solve this problem and compared the results of exact and approximate algorithms. Laser cutting, as a particular processing method, imposes multiple process requirements on path planning, which introduces new perspectives to researching such problems. Lin Lizong et al. [16] investigated the cutting order problem of nested contours in laser cutting, established a mathematical model of empty movement paths, added a graphic preprocessing stage, computed the minimum bounding rectangle of each contour, and used a two-level programming design genetic optimization algorithm under the laser cutting process conditions of internal cutting before external cutting for complex contour shapes and non-crossing cutting paths in the already-cut area. The contour line corner points and the boundary corner points of the bounding rectangle determine the positional relationship of the nested contour line. Considering practical processing issues such as plate heating during laser cutting, Song Lei et al. [17] formulated a multi-objective function mathematical model for laser cutting path planning. They employed a dual-chromosome genetic algorithm with a dual-chromosome coding method and suitably modified genetic algorithm steps such as crossover and mutation. To account for constraints such as "punching," "cutting along," and "no crossing cutting" paths, Wang Na et al. [18] established a constrained GTSP model. They utilized a bidirectional ant colony optimization algorithm to address the problem of optimizing the closed contour path. Wang Zheng et al. [19] constructed a generalized traveling salesman problem model to address the shortcut path optimization problem for multiple contours and employed a quantum evolutionary algorithm to obtain the processing sequence. Dynamic programming has the advantage of achieving the optimal decision; therefore, it is utilized to calculate individual fitness. The quantum replacement method also enhances the global search capability of the algorithm. Yang Jianjun et al. [20] proposed the concept of time distance. They utilized a dual encoding genetic algorithm to determine the contour processing sequence and the starting points of each contour while considering the thermal effect problem in laser cutting. Dewil et al. [21] viewed path planning as the division of a contour line, and the partition minimized the cost of connecting the rooted directed minimum spanning tree. They employed the Edmond–Liu algorithm to solve the tree problem and the improved Liu–Kernighan heuristic algorithm to solve the generalized traveling salesman problem. Hajad et al. [22] proposed a simulated annealing algorithm with an adaptive large neighborhood search to minimize the laser cutting path in a two-dimensional cutting process. The algorithm extracts cut profiles from the input image using image processing algorithms and assigns coordinates to the contours' pixels. Based on the generalized traveling salesman problem, the optimization algorithm considers all input image pixels as potential piercing locations. A laser beam makes a single visit and then does a complete cut of each profile consecutively. The simulation results showed that the proposed algorithm could successfully solve several datasets from the GTSP-Lib database with good solution quality. Additionally, the cutting path generated by the proposed method was shorter than that recommended by the commercial CAM software and other previous works.

The studies mentioned above examined the problem of optimizing cutting paths from different perspectives. However, the methodologies employed are often quite similar, with most converting the problem into a TSP by applying the constraint rule of sequential cutting of parts and subsequently employing relevant algorithms for optimization and solution. Nevertheless, the constraint rule of sequential cutting significantly impacts the practical effectiveness of cutting path optimization, leading to cases where the best path found is not a truly optimal solution.

This paper's primary focus is optimizing the numerical control laser cutting path for hull components without holes and with non-adjacent edges. Specifically, the relationship between the cutting constraints of the parts and the path of the laser head is explored, along with relevant optimization algorithms. This study addresses critical issues in optimizing hull components' numerical control laser cutting path.

3. Problem Description and Formulation

3.1. Problem Description

As shown in Figure 4, this paper considers a scenario where M parts without holes and with no common edges need to be cut from a plate. The cutting process begins at the lower left corner of the plate, denoted as the origin O, and the laser head cuts the contour shape of each part in turn. Finally, the laser head returns to the origin O to complete all parts' cutting and blanking processes. During the cutting process, the laser head is "laser on cutting" when moving along the part contour line and in the "laser off non cutting" state (i.e., empty travel) when moving between contours. Since the length of the part contour is fixed, minimizing the empty travel distance between contours is crucial in reducing the cutting path length when the laser head's movement speed is constant. Therefore, cutting path optimization aims to find the optimal path for the laser head, which minimizes the empty travel distance during the cutting process.

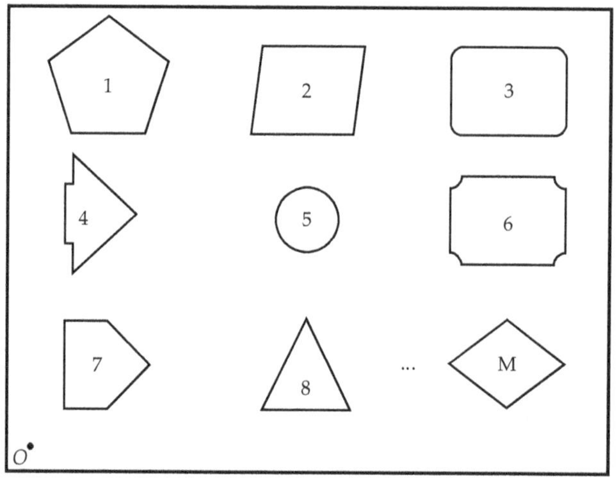

Figure 4. Layout of parts.

3.2. Model Establishment

3.2.1. Part-Cutting Constraint Rules

Many current studies in this field adopt the constraint rule of serial cutting of parts, where the laser head follows the part's contour for a complete loop before moving on to the next part. Thus, the shape of the part's contour may need to be considered when analyzing and optimizing the cutting path. A single point is selected from the contour as the starting and ending point of the contour (cutting point), which can replace the part's contour. Consequently, the contours of the parts to be cut can be simplified as a set of points on the plate plane, and the cutting path optimization problem can be transformed into the well-known traveling salesman problem (TSP) to find the shortest distance between point sets. Most studies initially utilize the nearest neighbor algorithm to determine the cutting points for each contour and then use various related algorithms to optimize and solve the TSP, thereby resolving the cutting path optimization problem. However, based on in-depth research, this method, while simplifying the problem, may have a specific impact on the optimization results. As shown in Figure 5a, two parts are waiting to be cut inside the sheet, with the vertices of part 1 being (P_1^1, P_2^1, P_3^1, P_4^1, P_5^1) and those of part 2 being (P_1^2, P_2^2, P_3^2, P_4^2). Using the nearest neighbor algorithm to determine the punching points as P_1^1 and P_1^2, the complete cutting path can be represented as:

$$O \to P_1^1 \to [\,P_2^1 \to P_3^1 \to P_4^1 \to P_5^1 \to P_1^1\,] \to P_1^2 \to [\,P_2^2 \to P_3^2 \to P_4^2 \to P_1^2\,] \to O$$

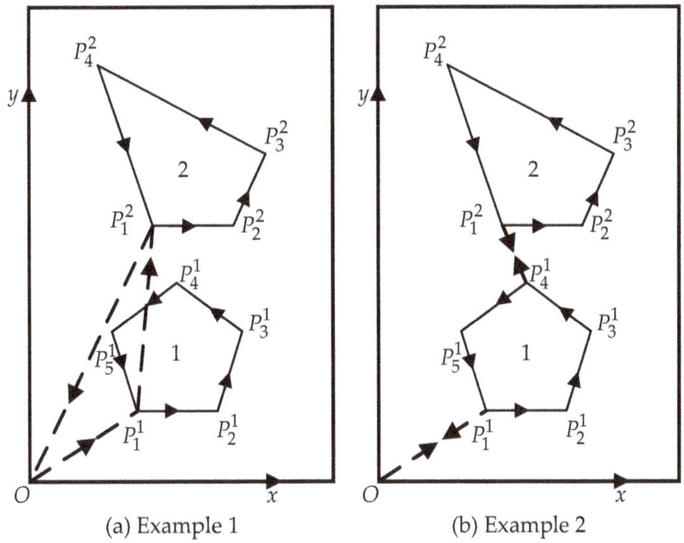

Figure 5. Example comparison of two cutting paths.

The empty travel distance is $S' = \overline{OP_1^1} + \overline{P_1^1P_1^2} + \overline{P_1^2O}$. If the laser head is not subject to the constraint of serial cutting, i.e., allowing it to change position to cut other parts before one part is completely cut, then the cutting path shown in Figure 5b can be obtained:

$$O \to P_1^1 \to [P_2^1 \to P_3^1 \to P_4^1] \to P_1^2 \to [P_2^2 \to P_3^2 \to P_4^2 \to P_1^2] \to P_4^1 \to [P_5^1 \to P_1^1] \to O$$

The empty travel distance, in this case, is

$$S = \overline{OP_1^1} + \overline{P_4^1P_1^2} + \overline{P_1^2P_4^1} + \overline{P_1^1O} = 2\overline{OP_1^1} + 2\overline{P_4^1P_1^2}$$

It is clear that $S' > S$. Hence, under the constraint of serial cutting, conventional algorithms may sometimes fail to find the optimal solution. To tackle this challenge, this paper proposes employing a part-cutting-based cutting constraint rule that permits the laser head to move freely between the contours of each part during the cutting process, without being constrained by the cutting rate of the part contours. Figure 5b provides an illustration of a cutting path that utilizes this part-cutting rule. This rule provides a beneficial means for optimization algorithms to discover superior cutting paths.

3.2.2. Model Establishment

Due to the differences between the partial cutting constraint rule and the serial cutting constraint rule, the approach and processing methods for optimizing the cutting path are entirely distinct. The equivalent TSP method used in traditional algorithms is no longer applicable. Therefore, this paper proposes a method of discretizing the contour segments of the parts. Let there be M parts, where the contour segment group of part 1 is $\{L_{11}, L_{12}, L_{13}, \ldots, L_{1m}\}$, with m being the number of contour segments of part 1, and the contour segment group of part 2 is $\{L_{21}, L_{22}, L_{23}, \ldots, L_{2n}\}$, with n being the number of contour segments of part 2. The contour segment group of part M is $\{L_{M1}, L_{M2}, L_{M3}, \ldots, L_{Mp}\}$, with p being the number of contour segments of part M. According to the cutting requirements of the parts, the laser head only needs to move along all the discretized contour segments on the plate, regardless of the order, to cut all parts. At this point, the cutting path optimization problem becomes a matter of planning the cutting sequence for each contour segment to minimize the laser head's travel distance without cutting, also known as empty travel distance. It is assumed that the cutting sequence of each part's contour segments has been determined according to some rule, as shown in Table 2.

Table 2. Cutting sequence of part contour segments.

Number	Contour Segments	Endpoint 1	Endpoint 2	x-Coordinate of Endpoint 1	y-Coordinate of Endpoint 1	x-Coordinate of Endpoint 2	y-Coordinate of Endpoint 2
1	L_1	P_{11}	P_{12}	x_{11}	y_{11}	x_{12}	y_{12}
2	L_2	P_{21}	P_{22}	x_{21}	y_{21}	x_{22}	y_{22}
3	L_3	P_{31}	P_{32}	x_{31}	y_{31}	x_{32}	y_{32}
4	L_4	P_{41}	P_{42}	x_{41}	y_{41}	x_{42}	y_{42}
…	…	…	…	…	…	…	…
N	L_N	P_{N1}	P_{N2}	x_{N1}	y_{N1}	x_{N2}	y_{N2}

The total number of contour line segments is denoted as N, and the coordinates of the origin O are $(0, 0)$. Thus, the cutting path of the laser head can be represented as $O \to [P_{11} \to P_{12}] \to [P_{21} \to P_{22}] \to \cdots \to [P_{N1} \to P_{N2}] \to O$.

The empty travel distance is

$$\begin{aligned} S_{Empty} &= \overline{OP_{11}} + \overline{P_{12}P_{21}} + \overline{P_{22}P_{31}} + \cdots + \overline{P_{(N-1)2}P_{N1}} + \overline{P_{N2}O} \\ &= \overline{OP_{11}} + \sum_{b=1}^{N-1} \overline{P_{b2}P_{(b+1)1}} + \overline{P_{N2}O} \end{aligned} \tag{1}$$

Therefore, the mathematical model for the optimization-of-cutting-path problem is expressed as follows:

$$\begin{aligned} \min S_{Empty} = \sqrt{x_{11}^2 + y_{11}^2} + \sqrt{x_{N2}^2 + y_{N2}^2} + \sum_{b=1}^{N-1} \sqrt{(x_{(b+1)1} - x_{b2})^2 + (y_{(b+1)1} - y_{b2})^2} \\ s.t.\ x_{ij} \geq 0 \quad i = 1, 2, \cdots, N; \ j = 1, 2 \\ y_{ij} \geq 0 \quad i = 1, 2, \cdots, N; \ j = 1, 2 \end{aligned} \tag{2}$$

In the mathematical model, the cutting path optimization problem of M hull parts is feasible; as shown in Formula (3), a feasible solution is divided into M segments, and each segment represents the cutting sequence of the line segment set in the contour of the part.

$$X = \left\{ \begin{array}{l} path_1 : [x_1^1, x_2^1, \ldots, x_m^1] \\ path_2 : [x_1^2, x_2^2, \ldots, x_n^2] \\ \cdots \\ path_M : [x_1^M, x_2^M, \ldots, x_p^M] \end{array} \right\} \tag{3}$$

4. Segmented Genetic Algorithm Based on Reinforcement Learning

4.1. Algorithm Overview

Based on the analysis in Section 3.2.2, it is evident that the crux of the cutting path optimization problem lies in selecting the sequence of part contour segments. This is a classic NP (non-deterministic polynomial) combinatorial optimization problem that can be solved through intelligent algorithms such as the genetic algorithm, simulated annealing algorithm, and ant colony algorithm.

In the above mathematical model, the encoding method of feasible solutions is complex, and conventional heuristic algorithms or path-planning algorithms applied to the model have poor optimization performance. In order to optimize the model, this paper proposes a segmented genetic algorithm based on reinforcement learning (RLSGA). In different diversity states, the population tries and accumulates to select the best crossover operator to obtain the shortest path of the tool.

The reasons for choosing a reinforcement-learning-based segmented genetic algorithm over other genetic algorithms are multifaceted. Firstly, it can effectively handle high-dimensional and complex problems, which traditional genetic algorithms often need help with due to their low efficiency and the vastness of the search space. Secondly, it can quickly find optimal global solutions by adapting the search space. Thirdly, it can achieve online learning, which enables continuous learning and optimization in real-time environments, whereas traditional genetic algorithms typically rely on offline learning. Lastly, it can better address nonlinear problems by modeling and solving them, which is impossible with traditional genetic algorithms that are typically only applicable to linear or convex optimization problems.

A genetic algorithm (GA) is a commonly used method for solving combinatorial optimization problems [23]. GAs have been applied to the solution of various combinatorial optimization models [24]. Giannakoglou [25] presents an approach to utilizing stochastic optimization and computational intelligence to optimize aerodynamic shapes to improve aircraft performance. The paper primarily focuses on using population-based search algorithms, particularly genetic algorithms, and explores methods to reduce the computational cost of these methods. The construction and use of surrogate or approximation models were also discussed as substitutes for the costly evaluation tool. The paper provides valuable insights into applying stochastic optimization in aerospace engineering design. Vosniakos et al. [26] proposed a systematic procedure for optimizing manufacturing cells by combining neural network simulation metamodels with genetic algorithms. The method optimizes design and operation parameters, overcoming the limitations of discrete event simulations. A neural network metamodel is used to calculate the fitness function, followed by a genetic algorithm to determine the best combination of parameters. The study discusses the conditions for the successful implementation of the proposed approach. Zhu et al. improved GAs using a multi-level method and achieved optimization of single-tool drilling path optimization (DPO) [27]. Although GAs have a fast convergence speed in solving problems, the diversity of the GA population often needs to be maintained as iterations proceed. In the later stage of evolution, the population search tends to slow down, and it is easy to fall into local optima. In order to improve the performance of GAs and overcome their limitations, GAs based on the theory of reinforcement learning have been widely used in the field of combinatorial optimization [28], such as by Li Runfo et al., who used reinforcement learning to automatically adjust GA parameters to solve the ship scheduling problem.

A GA regards the feasible solutions to a problem as chromosomes and multiple chromosomes form a population. The population continuously iterates and evolves through selection, crossover, and mutation until convergence. The core idea of reinforcement learning theory is "try" and "accumulate". The agent executes actions in the environment based on its state and accumulated rewards, obtains immediate rewards, and updates the accumulated rewards.

Three segmented crossover operators for GAs were designed, along with a migration operation, to solve the model proposed in this paper. Based on reinforcement learning, the population was treated as intelligent agents, and a state (s), action (a), and immediate reward (R) were designed. Different segmented crossover operators were selected based on the population's state to enhance the population's diversity while ensuring convergence.

In GAs, the quality of a chromosome is represented by its fitness value fit, with a higher fitness value indicating a better chromosome. Optimizing cutting paths minimizes the total length of empty travel (TL), which is the sum of the lengths of empty travel between all cutting segments. For a population containing n chromosomes, let L_i be the length of empty travel for the $i-th$ chromosome $(i = 1, 2, \cdots n)$; then, the fitness value of the $i-th$ chromosome is given by:

$$fit^i = \frac{1}{TL^i} \tag{4}$$

4.2. State

In order to increase the diversity of the GA population, at time t, the state of the population pop_t is determined by its diversity coefficient. The diversity coefficient div of the population is calculated by Equation (4). The diversity coefficient $div \in [0,1]$, and as $div \to 1$, the diversity of the population becomes better, and vice versa.

$$div(pop_t) \in (1 - 0.5\frac{t}{t_{\max}}) \cdot e^p \tag{5}$$

In the equation, $t_{\max}$ is the maximum number of iterations, and the coefficient p is calculated by Formula (5).

$$p = -\frac{(mean[fit_t])^2 \cdot \sum_{i=1}^{n} (D[i] - 1)}{n \cdot (best[fit_t] - mean[fit_t])} \tag{6}$$

The $mean[fit_t]$ and $best[fit_t]$ represent the population's mean and best fitness values at time t, respectively. $D[i]$ denotes the number of chromosomes with the same fitness value as the $i-th$ chromosome, including itself.

Three states of the population at time t are defined as follows: $s_t^1 : div \in [0, 0.25)$, $s_t^2 : div \in [0.25, 0.75]$, and $s_t^3 : div \in (0.75, 1]$.

4.3. Action

The encoding of feasible solutions determines the crossover method for chromosomes. As shown in Equation (2), feasible solutions are encoded using segmented vectors in the mathematical model. Therefore, when a GA is applied to solve this model, three-segmented crossover operators (SX) were designed in this paper, including positive segmented crossover (PSX), reverse segmented crossover (RSX), and intersect segmented crossover (ISX). These three crossover operators are actions that the intelligent agents (populations) can choose at any state.

4.3.1. Positive Segmented Crossover (PSX)

In RLSGA, the crossover is defined as follows: a chromosome randomly selects gene segments of a certain length from different positions on each path ($path_M$) and exchanges them with the corresponding gene segments of the other chromosome at the same positions and with the same length. The crossover consists of four steps: multiple segment exchange, repeated blanking, completeness checking and complementation, and blank filling. The positive segmented crossover operation on the chromosome is illustrated in Figure 6.

(a) Multiple segment exchange: Segmented crossover is used in RLSGA, where each chromosome comprises multiple parts of the contour line. The directions and orders of each line segment in different chromosomes are not always the same. A random section of each path is selected and exchanged to perform a forward segmented crossover operation on two chromosomes, as shown in Figure 2a. In order to ensure that the segments are of equal length and position, the start and end points for a segment are randomly selected from the shorter of the two chromosomes.

(b) Repeated blanking: After multiple segment swaps, the coding within the swapped segments remains fixed, and if there are any identical numbers outside the swapped segments, they are set to null and need to be filled. In Figure 2b, the empty positions are marked with Roman numerals on a white background. When Chromosome 1 receives the segment {5, 2} from Chromosome 2, the number 2 occurs twice outside the swapped segment, so the number outside the segment is set to null and marked as 'I.' Similarly, marks 'II,' 'III,' and 'IV' are assigned to other null positions outside the swapped segment.

(c) Completeness checking and blank transfer: Let l_m be the set of part contour segments on path m, so $|l_M|$ is constant for path m. Based on this property, the completeness of the path is defined as follows: under the blank state, if the number of remaining segments on a specific path plus the number of blank spaces is less than $|l_M|$, then it is an incomplete path; if the two are equal, then it is a complete path; and if the former is greater than the latter, it is an over-complete path. Incomplete paths cannot execute the fourth step of "blank completion" because the number of their blank spaces is less than the number of segments that need to be filled. At this time, random over-complete paths are continuously transferred to the non-complete paths until they become complete. Transfer blank spaces from over-complete paths to all incomplete paths until there are no incomplete paths in the chromosome.

(d) Empty slot completion: All the segments not on the chromosome are first filled with appropriate empty slots. Then, the remaining empty slots are filled with the remaining segments in a random order according to the principle of minimal increments.

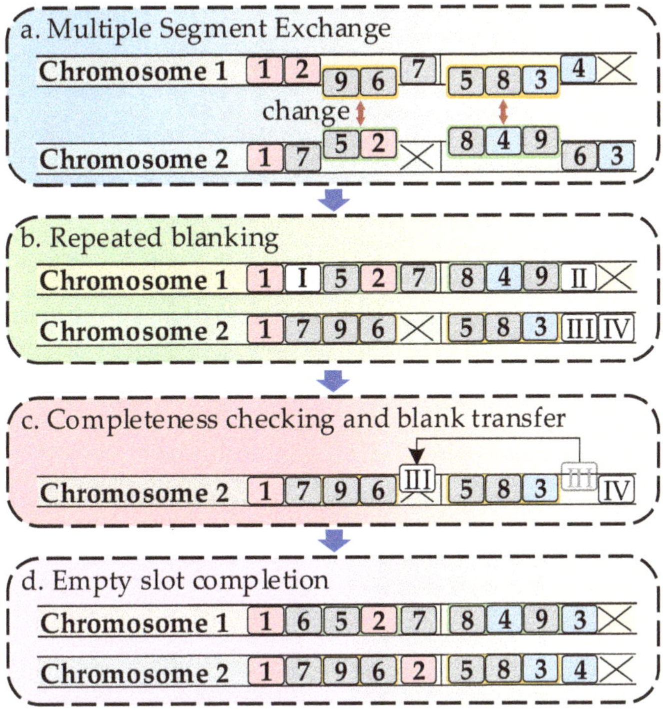

Figure 6. Positive segmented crossover operation step.

4.3.2. Reverse Segmented Crossover (RSX)

The difference between a reverse segmented crossover and a positive segmented crossover lies in the fact that, for the former, the extracted segment is reversed before being exchanged. In contrast, the latter does not reverse the extracted segment. All other steps are the same for both types of crossover.

4.3.3. Intersect Segmented Crossover (ISX)

The Intersect Segment Crossover (ISX) is designed to increase the diversity of the population. ISX only requires that the length of the segments is the same, but not their gene order. When performing ISX, after determining the segment start and end points on the shorter path, a segment of the same length is randomly selected from the long path with a satisfying starting point for the segment. Then, the two segments are exchanged. The remaining steps are the same as those for the Order Crossover.

4.4. Immediate Reward

The immediate reward feedback from the environment to the agent can reflect the rationality of its actions at the current moment in a specific state. In RLSGA, the immediate reward R of the agent is divided into three parts: the diversity reward R1, the best fitness value reward R2, and the average fitness value reward R3.

4.4.1. The Diversity Reward

In the later iterations of the GA algorithm, there is a high probability that all chromosomes in the population will converge to the same one, which quickly leads to the GA being trapped in local optima. To suppress the population assimilation rate and maintain the population's diversity, RLSGA adds a diversity reward. The diversity reward is expressed as the change in the population diversity coefficient div. At time t, the diversity reward is calculated by the following formula:

$$R_t^1 = div(pop_t) - div(pop_{t-1}) \tag{7}$$

4.4.2. The Best Fitness Value Reward

RLSGA aims to enhance the population's best fitness value $best[fit]$, thereby outputting it as the optimal solution for the current optimization when the iteration concludes. Therefore, positive feedback should be given to the population if the best fitness value is improved; otherwise, negative feedback will be provided. At time t, the reward for the best fitness value is calculated using the following formula:

$$R_t^2 = \begin{cases} +1.5 \; if \; best[fit_t] > best[fit_{t-1}] \\ -0.5 \; if \; best[fit_t] = best[fit_{t-1}] \\ -1.0 \; if \; best[fit_t] < best[fit_{t-1}] \end{cases} \tag{8}$$

4.4.3. The Average Fitness Value Reward

The average fitness value $mean[fit]$ of a population reflects the evolutionary state and trend of the entire population. At time t, the reward for the average fitness value is calculated using Equation (8). It should be noted that, if the average fitness value of the population at time t is equal to that at time $t-1$, it indicates that the population has likely not undergone any changes. Therefore, a substantial penalty will be imposed.

$$R_t^3 = \begin{cases} +1.0 \; if \; mean[fit_t] > best[fit_{t-1}] \\ -2.0 \; if \; mean[fit_t] = best[fit_{t-1}] \\ -0.5 \; if \; mean[fit_t] < best[fit_{t-1}] \end{cases} \tag{9}$$

The immediate reward of an agent at time t is the sum of three components:

$$R_t = R_t^1 + R_t^2 + R_t^3 \tag{10}$$

4.5. RLSGA Learning and Iterative Process

In summary, the learning and iteration process of the agent in RLSGA is illustrated in Figure 7.

In Figure 7, Q_t is the Q table at time t, which is used to store the accumulated rewards of different actions in each state. The Q_t table is updated using the $Q-learning$ algorithm, where the action sampled in a single step is independent of the action selected at time t. The Q_t table is updated based on Equation (11) in the $Q-learning$ algorithm, with $\alpha \in [0, 1)$ representing the learning rate and $\gamma \in [0, 1]$ representing the discount factor. The $Q-learning$ algorithm allows the agent to learn incrementally and update the accumulated rewards during the iteration process. As a result, the agent can more accurately determine what actions to take in the current state to maximize its return as the iteration progresses.

$$Q_{t+1}(s_t, a_t) = (1-\alpha)Q_t(s_t, a_t) + \alpha[R_t + \gamma Q_t(s_{t+1}, a_{t+1})] \tag{11}$$

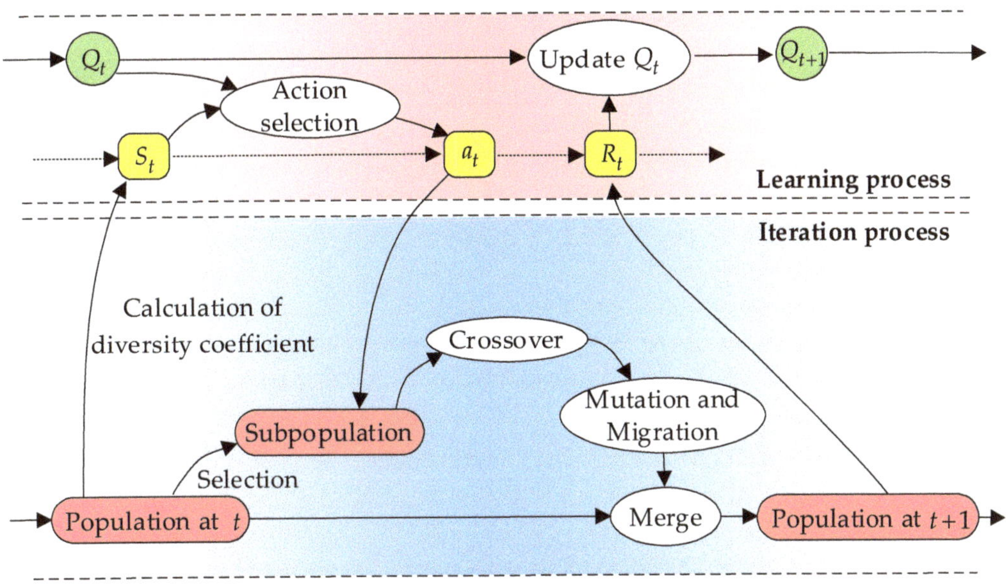

Figure 7. Warehouse layout of the robotic mobile fulfillment system.

The basic process of the RLSGA algorithm can be summarized as follows:

1. Selection: Use roulette wheel selection to choose n individuals from a population of size $Q - learning$ to form a subpopulation.
2. Action selection: The population obtains the state s_t based on its diversity coefficient and then selects an action (crossover operator) based on the state and the cumulative reward Q_t. The action selection method uses an ε-greedy approach, where an action is randomly selected with probability ε ($\varepsilon \in [0, 1]$), and the action with the maximum cumulative reward is selected with probability $1 - \varepsilon$.
3. Crossover: The individuals in the subpopulation undergo crossover with a probability of pC.
4. Mutation: Each individual in the subpopulation undergoes a mutation with a probability of pMu, which is defined as the exchange of two random encodings on a random path.
5. Migration: Each individual in the subpopulation undergoes migration with a probability of pMi, where migration is defined as transferring a random segment of a random path to a random position on another path.
6. Merge population: The subpopulation and parent population consist of $2n$ individuals. After merging the two populations, the top n individuals with high fitness values are selected as the next-generation population.
7. Obtain immediate reward: Obtain the immediate reward R_t based on the relationship between the fitness values and diversity coefficient of the current and previous generations.
8. Update Q_t to $Q_{t+1}.P_1^1$

5. Experimental Results and Discussions

For the proposed HPCPO model, two test questions are designed in this paper. On the test problem, the effectiveness of RLSGA is checked.

The methods have been implemented using the Python programming language, and the experimental environment has been a computer with Windows 10, an Intel(R) Core(TM) i7-10875H CPU @ 2.30 GHz, and 16.0 GB RAM (Lenovo, located in Shanghai, China).

5.1. Test Question Design

In this paper, a total of two test questions of two scales are designed, including five parts with 20 line segments (5–20) and ten parts with 45 line segments (10–45). The layout of the two test questions is shown in Figure 8.

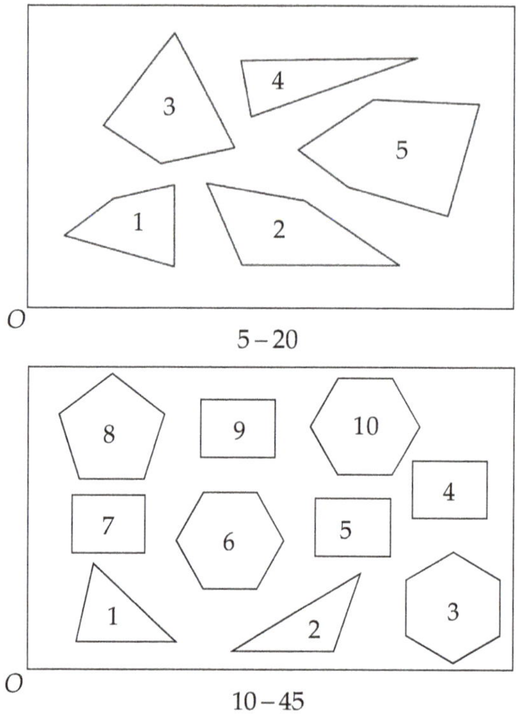

Figure 8. Layout diagram of 2 test cases.

5.2. Comparing Algorithms and Algorithm Parameters

The particle swarm optimization (PSO) algorithm [29], the optimal foraging algorithm (OFA) [30], and the whale optimization algorithm (WOA) [31] are excellent heuristic algorithms. In this paper, the above three heuristic algorithms are selected as the comparison algorithms for RLSGA. Specifically, to verify the impact of the reinforcement learning framework on genetic algorithms (GA), we included a segmented genetic algorithm (SGA) without a reinforcement learning framework in the comparison algorithms. This algorithm does not have attributes such as a population state or immediate rewards and randomly selects a crossover operator during chromosome crossover.

The algorithm parameters were set as follows. For PSO, we set the inertia weight ω to 0.3 and the acceleration coefficients c_1 and c_2 to 0.5, which are commonly used values in the literature [29]. For OFA, we set the parameter k to $0.9 - 0.5\frac{t}{t_{max}}$. This parameter setting enables OFA to dynamically adjust the exploration–exploitation tradeoff during the optimization process [30]. Regarding WOA, we used the default parameter values mentioned in [31], which were suitable for our problem domain. Specifically, we set $A = 2$, $A_{min} = 0$, $C = 1$, and $b = 1$, as recommended in [31]. It is worth noting that the parameters of the segmented genetic algorithm (SGA) without a reinforcement learning framework were set the same as those of the reinforced learning segmented genetic algorithm (RLSGA) during the iterative process. Since the parameter-tuning process in the reinforcement learning phase is a time-consuming task, the hyperparameter settings adopted in this study are based on existing research (Alipour et al.) and further tuned [32]. Specifically, we set

$pC = 0.6$, $pMu = 0.1$, and $pMi = 0.1$ during the iterative process, and $\alpha = 0.9$, $\gamma = 0.9$, and $\varepsilon = 0.1$ during the learning process.

The five algorithms used the same randomly initialized population as the initial population for optimization. The population size n for the five algorithms on test problems 5–20 and 10–45 were all set to 100 and 200, and the maximum iteration times were set to tmax = 1000. Each algorithm was independently run 30 times on each test problem.

5.3. Experiment and Analysis

Table 3 presents the average and best (indicated in bold) total path lengths obtained by the five algorithms, each independently run 30 times on the test problems. RLSGA achieved the best performance on the average and best values for each test problem. The gap between the average and best values obtained by RLSGA was significantly smaller than that of the other algorithms, indicating that RLSGA had lower randomness and more stable performance. For the more straightforward problem 5–20, all algorithms successfully obtained the optimal solution in every run. As the number of parts or part profiles in the test problem increased, the complexity of the problem also increased. RLSGA significantly outperformed the other algorithms on the 10–45 test problem. Overall, RLSGA had the best performance, followed by SGA and PSO, while OFA and WOA did not perform as well. The results indicate that RLSGA can effectively and reasonably optimize the HPCPO model.

Table 3. Algorithm-running results comparison.

Case	Average Total Path Lengths					Best Total Path Lengths				
	RLSGA	PSO	OFA	WOA	SGA	RLSGA	PSO	OFA	WOA	SGA
5–20	**49.68**	56.36	54.95	53.96	53.34	**49.68**	55.95	54.38	53.14	52.09
10–45	**84.31**	97.65	97.19	102.30	95.49	**81.79**	96.17	92.51	98.78	92.32

Figure 9 is the tool path corresponding to the optimal solution of RLSGA on the two test problems.

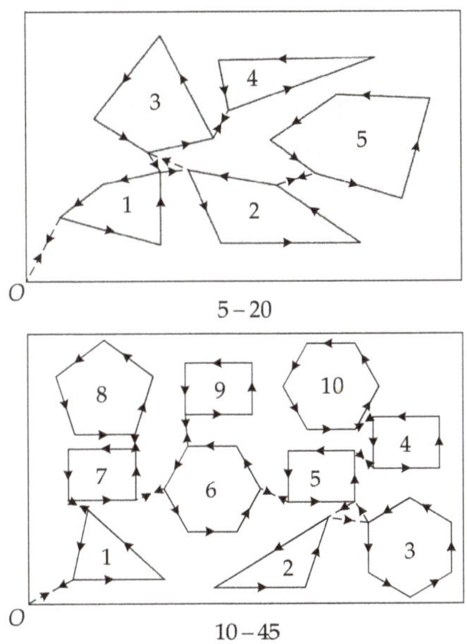

Figure 9. The optimal path graphs obtained by RLSGA on the test problems.

The convergence behavior of the five algorithms on the 5–20 and 10–45 test problems is shown in Figures 10 and 11, respectively. In general, all algorithms demonstrate a gradual convergence to the optimal solution, although there are variations in the convergence rate and the quality of the final solution.

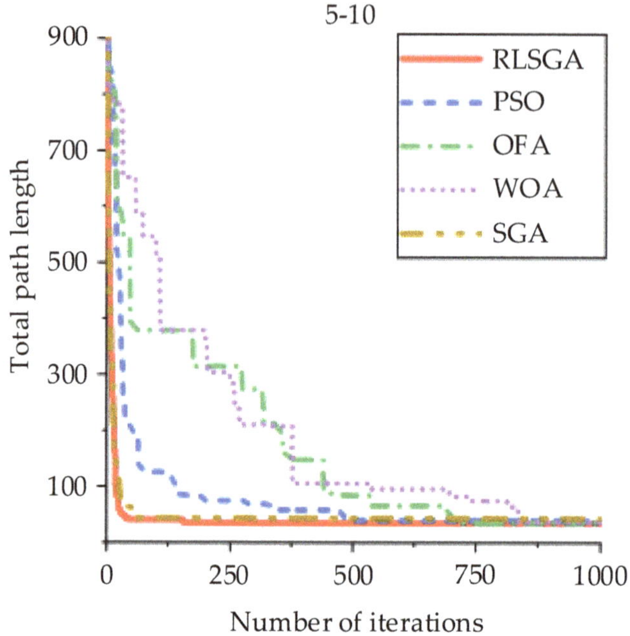

Figure 10. Comparison of rack travel distance between considering equilibrium and not considering.

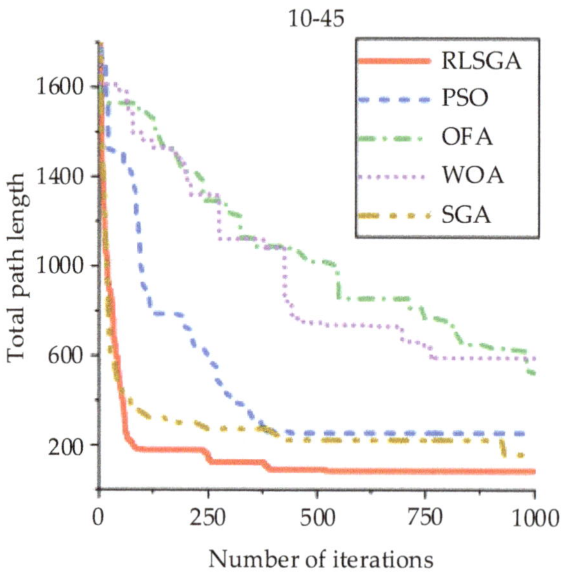

Figure 11. Comparison of rack travel distance between considering equilibrium and not considering.

On the 5–20 problem, RLSGA, SGA, and PSO show a relatively fast convergence, while OFA and WOA converge more slowly, as shown in Figure 10. Specifically, RLSGA and SGA converge almost simultaneously after 100 iterations, suggesting that both methods effectively explore the search space and exploit reasonable solutions. By contrast, the convergence curve of OFA exhibits more fluctuations, indicating that the algorithm may have difficulty escaping from local optima. Meanwhile, WOA converges steadily but more slowly than the other methods, indicating that its exploration strategy may need to be more effective in this problem.

On the 10–45 problem, RLSGA outperforms the other methods by a large margin, as shown in Figure 11. RLSGA and PSO converge to the optimal solution after about 1000 iterations, while SGA falls into a local optimum and only finds a better solution at the end of the iteration. This result suggests that the diversity reward in RLSGA effectively prevents premature convergence and maintains population diversity, enabling the algorithm to explore more promising regions of the search space.

To better understand the performance of the five algorithms, we compared their computational cost in terms of time complexity. Precisely, we measured the average running time of each algorithm on the 5–20 and 10–45 test problems. The results are presented in Table 4. As shown in Table 4, RLSGA and SGA have the lowest average running times on both test problems, while PSO, OFA, and WOA have higher running times. Although RLSGA's running time is slightly higher than SGA's, the difference is within an acceptable range, considering RLSGA's better performance in finding optimal solutions. These results indicate that RLSGA and SGA are the most computationally efficient algorithms for solving the HPCPO problem among the five compared algorithms. Specifically, RLSGA outperforms the other algorithms, including SGA, regarding solution quality and convergence speed.

Table 4. Comparison of average running times of five algorithms on test problems 5–20 and 10–45.

Algorithm	Test Problem 5–20	Test Problem 10–45
RLSGA	16.523	27.714
SGA	16.003	26.660
PSO	24.671	46.529
OFA	30.587	50.319
WOA	35.641	55.413

The data in the table are in seconds.

We can draw several conclusions based on the performance of RLSGA, SGA, and other comparison methods on the 5–20 and 10–45 problems. Firstly, RLSGA outperforms all other comparison methods, suggesting that the reinforcement learning framework with diversity reward effectively solves the HPCPO problem. Secondly, SGA performs better than OFA and WOA but outperforms RLSGA and PSO. This indicates that the genetic algorithm is a suitable method for this problem but may require additional improvements in the future.

To better understand the reasons for the superior performance of RLSGA, we further analyzed its behavior during the optimization process. Specifically, we observed that RLSGA sacrifices some convergence speed to maintain diversity. However, this tradeoff leads to a more robust and stable algorithm less likely to be trapped in local optima. The convergence curves of RLSGA on both problems are relatively smooth, indicating that the algorithm can consistently find reasonable solutions. This property is significant for real-world applications where the objective function may be noisy or non-differentiable.

Furthermore, we compared the results of RLSGA with those of SGA and found that RLSGA is superior in terms of convergence rate and solution quality. The success of RLSGA is due to the use of a reinforcement learning framework, which allows the algorithm to learn from past experiences and adapt to changes in the optimization landscape. Additionally, the diversity reward function in RLSGA encourages the algorithm to explore a wide range of solutions, which may help to avoid getting stuck in local optima.

In conclusion, our results demonstrate the effectiveness of the RLSGA algorithm for solving the HPCPO problem. The use of a reinforcement learning framework and a diversity reward function are key factors contributing to the superior performance of RLSGA. Further research can explore how to further improve the performance of genetic algorithms on this problem by incorporating additional techniques or optimizing the algorithm's parameters.

6. Conclusions

This paper proposes a hull parts cutting path optimization problem (HPCPO) model based on partial cutting rules for the laser cutting process of ship components in practical production activities. The HPCPO model represents the optimization of cutting paths for each component as the planning of the cutting sequence of each component's contour segments. To optimize the HPCPO, this paper proposes a reinforcement -learning-based segmented genetic algorithm (RLSGA), which enables the population to try and accumulate different segmented crossover operators at different diversity coefficient states to achieve maximum benefits. The performance of RLSGA is compared with four other algorithms on the designed test problem, and the results show that RLSGA outperforms the other algorithms and effectively solves the HPCPO problem.

The results show that the proposed reinforcement-learning-based segmented genetic algorithm (RLSGA) performs better than other algorithms in solving hull parts cutting path optimization problems based on partial cutting rules (HPCPO). However, there are still several directions for future work. Firstly, additional factors relevant to practical production activities, such as machine maintenance, should be considered to enhance the HPCPO model. Secondly, a more comprehensive comparison with other state-of-the-art algorithms in the field should be conducted to validate the RLSGA algorithm's effectiveness further. Lastly, it may be worthwhile to investigate the applicability of the RLSGA algorithm to other optimization problems in the field of laser cutting and other manufacturing domains.

Author Contributions: Conceptualization, X.L. and D.C.; methodology, X.L. and D.C.; writing—original draft preparation, X.L.; writing—review and editing, D.C. All authors have read and agreed to the published version of the manuscript.

Funding: This research received no external funding.

Institutional Review Board Statement: Not applicable.

Informed Consent Statement: Not applicable.

Data Availability Statement: Not applicable.

Conflicts of Interest: The authors declare no conflict of interest.

References

1. Stanić, V.; Hadjina, M.; Fafandjel, N.; Matulja, T. Toward shipbuilding 4.0-an industry 4.0 changing the face of the shipbuilding industry. *Brodogr. Teor. Praksa Brodogr. I Pomor. Teh.* **2018**, *69*, 111–128. [CrossRef]
2. Lu, J.; Ou, C.; Liao, C.; Zhang, Z.; Chen, K.; Liao, X. Formal modelling of a sheet metal smart manufacturing system by using Petri nets and first-order predicate logic. *J. Intell. Manuf.* **2021**, *32*, 1043–1063. [CrossRef]
3. Cebi, S.; Ozkok, M.; Kafali, M.; Kahraman, C. A Fuzzy Multiphase and Multicriteria Decision-Making Method for Cutting Technologies Used in Shipyards. *Int. J. Fuzzy Syst.* **2016**, *18*, 198–211. [CrossRef]
4. Oliveira, L.T.; Silva, E.F.; Oliveira, J.F.; Toledo, F.M.B. Integrating irregular strip packing and cutting path determination problems: A discrete exact approach. *Comput. Ind. Eng.* **2020**, *149*, 1–9. [CrossRef]
5. Hajad, M.; Tangwarodomnukun, V.; Jaturanonda, C.; Dumkum, C. Laser cutting path optimization with minimum heat accumulation. *Int. J. Adv. Manuf. Technol.* **2019**, *105*, 2569–2579. [CrossRef]
6. Madić, M.; Petrović, G.; Petković, D.; Antucheviciene, J.; Marinković, D. Application of a Robust Decision-Making Rule for Comprehensive Assessment of Laser Cutting Conditions and Performance. *Machines* **2022**, *10*, 153. [CrossRef]
7. Madić, M.; Gostimirović, M.; Rodić, D.; Radovanović, M.; Coteaţă, M. Mathematical modelling of the CO2 laser cutting process using genetic programming. *Facta Univ. Ser. Mech. Eng.* **2022**, *20*, 665–676. [CrossRef]
8. Sherif, S.U.; Jawahar, N.; Balamurali, M. Sequential optimization approach for nesting and cutting sequence in laser cutting. *J. Manuf. Syst.* **2014**, *33*, 624–638. [CrossRef]

9. Dewil, R.; Vansteenwegen, P.; Cattrysse, D. A review of cutting path algorithms for laser cutters. *Int. J. Adv. Manuf. Technol.* **2016**, *87*, 1865–1884. [CrossRef]
10. Lu, S. Research and Implementation of Laser Cutting Path Planning and Optimization. Master's Thesis, Hefei University of Technology, Hefei, China, 2020.
11. Li, N.N.; Chen, Z.W.; Chen, S.Z. Optimization of laser cutting path based on local search and genetic algorithm. *Comput. Eng. Appl.* **2010**, *46*, 234–236. [CrossRef]
12. Chen, T.; Yu, X.P.; Chen, J.F.; Han, X. Path Planning Based on Plane Unclosed Graphic Cutting. *Jisuanji Xiandaihua* **2019**, *8*, 39–43. [CrossRef]
13. Hu, S.H. Research on Optimizing Algorithm of Laser Processing Based on Graph Computational Geometry Technology. Master's Thesis, Huazhong University of Science and Technology, Wuhan, China, 2007. [CrossRef]
14. Zhou, R.; Ma, H.W. Path Planning of Laser Cutting Cooperative Operation Based on Improved Genetic Algorithm. *Logist. Sci.-Tech.* **2021**, *49*, 50–55. [CrossRef]
15. Chentsov, P.A.; Petunin, A.A.; Sesekin, A.N.; Shipacheva, E.N.; Sholohov, A.E. Heuristic algorithms for solving of the tool routing problem for CNC cutting machines. *Aip. Conf. Proceeding* **2015**, *1690*, 47–50. [CrossRef]
16. Lin, L.Z.; Li, M.Z.; Cheng, X.S. Laser Cutting Path Planning for Complex Contours Based on Mixed Enveloping Rectangles. *Forg. Stamp. Technol.* **2020**, *45*, 147–153. [CrossRef]
17. Song, L.; Wang, X.X.; Liu, X.Y. Optimization of double-chromosome genetic algorithm for laser cutting technology path. *Forg. Stamp. Technol.* **2021**, *46*, 119–125. [CrossRef]
18. Wang, N.; Wang, H.Y.; Jiang, Y.C. Optimization on laser cutting process path based on bidirectional ant colony algorithm. *Forg. Stamp. Technol.* **2020**, *45*, 30–35. [CrossRef]
19. Wang, Z.; Yang, W.B.; Wang, W.L. Path optimization for multi-contour based on quantum evolutionary algorithm. *Comput. Integr. Manuf. Syst.* **2017**, *23*, 2128–2135. [CrossRef]
20. Yang, J.J.; Liu, B.Y.; Ju, L.Y. Dual coding improved genetic algorithm for optimization of laser cutting path. *J. Army Eng. Univ. PLA* **2012**, *13*, 684–687. [CrossRef]
21. Dewil, R.; Vansteenwegen, P.; Cattrysse, D.; Laguna, M.; Vossen, T. An improvement heuristic framework for the laser cutting tool path problem. *Int. J. Prod. Res.* **2015**, *53*, 1761–1776. [CrossRef]
22. Hajad, M.; Tangwarodomnukun, V.; Jaturanonda, C.; Dumkum, C. Laser cutting path optimization using simulated annealing with an adaptive large neighborhood search. *Int. J. Adv. Manuf. Technol.* **2019**, *103*, 781–792. [CrossRef]
23. Karkalos, N.E.; Markopoulos, A.P.; Davim, J.P. *Computational Methods for Application in Industry 4.0*, 1st ed.; Springer: Cham, Switzerland, 2018; pp. 11–31. [CrossRef]
24. Wang, Y.; Wei, Y.; Wang, X.; Wang, Z.; Wang, H. A clustering-based extended genetic algorithm for the multidepot vehicle routing problem with time windows and three-dimensional loading constraints. *Appl. Soft Comput.* **2023**, *133*, 109922. [CrossRef]
25. Giannakoglou, K.C. Design of optimal aerodynamic shapes using stochastic optimization methods and computational intelligence. *Prog. Aerosp. Sci.* **2002**, *38*, 43–76. [CrossRef]
26. Vosniakos, G.C.; Tsifakis, A.; Benardos, P. Neural network simulation metamodels and genetic algorithms in analysis and design of manufacturing cells. *Int. J. Adv. Manuf. Technol.* **2006**, *29*, 541–550. [CrossRef]
27. Zhu, G.Y.; Chen, L.F. Holes machining process optimization with genetic algorithm. *Key Eng. Mater. Trans. Tech. Publ. Ltd.* **2011**, *460*, 117–122. [CrossRef]
28. Li, R.F.; Zhang, X.Y.; Li, J.J.; Chen, L. Application of Self-learning Genetic Algorithm Based on Reinforcement Learning in Ship Scheduling. *J. Dalian Marit. Univ.* **2022**, *48*, 20–30. [CrossRef]
29. Poli, R.; Kennedy, J.; Blackwell, T. Particle swarm optimization. *Swarm Intell.* **2007**, *1*, 33–57. [CrossRef]
30. Zhu, G.Y.; Zhang, W.B. Optimal foraging algorithm for global optimization. *Appl. Soft Comput.* **2017**, *51*, 294–313. [CrossRef]
31. Mirjalili, S.; Lewis, A. The whale optimization algorithm. *Adv. Eng. Softw.* **2016**, *95*, 51–67. [CrossRef]
32. Alipour, M.M.; Razavi, S.N.; Feizi-Derakhshi, M.-R.; Balafar, M.A. A hybrid algorithm using a genetic algorithm and multiagent reinforcement learning heuristic to solve the traveling salesman problem. *Neural Comput. Appl.* **2018**, *30*, 2935–2951. [CrossRef]

Journal of
Marine Science and Engineering

MDPI

Article

On the Fatigue Strength of Welded High-Strength Steel Joints in the As-Welded, Post-Weld-Treated and Repaired Conditions in a Typical Ship Structural Detail

Antti Ahola [1,*], Kalle Lipiäinen [1], Juuso Lindroos [2], Matti Koskimäki [1], Kari Laukia [2] and Timo Björk [1]

[1] Laboratory of Steel Structures, LUT University, Yliopistonkatu 34, FI-53850 Lappeenranta, Finland
[2] Aker Arctic Technology Oy, Merenkulkijankatu 6, FI-00980 Helsinki, Finland
* Correspondence: antti.ahola@lut.fi; Tel.: +358-50-475-2825

Abstract: Weld quality and life extension methods of welded details in ship structures made of high-strength and ultra-high-strength steels are of high importance to overcome the issues related to the fatigue characteristics of welded high-strength steels. The current work experimentally and numerically investigated the fatigue strength of a longitudinal stiffener detail, typically present in the bulkhead connections of ship hull. Two high-strength steel grades, namely EQ47TM and EQ70QT steels, were studied in welded plate connections using gas metal arc welding with rutile-cored wires. Fatigue tests were carried out on both small-scale specimens under axial and large-scale beam specimens under four-point bending loading. In addition to the joints tested in the as-welded condition, the high-frequency mechanical impact (HFMI) treatment was considered as a post-weld treatment technique in the fatigue test series. Furthermore, the large-scale beam specimens were pre-fatigued until substantial fatigue cracks were observed, after which they were re-tested after weld repairing and post-weld treatments to investigate the potential to rehabilitate fatigue-cracked ship details. The joints in the as-welded condition were performed in accordance with the current design recommendations. Due to the severe transition from the base material to the weld reinforcement in the joints welded with the rutile-cored wire, a successful HFMI treatment required geometrical modification of weld toe using a rotary burr to avoid any detrimental sub-cracks at the HFMI-treated region. Alternatively, the use of solid filler wires could potentially overcome these issues related to the welding quality. Repaired and post-weld-treated welds performed well in the re-tests, and the fatigue strength was almost twice higher than that of tests in the as-welded condition.

Keywords: fatigue; welded joint; ship structure; post-weld treatment; weld repair; high-strength steel

Citation: Ahola, A.; Lipiäinen, K.; Lindroos, J.; Koskimäki, M.; Laukia, K.; Björk, T. On the Fatigue Strength of Welded High-Strength Steel Joints in the As-Welded, Post-Weld-Treated and Repaired Conditions in a Typical Ship Structural Detail. *J. Mar. Sci. Eng.* **2023**, *11*, 644. https://doi.org/10.3390/jmse11030644

Academic Editor: Vincenzo Crupi

Received: 17 February 2023
Revised: 14 March 2023
Accepted: 16 March 2023
Published: 19 March 2023

1. Introduction

Fatigue is a decisive design criterion in ship and offshore structures to reach structural integrity and safety, as well as desirable lifecycle under cyclic load conditions. Of the structural elements, welded connections are usually susceptible to the fatigue failures due to their structural discontinuities, imperfections and misalignments [1–5], notch effects [6,7], high tensile residual stresses [8,9], and potential initial defects and flaws due to the welding [10,11]. Regarding the usage of advanced high-strength steel (HSS) materials, ship and offshore industries usually retained a certain level of conservatism but recent developments towards green technologies, as well as their potential regarding structural performance, led to an increasing interest in the applications of high-strength steels [12]. This particularly highlights the importance to consider the fatigue strength capacity of welded joints since, by default, an increase in the material strength does not contribute to an enhanced fatigue performance. High fatigue strength can be achieved by high-quality welds via the enhanced welding process preparation [13] and/or post-weld treatments [14].

In the case of occurring fatigue cracks, repair welding provides a solution to extend the service life of structural elements. Schubnell et al. [15] showed that the fatigue strength of repair-welded high-strength and low-strength repair-welded joints was similar or higher than that of the joints in the as-welded (AW) condition. Even though it was proven that PWT techniques are able to extend fatigue life of cracked or pre-fatigued welded components, they are usually limited to the small crack sizes [16,17]. Consequently, their applications in engineering structures, such as large ships with multiple critical details, are still limited without having continuous monitoring of potential cracks. Many experimental studies tended to focus on the repair welding of small-scale (laboratory) specimens and there is still a lack of experimental evidence of the applicability of repair welding in large-scale components.

The objective of this study was to demonstrate the potential and limitations of PWTs and repair welding to enhance fatigue strength of high-strength ship steels welded with the shipyard procedures. As a result, the fatigue design methodologies and best practices can be established and developed for welded high-strength steel structures. Within this scope, the present study investigated the fatigue strength of welded high-strength and extra-high-strength ship structural steel grades, namely EQ47TM and EQ70QT. Experimental fatigue tests were carried out on the small-scale tensile specimens and large-scale beam specimens with the longitudinal bulkhead stiffener joint. The specimens were investigated in the AW, HFMI-treated and repair-welded conditions. Numerical finite element (FE) analyses were conducted to study stress concentrations at the structural and notch levels.

2. Materials and Methods

2.1. Materials

Two different high-strength ship structural steels, EQ47TM and EQ70QT grades, were selected for the experimental study. EQ47TM is currently used in some ship constructions, e.g., in container ships, and it was selected as a reference grade since it is the highest strength grade that does not require any case-by-case actions from the classification societies when used in container ship structures. EQ70QT is a 690 MPa class extra-high-strength steel delivered in the quenched and tempered for ship and offshore applications. The EQ70QT steel became available for ship building due to the increased production capacity. Strength-matching positional rutile-cored wires (⌀1.2 mm): ESAB OK Tubrod 15.17 (T 46 3 1Ni P C1 2 H5) and ESAB OK Dual Shield 69 (T 69 6 Z P M21 2 H5) were used as filler metals for EQ47TM and 690QT steels, respectively, EN ISO 17632-A. Tables 1 and 2 presents the mechanical properties and chemical compositions of the studied materials, respectively.

Table 1. Mechanical properties of the studied materials. f_y is the yield strength of material, f_u is the ultimate strength of material, A is the elongation, and KV is the impact toughness (longitudinal, at plate top) at given temperature.

Material ID	Type	f_y (MPa)	f_u (MPa)	A (%)	KV at −40 °C (J)
EQ47TM	Nominal	460	570–720	17	46
	Mat. certificates	495–529	574–613	25–28	185
EQ70QT	Nominal	690	770-940	14	62
	Mat. certificates	841–855	874–879	16	195
OK Tubrod 15.17	Typical	544	613	26	124
Dual Shield 69	Typical	740	790	20	65
OK AristoRod 69 [a]	Typical	715	805	17	73

[a] Only applied for the second set of the HFMI-treated joints.

2.2. Specimens and Welding Preparation

Two types of fatigue test specimens were fabricated: coupon-shape specimens for fatigue testing under the uniaxial tension load condition (Figure 1a), and large-scale welded beam specimens for fatigue testing under four-point bending. Henceforth, in this paper, these two specimen types are referred as tensile specimens and beam specimens as per the load configurations. The purpose for using beam specimens was to replicate the manufacturing conditions of such structural details and, on the other hand, achieve welding residual stress state comparable with actual large-scale structures. With the tensile specimens, the test frequency could be increased, and a higher number of specimens were tested. In total, 6 beam specimens and 21 tensile specimens were tested. In the tensile specimens, the bulkhead plate transverse to the beam (see Figure 1b) was not attached to the specimen during fatigue testing but was attached to the specimen during welding to replicate the accessibility at the cope hole detail.

Table 2. Chemical composition of the studied materials (wt-%).

Material ID	Type	C	Si	Mn	Cr	Mo	V	Ni	Cu
EQ47TM	Nominal	0.2	0.55	1.7	46	–	–	–	–
	Mat. certificate	0.098	0.42	1.56	0.05	0.003	0.012	0.44	0.011
EQ70QT	Nominal	0.2	0.55	1.7	46	–	–	–	–
	Mat. certificate	0.159	0.19	1.42	0.25	0.498	0.021	0.04	0.012
OK Tubrod 15.17	Typical [a]	0.05	0.34	1.15	–	–	–	0.96	–
Dual Shield 69	Typical [a]	0.095	0.34	1.25	–	0.4	–	2.8	–
OK AristoRod 69	Typical [a]	0.089	0.53	1.54	0.26	0.24	–	–	–

[a] Undiluted weld metal.

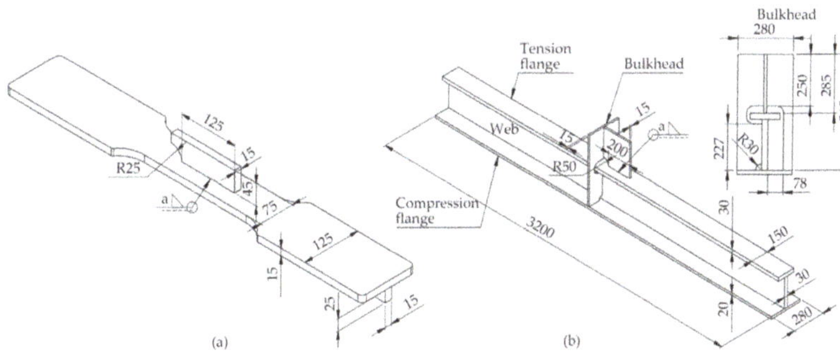

Figure 1. Shape and dimensions of the test specimens: (**a**) the tensile specimens, and (**b**) beam specimens (dimensions in mm).

The beam specimens were fabricated at shipyard with the normal welding procedure of the shipyard. The tensile specimens were fabricated in the laboratory conditions (at LUT University). First, the longitudinal fillet welds of the web-to-flange connections were welded with robotic welding. To compensate welding deformation due to the longitudinal seams and to have straight specimen for fatigue testing, pre-bending was applied in the welding of these welds. The fatigue-critical longitudinal gusset was welded with manual welding. Table 3 presents the welding parameters of the tensile specimens.

2.3. High-Frequency Mechanical Impact Treatment

In selected specimens, HFMI treatments were carried out using a commercial pneumatic HiFIT 'Basic' device [18] with a pin diameter of 4 mm. In the first set of the HFMI specimens (see Table 4, column 'HFMI'), the HFMI treatment was carried without any pre-processing of weld toe. However, as shown by the results (see Section 3), the use of rutile-cored caused severe transition from the base metal to the weld metal. As a result, the HFMI treatment unsuccessfully processed the fatigue-critical weld toe, no improvement in the fatigue strength was necessary found. Due to these reasons, the weld toes in the second set of HFMI-treated specimens were burr-ground (rotary burr with diameter of

6 mm) before the HFMI treatment (see Table 4, column 'BG+HFMI'). With the grinding, the transition was smoothened to enable accessibility to the toe position with the HFMI tool.

Table 3. Welding parameters of the tensile specimens. U is the welding voltage, I is the current, v_{travel} is the torch traveling speed, and v_{wire} is the wire feed rate.

Weld Description	Pass ID	Pos.	U (V)	I (A)	v_{travel} (mm/s)	v_{wire} (m/min)
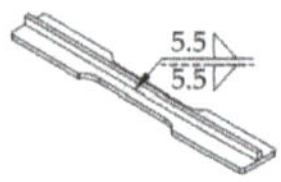	1–2	PB	25.9	310–320	4	13.5
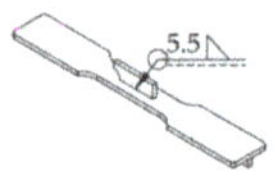	1–2 [a]	PB	28	310–320	4	12.0

[a] Welded with two passes with the start and stop positions at the gusset tips.

Table 4. Testing matrix. Detailed descriptions of the PWTs (burr grinding, BG, and HFMI treatment are given in Section 2.2).

Specimen Type		Material Grade	AW	HFMI	BG+HFMI	Repair-Welded
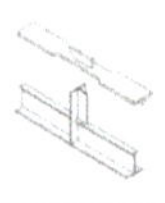	Tensile	EQ47	4	2	–	–
		EQ70	4	4	7	–
	Beam	EQ47	2	1	–	2
		EQ70	2	1	–	2

2.4. Repair Welding of the Beam Specimens

In the beam specimens, the repair welding was also studied to investigate the potential to rehabilitate pre-fatigued structures. The repair welding was applied to those four specimens tested in the AW condition. The fatigue tests of the beam specimens were stopped after crack reaching a depth of 10–15 mm. This depth corresponds to the crack size that can be clearly seen observed in real applications but is still repairable. The repair welding procedure was conducted as follows (Figure 2):

1. Fatigue crack was removed by plasma arc gouging and rotary burr;
2. The gouged dent was inspected with the liquid penetrant to ensure that the whole crack was removed before starting repair welding process;
3. Tungsten inert gas (TIG) welding was applied to re-melt and re-shape the deepest point of the dent;
4. Multipass TIG welding with the consumable of ESAB OK Tigrod 13.09 (⌀2.4 mm) and GMA welding with solid wire of ESAB OK 12.51 (for the EQ47 steel) and 13.29 (for the EQ70 steel) were used to fill the dent;
5. In the cope hole details, the weld reinforcement was ground to flush with a lamellar rotary grinder. At the opposite side of the gusset, HFMI treatment was applied.

2.5. Residual Stress Measurements

Before the fatigue testing (see Section 2.5), welding-induced residual stresses were characterized in both tensile and beam specimens. The surface residual stresses, in the longitudinal loading direction of the specimens, were measured using an X-Ray diffractometer (XRD, Stresstech X3000 G3) with a collimator spot diameter of 1 mm in the vicinity of the critical weld toes (see Figure 3). From the specimens in the AW condition, the residual stress distribution starting from the weld along the specimen was measured while in the HFMI-treated specimens, residual stresses were measured only from the HFMI groove.

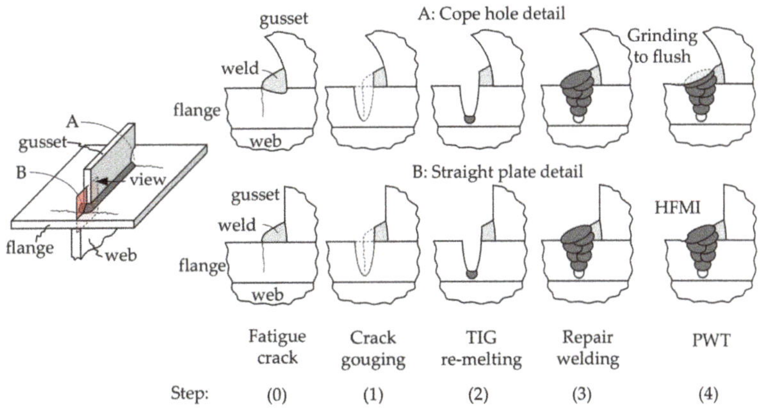

Figure 2. Schematic description of weld repair processing.

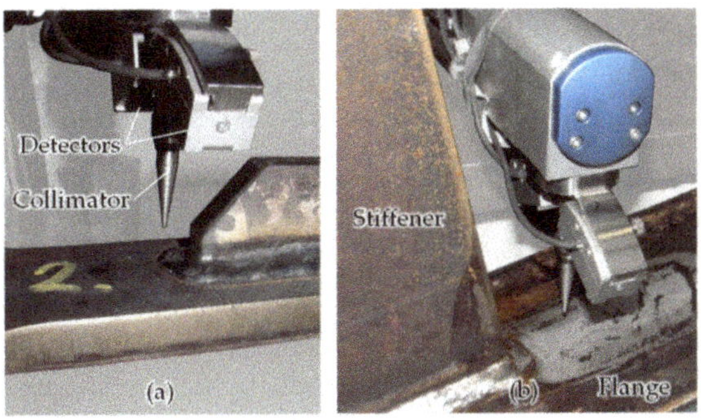

Figure 3. Residual stress measurements of (**a**) a tensile specimen and (**b**) beam specimen.

2.6. Fatigue Testing

Fatigue tests were carried out using two servo hydraulic fatigue test rigs. The tensile specimens were tested using uniaxial constant amplitude (CA) loading with the clamping system under applied stress ratios of $R = 0.1$ and $R = 0.5$ (Figure 4a). The beam specimens were fatigue-tested under CA four-point bending with a constant bending moment and zero shear force between the inner press rolls (Figure 4b). The distances between inner rolls and outer rolls were 700 mm and 3220 mm, respectively. The end criterion for fatigue testing in the tensile specimens was the total rupture of the specimen. As the beam specimens were pre-fatigued before the repair welding, the tests were stopped when substantial fatigue cracks were observed in the specimens–usually with the crack depth of a = 10–15 mm—enabling weld repairs in the flange. Subsequently, fatigue testing of these beam specimens was continued until cracks with substantial size in the flange were observed.

The test specimens were equipped with strain gages to capture both nominal strains (verifying test loads in the beam specimens) located far from the joint area and local structural strains at the hot-spot distances. In the tensile specimens (with the narrow web member), the HS strain gages were located at the 0.4t and 1.0t distances from the weld toe (type 'a' HS location). In the beam specimens, strain gage were located at 4 mm, 8 mm, and 12 mm distances from the weld toes as per the type 'b' HS location [19]. The HS strain measurement was conducted using strain gage chains. The strain gage instrumentation is exemplified in Figure 4c.

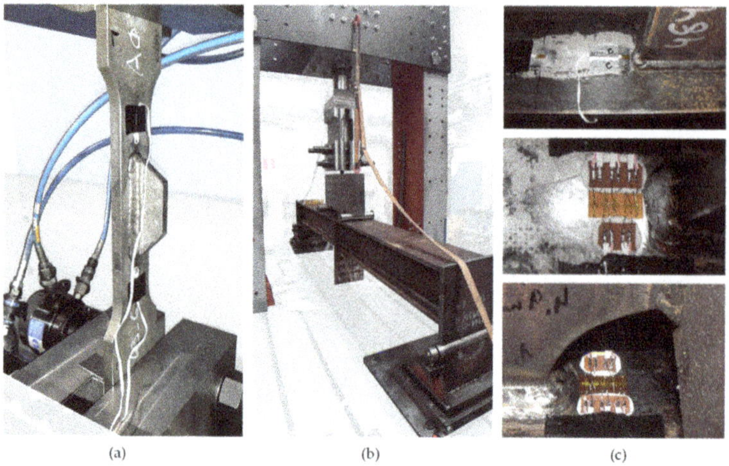

Figure 4. Fatigue testing equipment and instrumentation: fatigue tests of (a) the tensile specimens and (b) beam specimens, and (c) strain gage instrumentation.

2.7. Finite Element Analysis

To account for the effects of the studied specimen geometries and load configurations on the resulting stress concentrations, as well as to employ local approaches in fatigue assessments, numerical FE analyses were carried out using 8-noded brick shape linear solid elements with FEMAP 2022.2 (Siemens PLM) software. The geometry models for the tensile and beam specimens were created to include fictitious notch radii at the critical weld toes as per the concept of the effective notch stress (ENS) [20] with the reference radius of r_{ref} = 1.0 mm. Half symmetry was utilized in the model for the tensile specimens. At the critical toe position, element side length (tangential direction) was 0.10 mm with an aspect ratio of 2.5:1 (tangential/radial). The boundary and load conditions were modeled according to the test load conditions, i.e., under uniaxial tension and four-point bending loads in the tensile and beam specimens, respectively. Figure 5 presents the element mesh models together with the load and boundary conditions.

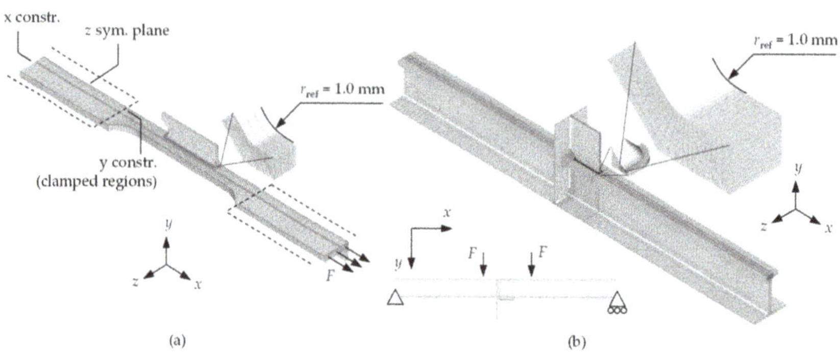

Figure 5. FE meshes for (a) the tensile specimens and (b) beam specimens (not in scale).

3. Results

3.1. Fatigue Tests

Figure 6 presents the results of the tensile specimens in the nominal stress system. The nominal stress refers to the value based on the axial load and cross-sectional area. For the longitudinal gusset length of 125 mm, FAT63/Class F1 (with a slope parameter of m = 3) can be selected as a characteristic design curve for joints in the AW condition [21–23]. For the HFMI-treated joints, FAT140 and FAT125 were recommended under the $R < 0.15$ load conditions for yield strengths of 550 MPa $\leq f_y <$ 750 MPa and 355 MPa $\leq f_y <$ 550 MPa, respectively [24]. In this case, a slope parameter of $m = 5$ was applied. For $R = 0.5$, three FAT class decrease in fatigue strength was suggested [24]. Here, FAT100 and FAT90 can be claimed for these joints [24]. These FAT classes can be observed in Figure 6. In the AW condition, R ratio did not show distinguishing results; both load configurations were clearly above the design curve.

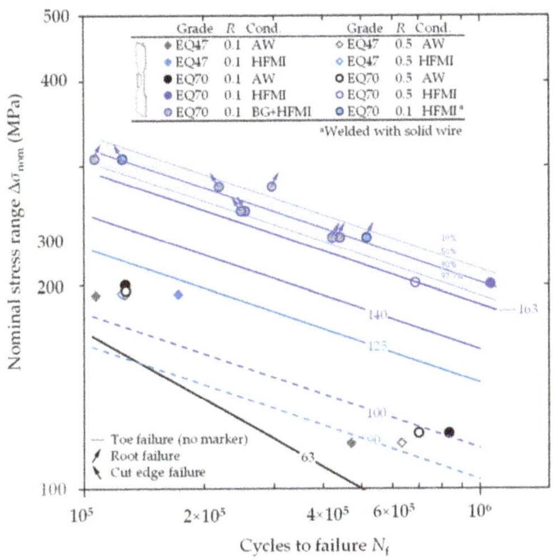

Figure 6. Fatigue test results of tensile specimens in the nominal stress system. BG+HFMI refers to the specimens welded specimens (rutile-cored wire) prepared by burr grinding followed by the HFMI treatment.

In the HFMI-treated conditions, the obtained fatigue strengths were highly related to the observed welding quality. The HFMI-treated EQ47 specimens suffered from the low welding quality, and no improvements were found compared to the AW specimens. The failure observations are presented in Section 3.2. In the EQ70 specimens, the HFMI treatments were more successful and even without BG processing, fatigue strengths above the FAT140 curve were obtained. With BG processing followed by the HFMI treatment (see BG+HFMI specimens in Figure 6), the successful HFMI treatment was confirmed but the fatigue strength of these specimens was comparable to the one of specimens without BG treatment. In the successful HFMI-treated specimens, with few exceptions, failures occurred either from the weld root or cut edge (see schematic presentations of failure modes in Figure 7). As the test results of these $R = 0.1$ HFMI specimens commeasured into a single scatter band, an S-N curve was statistically fitted into these points, employing the conventional statistical approach [22]:

$$\log N_{\mathrm{f}} = m \log \Delta\sigma + \log C \tag{1}$$

where N_f is the cycles to failure, m is the slope parameter, $\Delta\sigma$ is the stress range and C is the fatigue capacity. The mean curve, corresponding to the survival probability of P_s = 50%, was fitted and the lower and upper bounds, and the characteristic (P_s = 97.7%) curve was obtained as follows:

$$\log C_{97.7\%} = \log C_{50\%} - k \cdot \mathrm{Stdv} \tag{2}$$

where k is the characteristic value and Stdv is the standard deviation. With this approach, characteristic fatigue strength capacity of $\Delta\sigma_{c,97.7\%}$ = 163 MPa was obtained for the EQ70 specimens in the HFMI-treated condition under the R = 0.1 load conditions (see Figure 6).

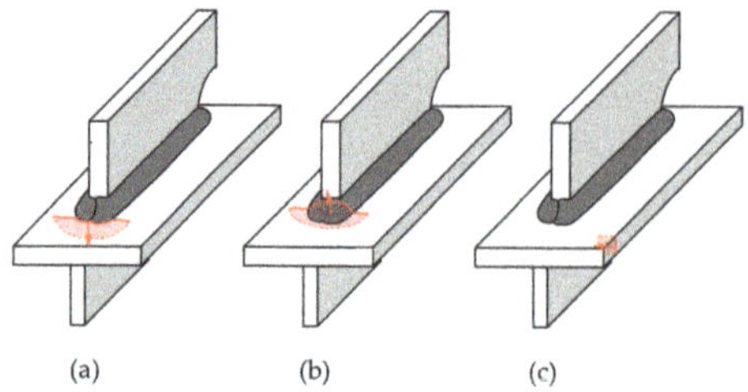

Figure 7. Observed failure locations in the specimens: from (**a**) weld toe, (**b**) weld root, and (**c**) thermally cut edge.

Figure 8 presents the fatigue test results for the beam specimens. For these specimens, the nominal stresses were obtained based on the acting bending moment and section modulus of the beam. The fatigue strength of the beam specimens was slightly lower than that of the tensile specimens. However, FAT63 was conservative in comparison to the test results.

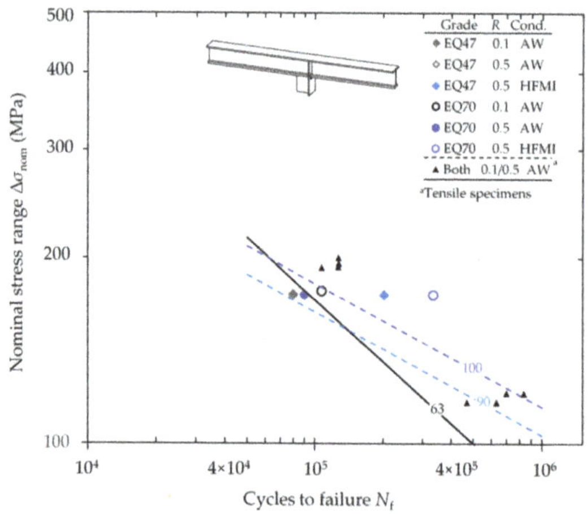

Figure 8. Fatigue test results of the beam specimens in the nominal stress system ([a] Tensile specimens in Figure 5).

Figure 9 presents the fatigue test results of the repair-welded beam specimens in comparison to the tensile and beam specimens tested in the AW condition. The repair-welded specimens were also tested at the higher load levels to reach fatigue failures at similar number of cycles (close to 10^5 cycles) due to the lower test frequency. As can be seen, the repair-welded specimens performed better than the pre-fatigued specimens. Due to the PWTs, the failure mode changed from the weld toe failure to both weld root failure and toe failures from the HFMI-treated groove. In the S-N data plots of repair-welded specimens in Figure 9, the pre-fatigue test cycles were not accounted in the re-test; although, the weld root experienced the fatigue loading from the previous test. Considering these aspects, the fatigue lives would more clearly exceed the design lives of the corresponding design curves for the HFMI treatments.

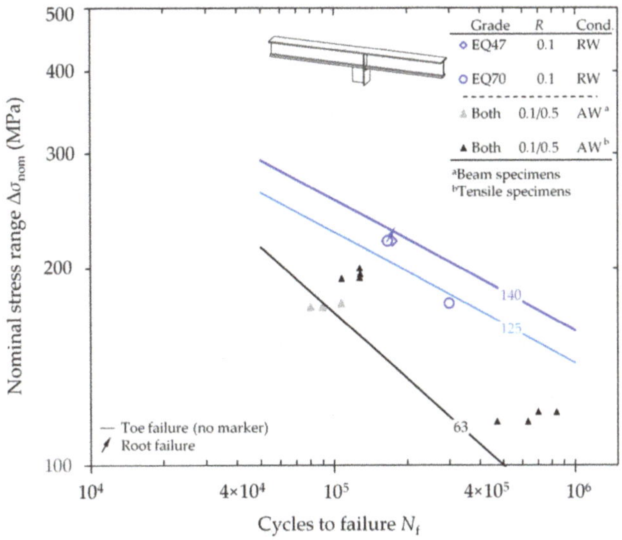

Figure 9. Fatigue test results of the repair-welded (RW) beam specimens in the nominal stress system ([a] Beam specimens in Figure 7; [b] Tensile specimens in Figure 5).

3.2. Failure Observations

As mentioned in Section 3.1 (see also Figure 7), different failure modes were observed in the specimens. As expected, all specimens in the AW condition failed from the weld toe position; an exemplifying crack path was shown in Figure 10a. Those HFMI EQ47 specimens not reaching the design curve of FAT125 (see Figure 6) failed from the weld toe but outside the HFMI-treated groove, as demonstrated in Figure 10b. Figure 11a,b exemplified the weld toe and root failure modes, respectively, in the repair-welded specimens. The repair-welded regions were free of new fatigue cracks in all four studied specimens.

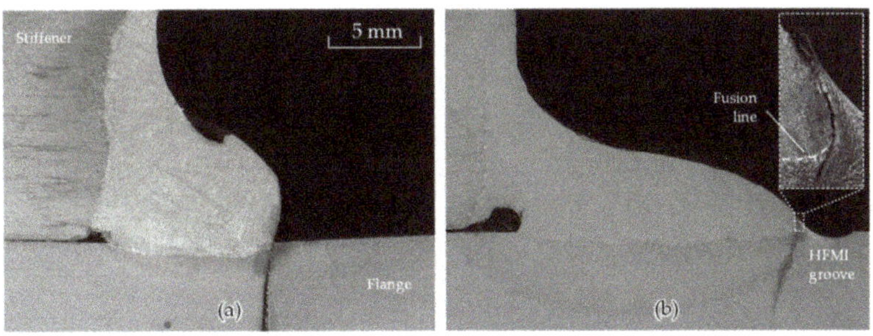

Figure 10. Macro graphs on the failures in the tensile specimens: (**a**) an EQ70 specimen failing from the weld toe, and (**b**) a HFMI-treated EQ47 specimen failing from the edge of the HFMI groove in the vicinity of the fusion line.

Figure 12a,b, respectively, demonstrates the failure modes in the successfully HFMI-treated prepared using rutile-cored wire (burr grinding and HFMI treatment) and solid wire (only HFMI treatment). The use of solid wire produced better pre-conditions for the HFMI treatment and failures occurred from the weld root. Similarly, burr grinding removed the material to enable accessibility to the original weld toe.

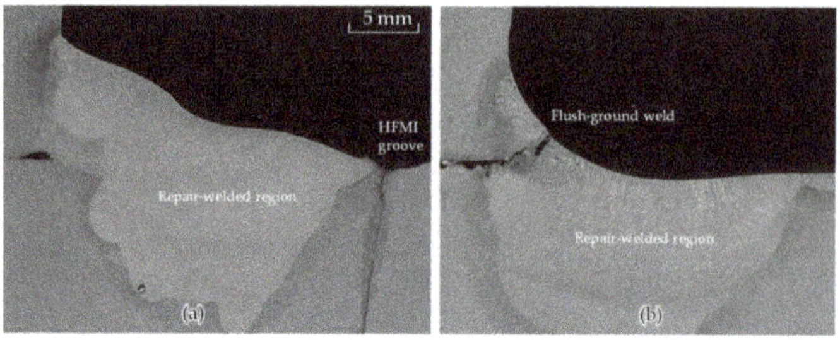

Figure 11. Macro graphs on the failures in the repair-welded beam specimens: (**a**) failure from the HFMI groove, and (**b**) failure from the weld root at the side of flush-ground toe (the cope hole side of the gusset).

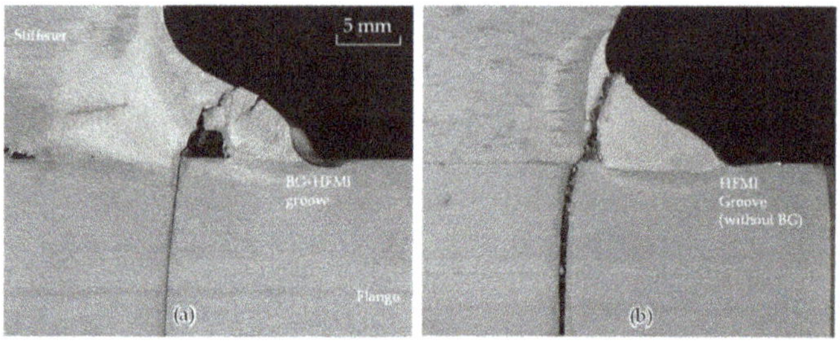

Figure 12. Macro graphs on the root failures in the successful HFMI treatments: (**a**) burr-ground and HFMI-treated specimen, and (**b**) HFMI-treated specimen (prepared using the solid wire).

3.3. Residual Stresses

Figure 13 presents the XRD measurement results on the surface residual stresses along the specimen's flange starting from the weld toe position. As expected, the welding residual stresses were highly in tension in both investigated specimen types in the AW condition–roughly 50% from the yield strength of the material. In the beam specimens, plate surface conditions potentially are a cause for the varying and highly scattering results, i.e., showing both tensile and compressive residual stress in the vicinity of the weld. However, the results provide an overview on the magnitude of the residual stresses in the specimens, confirming that the specimens represent real ship conditions with high tensile residual stresses. By applying the HFMI treatment, an introduction of compressive residual stresses was confirmed. Furthermore, a strain hardening phenomenon [25] could be identified by the measured full width at half maximum (FWHM) values, as seen in Table 5. The measurements were performed on the individual specimens of the series, and the bulkhead stiffener caused some geometrical inaccessibility to measure weld toe position (low tilting angles were applied in the XRD measurements). Due to these reasons, the results were slightly scattered, particularly in the case of the EQ47 specimens. In addition, compared to the yield strength of material, the EQ70 specimens had slightly lower residual stresses. Potentially, this can be explained by the microstructural (volume) changes at the heat-affected zone with the EQ70 steel with a higher carbon content.

Table 5. Comparison of the measured residual stresses (at toe/groove position) and FWHM values in the AW and HFMI-treated tensile specimens.

Material Grade	σ_{res} (MPa)		FWHM (°)	
	AW	HFMI	AW	HFMI
EQ47	200 ... 500	−200 ... −500	1.7 ... 2.0	3.1 ... 3.5
EQ70	70 ... 130	−150 ... −350	2.2 ... 2.5	3.5 ... 4.2

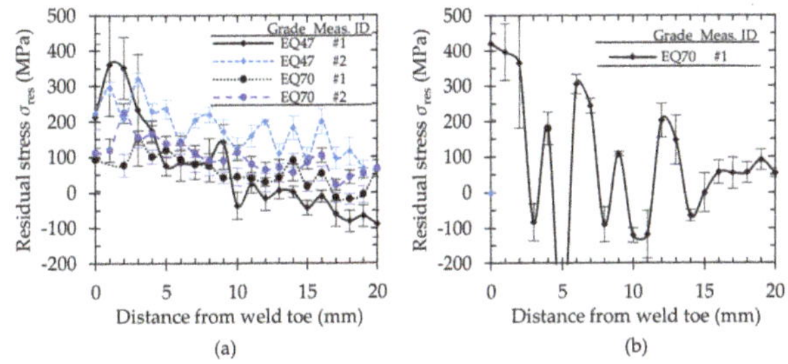

Figure 13. Residual stress measurement results in the specimens in the AW condition: (**a**) tensile specimens, and (**b**) beam specimen.

3.4. Stress Concentrations and Fatigue Assessments Using Structural and Notch Stresses

Figure 14 presents the results of FE analyses in terms of the stress concentration factors (SCFs), obtained for the structural stress and ENS systems. By using the values shown in Figure 14, the fatigue notch factors of K_f = 3.16 (=5.55/1.76) and K_f = 2.27 (=4.00/1.70) were obtained for the beam and tensile specimens, respectively. These observations showed that the specimen type influenced the resulting SCF in both structural and notch stress levels. By employing these values, Figure 15 presents the *S-N* curves plotted using the local approaches. Repair-welded specimens were excluded from this evaluation. FAT100 for the structural HS stress approach and FAT225 for the ENS system (maximum principal stress criterion with r_{ref} = 1.0 mm) are the characteristic design curves (a slope parameter of m = 3). For the mean fatigue strength of butt-welded specimens in the ENS system, Nykänen and Björk [26] obtained 309 MPa. This is in line with the recommended safety factor of j_σ = 1.37 [27] and FAT225, giving 1.37 × 225 MPa = 308 MPa. For the HFMI-treated joints, FAT200 and FAT180 (m = 5) were suggested for yield strength of 550 MPa $\leq f_y <$ 750 MPa and 550 MPa $\leq f_y <$ 750 MPa, respectively. For the ENS system, similar classes are FAT400 and FAT360 (m = 5) [24]. Compared to the nominal stress results (see Figures 6 and 8), the local approaches produced more conservative results. In addition, with the ENS system, the tensile and beam specimens seemingly fit better into a single scatter band. Compared to the mean fatigue strength obtained in the previous study [26], the results fit to these reasonably well and the ENS system produced conservative assessments.

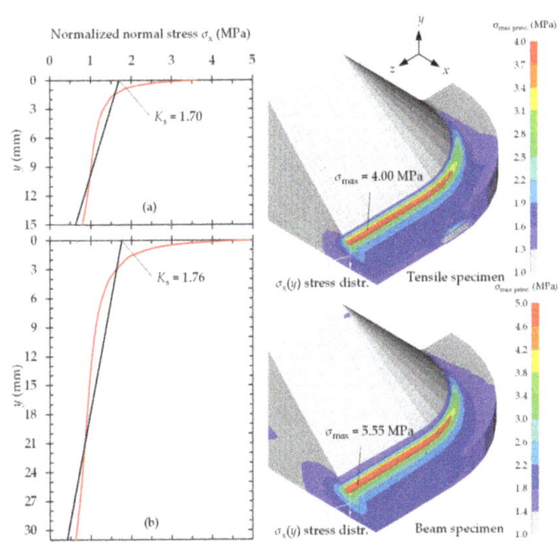

Figure 14. Results of FE analyses: stress distributions and obtained structural SCFs and ENSs at the toe positions using the maximum principal stress criterion: (**a**) the tensile and (**b**) beam specimens.

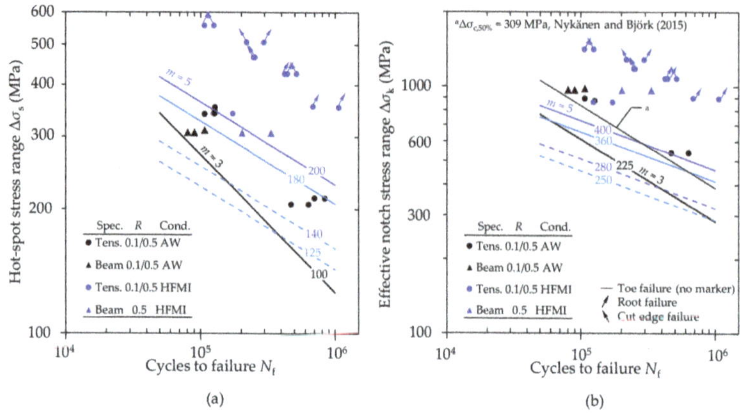

Figure 15. Fatigue test results in terms of (**a**) structural HS stress and (**b**) ENS systems [26].

4. Discussion

The present study experimentally and numerically studied the fatigue strength of the welded bulkhead stiffener connection made of high-strength and extra-high-strength steels and susceptible to fatigue failures under cyclic loads. The fatigue test results (Figures 6 and 8) showed that, in the AW condition, the fatigue strength of these details were in line with the current recommendations. However, the number of tested AW specimens were limited, and an establishment of separate *S-N* curves would require more extensive testing. The produced welding quality, albeit not showing the optimal geometrical quality in the gusset tips, seemed to have sufficient fatigue strength. The beam specimens had slightly lower fatigue strength in the nominal stress system. The tensile specimens had lower plate thickness and shorter longitudinal stiffener than the beam specimens. Furthermore, the measured tensile residual stresses were slightly higher in the beam specimens than in the tensile specimens (Figure 13). However, the results showed the *R* ratio did not have a significant influence on the fatigue strength and, thus, the differences in the residual stresses cannot fully explain these differences. This is in line with the recommended practice [22] and experimental observations on the *R* ratio effects under different residual stress states [8]. In the beam specimens, the tests were interrupted after observing the crack, and thus, the obtained cycles to failure did not correspond to the total life, i.e., crack propagation life was not fully considered.

The initially fatigue-tested HFMI-treated specimens suffered from the sharp transition with low notch opening angle, and HFMI tool pin was inaccessible to treat the weld toe. The depth of the HFMI groove was acceptable and groove visually showed acceptable conditions; however, the pin did not reach the original toe position, as demonstrated by the crack paths. Based on the fatigue test observations on the HFMI-treated joints manufactured with the solid wire (Figures 6 and 7), the unmanufacturable HFMI treatment was related to the applied rutile-cored wires. These wires are specifically intended for all welding positions with good productivity, which is meaningful in shipyard conditions. However, from the HFMI treatment viewpoint, weld toe grinding, or other technique would be needed to confirm successful HFMI treatments. This was also outlined in the IIW Recommendations for the HFMI treatment [24]. In burr-ground and, subsequently, HFMI-treated specimens, two fatigue failure modes were observed: fatigue cracks originated from both weld root and cut edges. The change between these two failure modes is expectedly related to the weld root quality in the gusset tips. In the case of lack of fusion in the frontal surface of stiffener, this initial defect, transverse to the loading direction, acts as a crack initiation/propagation location. If the successful fusion of the frontal surface is reached, a competing crack initiation from the cut edge is present.

The results of the successfully HFMI-treated joints showed conservative results with respect to the design curves (FAT140 with $m = 5$), with the characteristic fatigue strength of $\Delta\sigma_{c,97.7\%} = 163$ MPa, which indicates that HFMI treatment is a viable option to enhance fatigue strength of welded ship details made of (extra-)high-strength steels. The design recommendation for this detail category is FAT140 and, in the light of these results, one FAT class higher (FAT160) could be even suggested. In general, these findings are in line with the recent works conducted on the statistical re-evaluations of HFMI-treated joints made of high-strength steels [28]. The main focus of the HFMI experiments was on the EQ70 steel and further investigation would be needed for the 500 MPa class grades.

In the beam specimens, repair welding was applied after pre-fatiguing specimens resulting in an observable fatigue crack. After repair welding, the weld toes were post-weld-treated with the HFMI treatment and weld profiling (grinding to flush). With these procedures, the failures from the weld toe positions were avoided. The re-tested beam specimens failed from the weld root, instead, similar to some of those HFMI-treated tensile specimens. In fact, the fatigue strength capacity of these joints in the HFMI-treated condition is limited by the weld root fatigue capacity. To avoid such undesirable root failures, fully penetrating welds at the gusset tips should be used. From shipyard viewpoints, this would significantly increase workload related to the groove preparation. On the other hand, as failures were observed from the cut edges of flanges, it is evident that PWTs can enhance the fatigue strength of extra-high-strength steel weldments to the level of thermally cut edges, and thus, the full potential regarding the fatigue capacity of these steels can be claimed. The fatigue test results of those specimens failing from the cut edges were in line with the suggested fatigue class of FAT150 ($m = 3$) for thermally cut edges [29].

The structural SCFs were 1.70 and 1.76 for the tensile and beam specimens, respectively (Figure 14). Consequently, the structural HS stress method gave more conservative results than the nominal stress approach since SCFs were higher than the ratios FAT100/FAT71 (=1.41) and FAT100/FAT63 (=1.59). Moreover, the fatigue notch factors were different for the studied geometries, and an application of ENS commeasured the data points into a same scatter band (Figure 15b). Since the weld geometry was similar in both tensile and beam specimens, the changes in the fatigue notch factors (K_f) can be explained by the thickness effect as well as, partially, by the stress gradient and increased share of bending stress in the nominally applied loading, which was identified as a key factor influencing notch SCFs in [30,31].

5. Conclusions

Based on the experimental and numerical work carried out within this work, the following conclusions can be drawn:

- Fatigue strength of welded bulkhead stiffener connection prepared with the rutile-cored wire was in line with the detail category of FAT63. The sharp transitions from the base metal to the weld reinforcement were observed, but based on the experimental results, sufficient fatigue strength was found.
- Pre-conditions for the successful HFMI treatment must be confirmed, either using burr grinding or equivalent geometry modifications, or employing solid filler wires. In this study, it was specifically related to the welding quality produced by the rutile-cored wires. With the successful treatment, the results were conservative compared to the design curves, and characteristic ($P_s = 97.7\%$) design fatigue capacity was 173 MPa at two million cycles. In the majority of the HFMI-treated specimens, failures occurred at the weld root or thermally cut edges and, consequently, it can be concluded that HFMI treatment is a viable option to enhance the fatigue performance of welded ship details made of extra-high-strength steels. If HFMI processes are applied, careful attention should be paid on the weld root quality; the gusset end face should be welded without any lack of penetration.
- Repair welding can extend the fatigue life of structural components. In this study, the specimens experienced failures outside the repair-welded regions (mostly from the

weld root), and fatigue lives were higher than that of pre-fatigued specimens. In repair welding, it is, thus, important to consider different failure modes from the different crack initiation points.
- With the investigated joint type, a use of the local approaches (structural HS stress method and ENS concept) produced more conservative results than the nominal stress method. Differences in the specimen geometries and load configuration gave different SCFs and can, thus, explain the differences found in the fatigue capacities when using the nominal stress system.

Author Contributions: Conceptualization, A.A. and M.K.; methodology, A.A. and T.B.; software, J.L. and A.A.; validation, M.K., A.A. and K.L. (Kalle Lipiäinen); formal analysis, M.K, A.A. and K.L. (Kalle Lipiäinen); investigation, A.A., K.L. (Kalle Lipiäinen) and M.K.; resources, M.K., A.A. and T.B.; data curation, M.K.; writing—original draft preparation, A.A.; writing—review and editing, A.A., K.L. (Kalle Lipiäinen), J.L., M.K., K.L. (Kari Laukia) and T.B.; visualization, A.A., K.L. (Kalle Lipiäinen); supervision, T.B. and K.L. (Kari Laukia); project administration, T.B. and K.L. (Kari Laukia); funding acquisition, T.B. and K.L. (Kari Laukia). All authors have read and agreed to the published version of the manuscript.

Funding: This research was funded by Business Finland in the Intelligent Steel Applications (ISA) project, grant number 7386/31/2018. The research work and manuscript were finished in the Fossil-Free Steel Applications (FOSSA) project funded by Business Finland (grant number 5498/31/2021). The APC was funded by LUT University.

Data Availability Statement: The research data are available upon request to the authors.

Acknowledgments: The authors wish to thank SSAB Europe Oy for providing the EQ steel materials for the experimental testing.

Conflicts of Interest: The authors declare no conflict of interest.

References

1. Björk, T.; Samuelsson, J.; Marquis, G. The Need for a Weld Quality System for Fatigue Loaded Structures. *Weld. World* **2008**, *52*, 34–46. [CrossRef]
2. Eggert, L.; Fricke, W.; Paetzold, H. Fatigue Strength of Thin-Plated Block Joints with Typical Shipbuilding Imperfections. *Weld. World* **2012**, *56*, 119–128. [CrossRef]
3. Lillemäe, I.; Lammi, H.; Molter, L.; Remes, H. Fatigue Strength of Welded Butt Joints in Thin and Slender Specimens. *Int. J. Fatigue* **2012**, *44*, 98–106. [CrossRef]
4. Lotsberg, I. Stress Concentrations Due to Misalignment at Butt Welds in Plated Structures and at Girth Welds in Tubulars. *Int. J. Fatigue* **2009**, *31*, 1137–1345. [CrossRef]
5. Ahola, A.; Björk, T. Fatigue Strength of Misaligned Non-Load-Carrying Cruciform Joints Made of Ultra-High-Strength Steel. *J. Constr. Steel Res.* **2020**, *175*, 106334. [CrossRef]
6. Niraula, A.; Remes, H.; Lehto, P. Local Weld Geometry—Based Characterization of Fatigue Strength in Laser—MAG Hybrid Welded Joints. *Weld. World* **2023**. [CrossRef]
7. Braun, M.; Ahola, A.; Milaković, A.-S.; Ehlers, S. Comparison of Local Fatigue Assessment Methods for High-Quality Butt-Welded Joints Made of High-Strength Steel. *Forces Mech.* **2022**, *6*, 100056. [CrossRef]
8. Baumgartner, J.; Bruder, T. Influence of Weld Geometry and Residual Stresses on the Fatigue Strength of Longitudinal Stiffeners. *Weld. World* **2013**, *57*, 841–855. [CrossRef]
9. Schroepfer, D.; Kromm, A.; Kannengiesser, T. Formation of Multi-Axial Welding Stresses Due to Material Behaviour during Fabrication of High-Strength Steel Components. *Weld. World* **2019**, *63*, 43–51. [CrossRef]
10. Jonsson, B.; Samuelsson, J.; Marquis, G. Development of Weld Quality Criteria Based on Fatigue Performance. *Weld. World* **2011**, *55*, 79–88. [CrossRef]
11. Hobbacher, A.; Kassner, M. On Relation between Fatigue Properties of Welded Joints, Quality Criteria and Groups in Iso 5817. *Weld. World* **2012**, *56*, 153–169. [CrossRef]
12. Billingham, J.; Sharp, J.V.; Spurrier, J.; Kilgallon, P.J. Review of the Performance of High Strength Steels Used Offshore. In *Health & Safety Executive, Research Report 105*; HSE Books: Norwich, UK, 2003.
13. Stoschka, M.; Leitner, M.; Posch, G.; Eichlseder, W. Effect of High-Strength Filler Metals on the Fatigue Behaviour of Butt Joints. *Weld. World* **2012**, *57*, 85–96. [CrossRef]
14. Haagensen, P.J. Fatigue Strength Improvement Methods. In *Fracture and Fatigue of Welded Joints and Structures*; MacDonald, K.A., Ed.; Woodhead Publishing: Cambridge, UK, 2011; pp. 297–329.

15. Schubnell, J.; Ladendorf, P.; Sarmast, A.; Farajian, M.; Knödel, P. Fatigue Performance of High- and Low-Strength Repaired Welded Steel Joints. *Metals* **2021**, *11*, 293. [CrossRef]
16. Leitner, M.; Barsoum, Z.; Schäfers, F. Crack Propagation Analysis and Rehabilitation by HFMI of Pre-Fatigued Welded Structures. *Weld. World* **2016**, *60*, 581–592. [CrossRef]
17. Al-Karawi, H.; von Bock und Polach, R.U.F.; Al-Emrani, M. Fatigue Life Extension of Existing Welded Structures via High Frequency Mechanical Impact (HFMI) Treatment. *Eng. Struct.* **2021**, *239*, 112234. [CrossRef]
18. HiFIT. The HiFIT Process. Available online: https://www.hifit.de/en/ (accessed on 9 June 2022).
19. Niemi, E.; Fricke, W.; Maddox, S.J. *Structural Hot-Spot Stress Approach to Fatigue Analysis of Welded Components*, 2nd ed.; Springer Singapore: Singapore, 2017; ISBN 978-981-10-5567-6.
20. Fricke, W. *IIW Recommendations for the Fatigue Assessment of Welded Structures by Notch Stress Analysis*; Woodhead Publishing: Cambridge, UK, 2012.
21. *EN 1993-1-9 Eurocode 3: Design of Steel Structures—Part 1–9: Fatigue*; European Committee for Standardization: Brussels, Belgium, 2005.
22. Hobbacher, A. *Recommendations for Fatigue Design of Welded Joints and Components*, 2nd ed.; Springer International Publishing: Cham, Switzerland, 2016; ISBN 978-3-319-23757-2.
23. DNVGL-RP-C203 Fatigue Design of Offshore Steel Structures 2016. Available online: http://www.dnvgl.com (accessed on 16 February 2023).
24. Marquis, G.B.; Barsoum, Z. *IIW Recommendations for the HFMI Treatment—For Improving the Fatigue Strength of Welded Joints*; Springer Singapore: Singapore, 2016; ISBN 978-981-10-2503-7.
25. Leitner, M.; Khurshid, M.; Barsoum, Z. Stability of High Frequency Mechanical Impact (HFMI) Post-Treatment Induced Residual Stress States under Cyclic Loading of Welded Steel Joints. *Eng. Struct.* **2017**, *143*, 589–602. [CrossRef]
26. Nykänen, T.; Björk, T. Assessment of Fatigue Strength of Steel Butt-Welded Joints in as-Welded Condition—Alternative Approaches for Curve Fitting and Mean Stress Effect Analysis. *Mar. Struct.* **2015**, *44*, 288–310. [CrossRef]
27. Radaj, D.; Sonsino, C.M.; Fricke, W. *Fatigue Assessment of Welded Joints by Local Approaches*, 2nd ed.; Woodhead Publishing: Cambridge, UK, 2006; ISBN 9781855739383.
28. Brunnhofer, P.; Buzzi, C.; Pertoll, T.; Rieger, M.; Leitner, M. Fatigue Design of Mild and High-Strength Steel Cruciform Joints in as-Welded and HFMI-Treated Condition by Nominal and Effective Notch Stress Approach. *Procedia Struct. Integr.* **2022**, *38*, 477–489. [CrossRef]
29. Lipiäinen, K.; Kaijalainen, A.; Ahola, A.; Björk, T. Fatigue Strength Assessment of Cut Edges Considering Material Strength and Cutting Quality. *Int. J. Fatigue* **2021**, *149*, 106263. [CrossRef]
30. Ahola, A.; Björk, T. Bending Effects on the Fatigue Strength of Non-Load-Carrying Transverse Attachment Joints Made of Ultra-High-Strength Steel. In Proceedings of the 8th International Conference on Structural Engineering, Mechanics and Computation (SEMC 2022), Cape Town, South Africa, 5–7 September 2022; pp. 1130–1135.
31. Ahola, A. Stress Components and Local Effects in the Fatigue Strength Assessment of Fillet Weld Joints Made of Ultra-High-Strength Steels. Ph.D. Thesis, LUT University, Lappeenranta, Finland, 2020.

Journal of
Marine Science and Engineering

MDPI

Article

Fatigue Strength of AH36 Thermal Cut Steel Edges at Sub-Zero Temperatures

Marten Beiler, Jan-Hendrik Grimm *, Trong-Nghia Bui, Franz von Bock und Polach and Moritz Braun *

Institute for Ship Structural Design and Analysis, Hamburg University of Technology, 21073 Hamburg, Germany
* Correspondence: jan.grimm@tuhh.de (J.-H.G.); moritz.br@tuhh.de (M.B.); Tel.: +49-40-42878-4648 (J.-H.G.); +49-40-42878-6091 (M.B.)

Abstract: Cruise ships or yachts that operate in areas with seasonal freezing temperatures have large openings in the outer shell. Those are thermal cut edges, and they are exposed to very low temperatures. From fatigue crack growth testing of base materials, it is known that low temperatures can accelerate the crack growth, which may reduce the fatigue life of a structure. However, the current guidelines and rules of classification societies do not provide design curves for the fatigue assessment of thermal cut steel edges at sub-zero temperatures. Therefore, fatigue tests of thermal cut edges are conducted at −20 °C and −50 °C as well as at room temperatures for reference. The specimens are plasma-cut and tested in a temperature chamber under uniaxial loading with a resonant testing machine. The test results are statistically evaluated using linear regression and the maximum likelihood method. The results show that the fatigue strength at sub-zero temperatures is significantly higher compared to room temperature. The test results of this study indicate that sub-zero temperatures down to −50 °C do not seem to cause a reduced fatigue life of thermal cut steel edges.

Keywords: ductile–brittle transition; fatigue testing; high-strength steel; low temperature; temperature dependence of fatigue curves; thermal cut steel edges

Citation: Beiler, M.; Grimm, J.-H.; Bui, T.-N.; von Bock und Polach, F.; Braun, M. Fatigue Strength of AH36 Thermal Cut Steel Edges at Sub-Zero Temperatures. *J. Mar. Sci. Eng.* **2023**, *11*, 346. https://doi.org/10.3390/jmse11020346

Academic Editor: Joško Parunov

Received: 25 November 2022
Revised: 13 January 2023
Accepted: 31 January 2023
Published: 4 February 2023

1. Introduction

Sub-zero temperatures influence the fatigue strength of ship structures. Apart from the root and toe failure of welds, free thermally cut steel edges are also critical details in the ship structure for crack initiation under cyclic loading. Especially the ship hull of modern cruise ships and large yachts, which hold a significant number of openings for windows and balcony doors (Figure 1). When these kinds of ships operate in polar and subpolar conditions, those cut outs may be exposed to very low temperatures, which may lead to a change in their fatigue behavior. Low temperatures refer to a combination of air temperature and the wind-chill effect (e.g., the temperature–humidity–wind index, THW index), which may lower the temperature by 10 °C or even 20 °C.

The fatigue performance of free plate edges [1–3] is researched for normal temperatures and also the effect of high temperatures is considered in the standards for the fatigue assessment of steel structures. Accordingly, current standards such as the recommendations of the International Institute of Welding (IIW) [4] provide a factor for fatigue assessment at high temperatures, which depends on the modulus of elasticity; however, studies on the impact of sub-zero temperatures are scarce.

Experiments of crack growth in the base material revealed that the fatigue of ferritic structural steels with a body-centered cubic crystal structure undergoes a ductile to brittle transition. This is similar to of the behavior of the fractur toughness at low temperatures. This transition is called the fatigue ductile–brittle transition (FDBT) [5]. This point of transition is defined by the fatigue transition temperature (FTT) below, where the crack growth is accelerated. In the temperature range below room temperature (RT) but above the FTT, the crack growth rate decreases. In different studies, it is found that the FTT is close to the temperature of the ductile to brittle transition of the fracture toughness. Furthermore, the FTT is generally located below the fracture–appearance transition temperature but above the T_{27J} temperature, at which the specimen yields an energy of 27 J in the Charpy test [6]. In the study by Walters et al. [5], the authors confirm a relationship between the T_{27J} and FTT in their tests of the S460 base material, which was originally proposed by Wallin et al. [7]. Dzioba and Lipiec [8] investigated the change in the fracture surface with decreasing temperatures for plain steel SENB specimens of ferritic S355 steel in uniaxial tensile tests. They found that at −50 °C the fracture surface shows only ductile fractures, but at −80 °C, the fracture surface featured a mixed type of fractures, ductile fractures, and brittle fractures. At the lowest test temperature of −120 °C the total fracture surface was dominated by a brittle fracture. This indicates that the temperature range for a change from ductile to brittle fracture lies approximately between −50 °C and −100 °C for ferritic S355 steel.

Figure 1. Stress concentration in a thermal cut radius of a typical opening in a ship structure.

Thermal cut edges represent an additional weakness, such as welds compared to the base material. For different welds, a significant increase in fatigue strength of more than 20% is found for temperatures down to −50 °C in comparison to room temperature [9,10].

This indicates that some standards are very conservative and that welded joints can be safely applied to sub-zero temperatures down to −50 °C, if the fracture toughness requirements are met.

Compared to welds, the surface defect size or notch size of the thermal cut edges are reduced, and their failure mode is governed by crack initiation instead of crack growth [11]. Due to this reason, the surfaces and qualities of cut edges have to be considered in the assessment [3]. In order to bridge the knowledge gap concerning the impact of sub-zero temperatures on fatigue performance of critical details in ship structures, free thermal cut steel edges are tested in this study. The focus is on sub-zero temperatures close to the fatigue ductile to brittle transition temperature. For this purpose, plasma-cut specimens are tested at room temperature (RT), −20 °C, and −50 °C. The test results are statistically evaluated to find the best fit S-N (stress–lifecycles) curves for the results of each test temperature.

2. Materials and Methods

To analyze the fatigue performance of thermal cut steel edges at sub-zero temperatures, a specimen design for all test series at the different temperatures is considered. All specimens are made of higher strength structural steel S355 (AH36) and are cut out of one plate via plasma cutting. The chemical composition of the used steel and the mechanical properties are listed in Table 1. The plasma cutting was performed in a shipyard and corresponds to typical ship building practice. The specimens are 500 mm long, 80 mm wide at both ends, and have a thickness of 12 mm. The specimens are hourglass shaped and have a reduced width of 30 mm by a radius of 500 mm in the center (see Figure 2).

Table 1. Chemical composition of alloying elements in % and observed mechanical properties of AH36 steel [12].

C	Si	Mn	P	S	Al	N
0.09	0.25	1.16	0.017	0.006	0.028	0.007
Cr	**Cu**	**Ni**	**Ti**	**V**	**Nb**	**Mo**
0.09	0.22	0.12	0.001	0.001	0.024	0.001
CEV	**Tensile Strength R_m (MPa)**	**Yield Strength R_{eh} (MPa)**	**Failure Strain (%)**			
0.32	538	423	29.2			

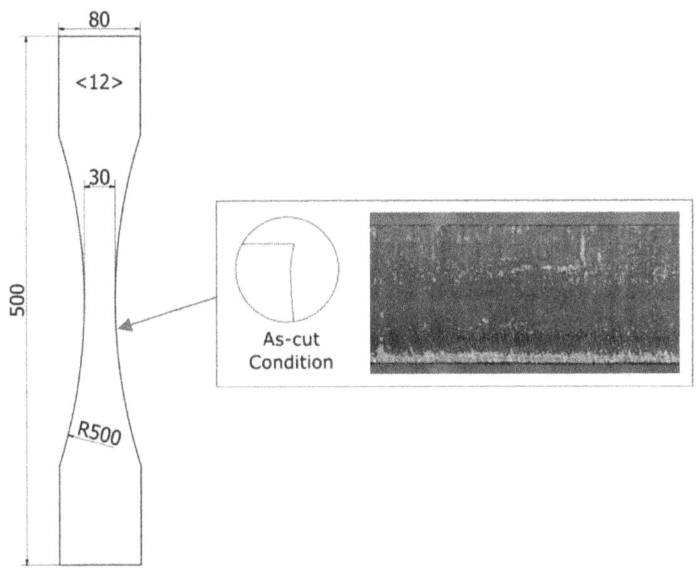

Figure 2. Specimen dimension and cut condition, modified from [12].

Before testing, the specimen specifics are recorded. For this, the geometry of the smallest cross section in the center is measured as well as the angular misalignment. Following the specifications of ISO 9013 [13], the specimens show a good surface quality (quality range 2). In case of plasma cutting, surfaces with low roughness values are produced, so that the roughness is in generally low, and the fatigue performance depends more on the cut edge defects [2,3]. The cut edges were not treated after cutting. Hence, the fatigue tests were carried out in as-cut condition. Figure 3 shows the microstructure of the polished inclined cross section of one of the specimens in the heat-affected zone.

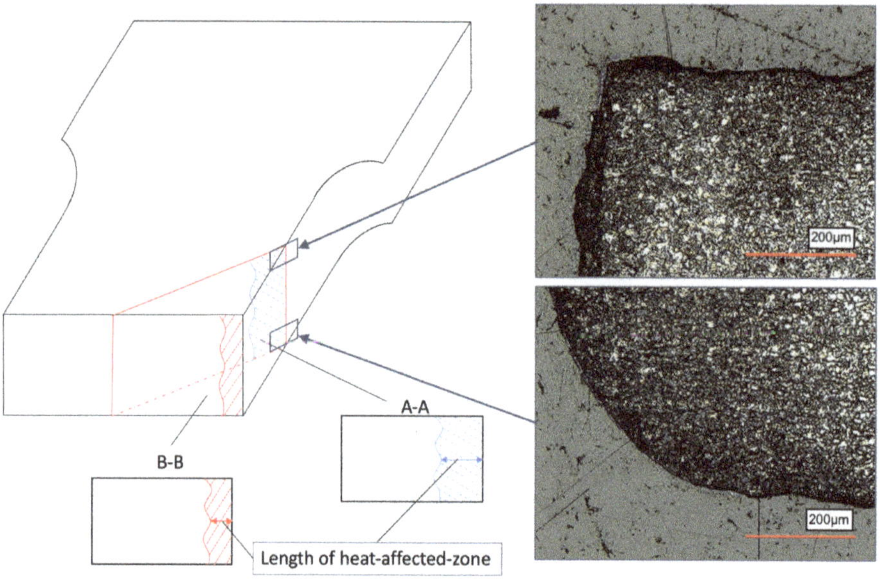

Figure 3. Microstructure of polished inclined cross section. Top side edge with sharp corner (flame entry) and more round edge on bottom side.

The fatigue testing of the specimens was carried out on a Schenk horizontal resonance testing machine. All tests were conducted with a stress ratio (ratio between the minimum stress value and the maximum stress value) of $R = 0.1$ and the failure criterion was a clearly visible crack. Experimental tests were conducted under uniaxial load in longitudinal direction of the test bodies. During testing, cooling was achieved by vaporized nitrogen, which was blown into a temperature chamber. For an almost constant temperature of $-20\ ^\circ\text{C} \pm 1\ ^\circ\text{C}$ and $-50\ ^\circ\text{C} \pm 1\ ^\circ\text{C}$ during the tests, the amount of nitrogen was controlled by a temperature gauge [9]. Additionally, the temperature of both the specimen and the air in the chamber was measured. An illustration of the clamping jaws of the resonance testing machine with the temperature chamber is shown in Figure 4.

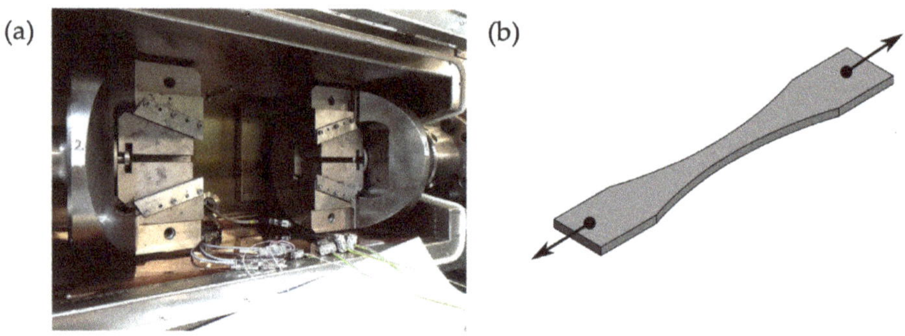

Figure 4. Temperature chamber that covers the clamping jaws of the horizontal testing machine (**a**) and uniaxial loaded specimen (**b**).

Each specimen was clamped at both ends of the test body in the jaws of the horizontal testing machine. Due to a limited number of specimens for each temperature, pearl-string method was applied to cover a wide range of stress ranges. At least ten specimens were tested at each temperature ensuring a statistical significance of the data recorded. Stress ranges refer to the nominal stress in the specimens' center cross section.

3. Results

3.1. Statistical Evaluation of Test Results

The fatigue test results are statistically evaluated with two different methods to obtain the mean stress–life (S-N) curve. On the one hand, a linear regression is used to evaluate the S-N curve minimizing the residual sum of squares. In addition to this method, the evaluation is also carried out with the maximum likelihood method. In contrast to a linear regression, the maximum likelihood method takes the runouts into account. Using the maximum likelihood method, a bilinear S-N curve with the highest probability of occurrence is fitted to the data [14].

The method of linear regression accounts for all test data before the knee point. Because of that, the knee point is specified according to the recommendations of the German Welding Society, DVS [14]. The high cycle fatigue regime before the knee point of the S-N curves is represented by the Basquin equation (Equation (1)), which is logarithmized for the common representation of the S-N curves:

$$\lg N = \lg N_k \cdot \Delta\sigma_k{}^m + m \cdot \lg \Delta\sigma \qquad (1)$$

With N as the number of cycles, N_k as the number of cycles at the knee point, m as the slope, $\Delta\sigma$ as the stress range, and $\Delta\sigma_k$ as the stress range at the knee point. By using linear regression [14], the slope in the high cycle fatigue regime was calculated as well as the stress range at $2 \cdot 10^6$ cycles for 50% and 97.7% survival probability. This number of cycles reflects the FAT-class values (see, e.g., IIW [4]). Moreover, the band width of the scatter between 97.7% and 2.3% is determined. Within this method, it is assumed that the scatter is represented by the logarithmic normal distribution and is independent of the load level.

In case a specimen did not fail before $2 \cdot 10^6$ cycles were reached, those runouts are marked with an arrow in the S-N diagrams. The results for all three test temperatures are shown in the following Figure 5a–c.

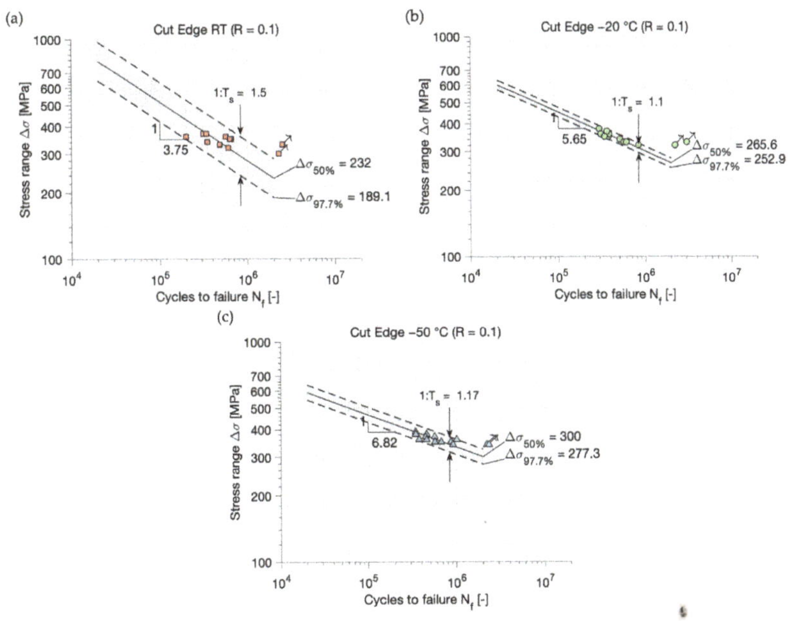

Figure 5. S-N curves of AH36 cut edges from linear regression for RT (**a**), −20 °C (**b**) and −50 °C (**c**).

When comparing the S-N curves for different temperatures, it can be observed that the fatigue strength increases with decreasing temperature. In addition, the scatter is much higher at RT compared to the sub-zero temperatures. As the slope for RT differs significantly, a linear regression does not seem to be well-suited for data with a high scatter when only a small number of specimens is tested.

Because of that, and the fact that linear regression does not take the runouts into account, the maximum likelihood method was applied to the test results. This method is also recommended by the DVS [14]. With this method, the runouts can be integrated into the evaluation, which can lead to better suited S-N curves for the given data, especially when the number of specimens is small. The method is about to find the bilinear S-N curve with the highest probability of occurrence from the data. As it is shown by Störzel et al. [15], the maximum likelihood method for fatigue test evaluation leads to reliable parameters of the S-N curve. The S-N curve in the maximum likelihood method is represented by a bilinear curve, where both parts of the curve are represented by the Basquin equation (Equation (1)). The curve parameters for the bilinear S-N curve are:

- Slope in the high cycle fatigue regime m_1;
- Stress range at the knee point S_k;
- Standard deviation stdv, as a measure of the scatter in the direction of the stress range;
- Cycles to failure at the knee point N_k;
- Slope in the long-life fatigue regime m_2.

The knee point and the slope in the long-life fatigue regime are specified for the calculation and the other three parameters are determined by optimization. The optimization is about to find the highest probability of occurrence for the test points I (S_i, N_i). [15]

$$P_{tot,max} = \max(\prod_{i=1}^{n} P_i) \tag{2}$$

In Equation (2), P_i represents the probability of a test result that failed or is a runout. As in this method, the scatter is described by the logarithmic normal distribution, and the probability for a failure and a runout are shown in Equations (3) and (4).

$$P_{i=1,fail} = \frac{1}{stdv \cdot \sqrt{2 \cdot \pi}} \cdot e^{-0.5 \cdot \left(\frac{\lg(y_i) - \log(y(x_i))}{stdv}\right)^2} dy \tag{3}$$

$$P_{i=2,runout} = \frac{1}{\sqrt{2 \cdot \pi}} \int_{t_i}^{\infty} e^{-0.5 \cdot t^2} dt \tag{4}$$

$$with:\ t_i = \frac{(\lg(y_i) - \lg(y(x_i)))}{stdv} \tag{5}$$

It is further assumed that the scatter is independent of the cycles to failure and that the width of the scatter band is constant at the entire S-N curve. Furthermore, it is a prerequisite for the application of the maximum likelihood method that lines of the same probability meet in the knee point.

In order to find the highest probability of occurrence for the parameter of the S-N curve its natural logarithm is used in the optimization in accordance with [15]. These functions are then called support functions (sup). However, not only the maximum of one support function leads to the best fit S-N curve. The data can be distinguished into three different groups, which depends on the relative position to the knee point:

1. Specimens that failed before the knee point ($N_i < N_k$);
2. Specimens that failed behind the knee point ($N_i > N_k$);
3. Specimens that have not failed (runouts).

For all of the three groups, the support function is different. Because of this, the maximization of the probability of occurrence has to take all support functions into account to find the best fit S-N curve, which is represented by Equation 6.

$$\ln P_{tot,\max} = \max(sup_1 + sup_2 + sup_B) \tag{6}$$

The support functions evaluate the distance between the test point and the S-N curve from the regression line in the direction of the stress range. As the slope of the S-N curve is different before and behind the knee point, this has to be considered for the support functions. Then, the equation for the support functions that are optimized have the following form [14].

$$For\ N_i < N_k \quad sup_1 = \sum\nolimits_{m_1} \left(\frac{(-0.5 \cdot (\lg(y_i) - \lg(y(x_i)))^2}{stdv^2} - \ln(m_1 \cdot stdv) \right) \tag{7}$$

$$For\ N_i > N_k \quad sup_2 = \sum\nolimits_{m_2} \left(\frac{(-0.5 \cdot (\lg(y_i) - \lg(y(x_i)))^2}{stdv^2} - \ln(m_2 \cdot stdv) \right) \tag{8}$$

$$For\ runouts \quad sup_B = \sum\nolimits_{m_B} \ln \frac{1}{\sqrt{2 \cdot \pi}} \cdot \int_{t_i}^{\infty} e^{-0.5 \cdot t^2} dx \tag{9}$$

The parameters that are optimized are $stdv$, m_1, and S_k. This is conducted by variating these parameters and then calculating the maximum of the support functions according to Equation 6. Due to discontinuity at each test point, the knee point cannot be adopted together with the other parameters [15]. The only way to do this is to optimize the knee point before optimizing the other parameters through finding the knee point with the highest probability of occurrence.

The knee point is determined by variating N_k from $0.5 \cdot 10^6$ up to $2 \cdot 10^6$ in steps of $0.1 \cdot 10^6$. The knee point is selected for which the test results have the greatest probability of occurrence. The fixed slope in the high cycle fatigue regime was set to m_2 = 22, according to the recommendation by DVS [14]. Besides the scatter index between a 97.7% and 2.3% survival probability, the resulting FAT-class value, the stress range for a 50% survival probability at the knee point, and the slope m_1 are calculated. The results of the statistical evaluation with the maximum likelihood method are shown in Figure 6a–c.

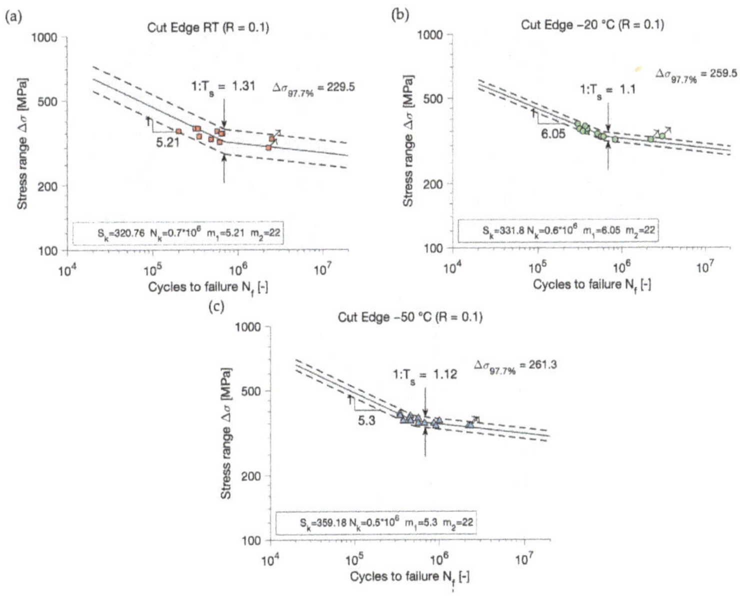

Figure 6. S-N curves of AH36 cut edges from maximum likelihood method for RT (**a**), −20 °C (**b**) and −50 °C (**c**).

The evaluation with the maximum likelihood method confirms the increase in fatigue strength with decreasing temperatures, but the change in the slope is not as significant as in the evaluation with linear regression. Comparing the FAT-class values of the S-N curves, a clear difference was observed between the RT and subzero temperatures. However, the difference in the FAT-class values between −20 °C and −50 °C has become very small using the maximum likelihood method. By comparing the knee point stress range S_k, a difference between the test results at sub-zero temperatures is observed. Moreover, only very small changes in the slope and also in the band width of the scatter can be observed for sub-zero temperatures, in comparison to the results with the method of linear regression. This was expected because the failure mechanism is the same as before. In comparison, the slope at RT is less steep and the band width of the scatter is smaller when using the maximum likelihood method.

3.2. Fracture Surface

As von Bock und Polach et al. (2022) [16] showed, the position of the crack initiation can have an influence on the fatigue strength of thermal cut edges. The crack initiation at untreated thermal cut edges starts, in most cases, at the corner of the cut edge [17]. A distinction is made between cracks that appear in the grooves on the thermal cut steel surface and cracks that appear in the corner of the cut edges, in case of post-treated specimens in the chamfer [1]. The crack growth can be traced by looking at the fracture surface. Thereby, it is found that all fracture surfaces (independent of the test temperature) show the same position for crack initiation. Figure 7 shows the fracture surface of specimens, which are tested at RT (a), −20 °C (b), and −50 °C (c). The crack initiation starts at the topside corner of the cut edges and propagates to the middle of the fracture surface. When the crack length reaches a certain size, the stiffness of the specimen decreases, and the longitudinal displacement becomes larger due to the constant load amplitude. At pre-set threshold values for displacement and frequency change, the resonance testing machine stops the loading procedure.

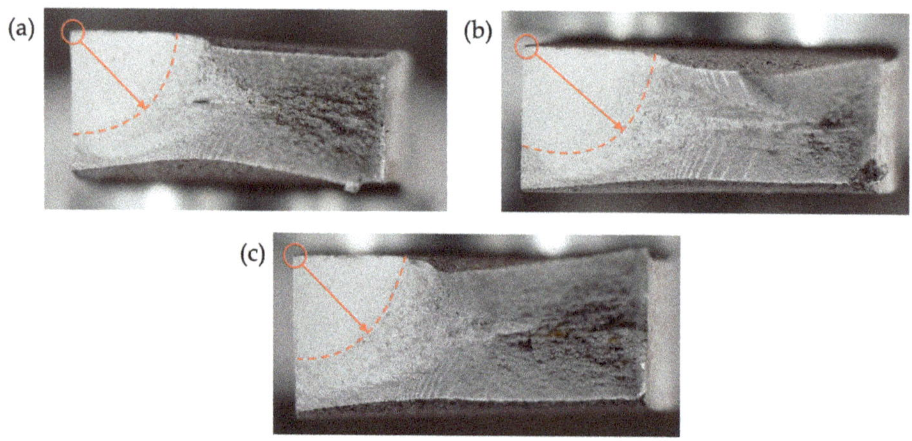

Figure 7. Fracture surface of AH36 cut edges tested at temperatures of RT (**a**), −20 °C (**b**) and −50 °C (**c**), with location of crack initiation and direction of crack propagation.

3.3. Influence of the Crack Position.

Corners of the thermal cut steel surface led to local stress concentrations. As those stress concentrations can reduce the lifetime of a specimen significantly, a failure does not only occur at the midsection of a specimen, where the nominal stress is the highest [1]. A specimen where the crack did not occur at the smallest cross section is shown in Figure 8.

The crack positions are measured and the nominal stress according to the local cross section is corrected. This procedure can lead to a sharp drop in the stress range when the crack location is relatively far away from the smallest cross section. The total stress at the crack location is a combination of the nominal stress and the stress concentration and must be higher than the total stress at the smallest cross section. Hence, it can be discussed if the nominal stress at the smallest cross section might not be more representative than the nominal stress at the actual crack location. However, for this study, a statistical evaluation is also carried out for the recalculated test series, where the nominal stresses at the actual crack location are used. The statistical evaluation can become a problem in case single values deviate too much from the others. This can be due to a difference in shape but also to changes in the test conditions. Such outliers lead to a very wide scatter band, which represents the test results incorrectly [18]. To deal with such outliers a graphical analysis is used as there are no methods for a mathematical evaluation [18]. In the test results at −50 °C, one outlier is found. In this case, the crack originates 48 mm out of the middle of the specimen. Plotting the modified stress range of this test next to the S-N curve of the remaining 13 data points, it is seen that the value does not fit in the scatter band. This is shown in Figure 9. Thereby, the statistical evaluation was conducted with the maximum likelihood method with an optimization of the knee point. In case of the tests at −20 °C and RT, no outliers were observed.

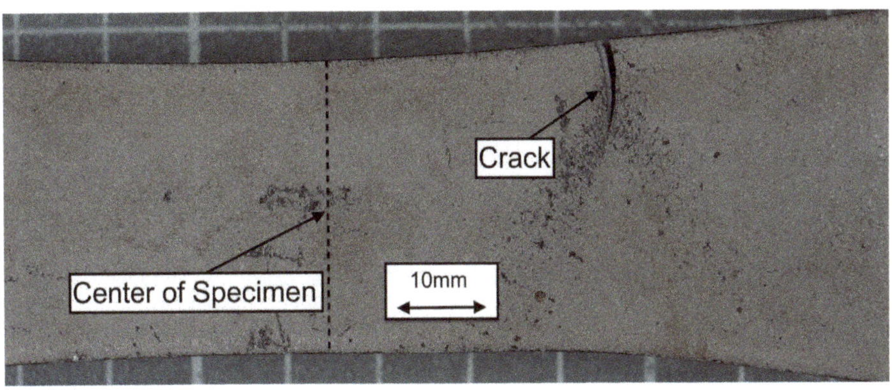

Figure 8. Specimen where the crack position is not at the smallest cross section.

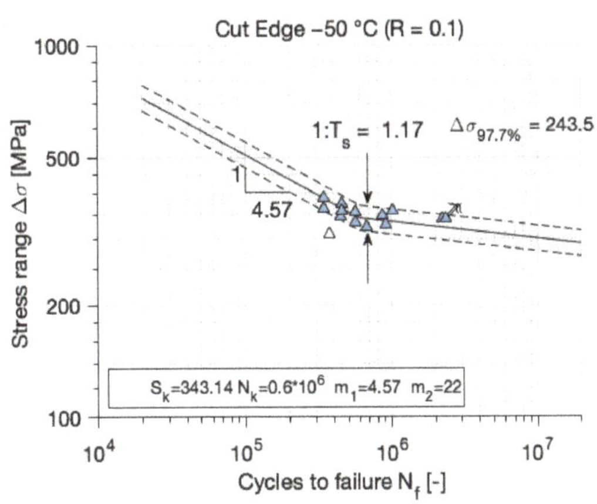

Figure 9. S-N curve of AH36 cut edges for test results at −50 °C with modified stress ranges evaluated with the maximum likelihood method. The white triangle shows the outliner.

In all three of the modified S-N curves, a decrease in the slope exponent is observed. Furthermore, the knee point stress range for a 50% survival probability is lower than in the evaluation with the initial test data. This was expected because the stress ranges are more or less reduced for every data point. However, in total, the difference in the fatigue strength between the sub-zero temperatures remains almost the same. In addition to the evaluation with the maximum likelihood method, a linear regression is used here as well. Thereby, it is observed that the slope differs even more than the results of the initial test data. For example, the slope of the results of the test series at −50 °C has decreased to $m = 3.84$, which is close to the recommended slopes in the guidelines of DNV ($m = 4$) [19]. All S-N curve parameters of all test series for both statistical methods are summarized in Table 2.

Table 2. S-N curve parameter evaluated with linear regression and with maximum likelihood method for all test series. "mod." refers to the actual crack position, i.e., the modified stress.

Statistical Method	Linear Regression			Maximum Likelihood Method					
Parameter	m	$1{:}T_s$	FAT (MPa)	m_1	m_2	$1{:}T_s$	N_k	S_k (MPa)	FAT (MPa)
RT	3.75	1.5	189.1	5.21	22	1.31	$0.7 \cdot 10^6$	320.76	229.5
−20 °C	5.65	1.1	252.9	6.05	22	1.1	$0.6 \cdot 10^6$	331.8	259.5
−50 °C	6.82	1.17	277.3	5.3	22	1.12	$0.5 \cdot 10^6$	359.18	261.3
RT mod.	3.57	1.51	179.3	4.8	22	1.33	$0.7 \cdot 10^6$	310.11	216.5
−20 °C mod.	5.12	1.12	237.9	5.26	22	1.13	$0.6 \cdot 10^6$	323.92	242.2

3.4. Linear Regression with Constant Slope

In order to obtain the fatigue performance of thermal cut edge steel at sub-zero temperatures, the best fit S-N curves are evaluated using both the previously described statistical methods. This procedure leads to different slopes for the different test series, which makes the comparison of values complicated, because the difference between the test series depends on the cycles to failure. In order to obtain information about the difference in the stress range, a linear regression is conducted with fixed slopes. Thus, the slopes are set to $m = 3$, as recommended by the IIW [4], and also $m = 4$ and $m = 3.5$, as in the guidelines of DNV [19]. Then, the FAT-class values of each test series are calculated with fixed slopes. The results of the calculation and the corresponding FAT-class values from the guidelines for similar cases are shown in Table 3.

Table 3. FAT-class values for test series RT, −20 °C, and −50 °C, and for test series with modified stress range values with different fixed slopes and FAT-class values from guidelines [4,19].

Test Series	RT (MPa)	−20 °C (MPa)	−50 °C (MPa)	RT Mod. (MPa)	−20 °C Mod. (MPa)	−50 °C Mod. (MPa)	Guideline (IIW Case No. 122, DNV Case B2 and C)
$m = 3$	162	181.9	190.8	158.7	178.5	185.2	125 (IIW)
$m = 3.5$	181	202.0	210.4	177.0	198.0	203.4	125 (DNV)
$m = 4$	196.5	218.2	226.4	191.9	213.4	217.9	140 (DNV)

The increase in fatigue strength is represented by the increase in FAT-class values. In Figure 10, the percentage rises to the FAT-class value at RT for the different slopes are shown. Thereby, the increase is the greatest for a fixed slope of $m = 3$ and smallest for $m = 4$. However, for all used slopes, a significant increase is observed.

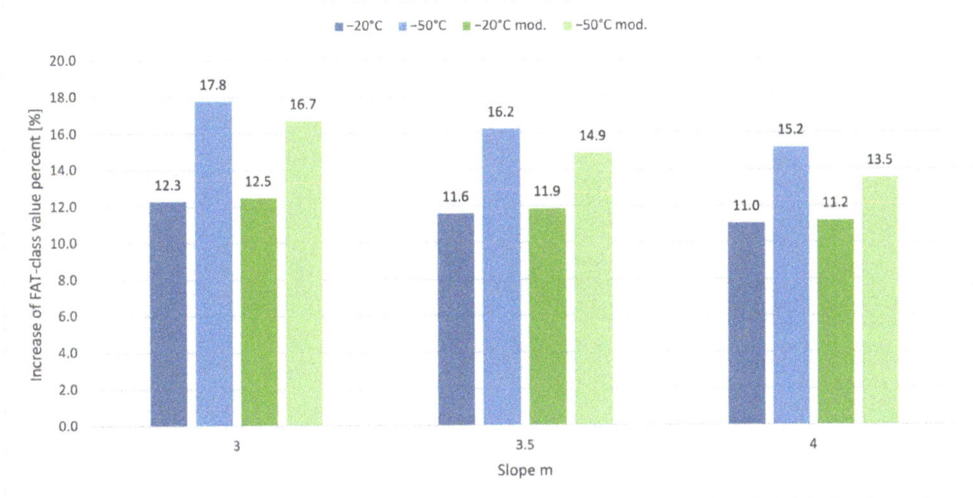

Figure 10. Increase in the FAT-class value in percent of tested AH36 cut edges at sub-zero temperatures compared to RT and of modified test series compared to RT modified. In the modified test series, the nominal stresses in the cross sections at the actual crack locations are used instead of the nominal stresses at the smallest cross sections.

4. Discussion

Fatigue tests of thermal cut steel edges at sub-zero temperatures are conducted and statistically evaluated. Independent to the statistical method, an increase in the fatigue strength of plasma-cut specimens made of AH36 at sub-zero temperatures compared to RT is observed. Despite this, the results evaluated with the linear regression method differ from those evaluated with the maximum likelihood method. The slope of the RT test series evaluated with the method of linear regression is much larger than the slope of the test results from the sub-zero temperatures. The evaluation with the maximum likelihood method shows a smaller slope at RT, which is closer to those at sub-zero temperatures. This is probably caused by the optimized knee point position and the fact that the maximum likelihood method takes the runouts into account. In addition, the number of tested specimens is very limited.

The difference in the statistical methods can also be observed by looking at the band width of the scatter. The tests at RT that were assessed with the linear regression method led to a much wider scatter band than for the maximum likelihood method. This might be caused by the changed knee point position. In comparison to this, the results of the sub-zero tests are not as much influenced by the type of statistical method. So, the difference in the band width of the scatter is relatively small between both methods, like it was also observed in case of the slope. However, with the maximum likelihood method, the difference between the test series is smaller compared to the RT test results with the method of linear regression.

The evaluation with the maximum likelihood method shows a difference in the fatigue strength of the three test temperatures at the knee point, but for the comparison of the fatigue strength, the position of the knee point and the slopes of the curves have to be considered. This can be completed by calculating the stress range of the −20 °C test series at the number of cycles where the −50 °C test series has its knee point. The stress range of the −50 °C test series at $0.5{\cdot}10^6$ cycles to failure is then about 18 MPa greater than the stress range of the −20 °C test series, which demonstrates an increase of about 5%. A similar difference in the stress range can be observed at the knee points of the modified S-N curves between the tests at −20 °C and −50 °C. However, this kind of comparison is still full of uncertainties. The difference in the stress range between the test series depends on the cycles to failure, as the curves have different slopes. For example, the FAT-class values of the test series −20 °C and −50 °C only differ by about 2 MPa. Because of that, this way of comparison can only give information about the trend of the curves.

A better way to compare the stress ranges is to use fixed slopes for the evaluation with the linear regression method. Through this, the difference in fatigue strength can be evaluated at any position of the S-N curve. Using fixed slopes of $m = 3$, $m = 3.5$, and $m = 4$, which are in the range given by common guidelines such as IIW or DNV, an increase up to 17.8% for a fixed slope of m = 3 for the −50 °C test series without a stress range correction in comparison to the RT test series is observed. However, the slopes that are evaluated with the maximum likelihood method are less steep than $m = 3$. Because of that, a fixed slope of $m = 4$ seems to be better suited for this test series. In that case, the increase from RT to −20 °C (11%) and to −50 °C (15.2%) is smaller, but still significant. Considering the modified S-N curves, where the stress range values are related to the cross section at the crack position, the increase for the tests at −50 °C is smaller, and remains almost the same for the −20 °C test series. In that case, the comparison of the FAT-class values is especially well-suited for the test series at RT and at −50 °C with fixed slopes of $m = 3.5$ and $m = 4$, as the evaluated slopes with the linear regression method are for both between these values. Then, an increase from RT to −50 °C of about 13.5% for a slope of $m = 4$ and 14.9% for a slope of $m = 3.5$ is calculated.

The FAT-class values at RT, independent of the slope, surpass the FAT-class values of the guidelines. As the test series of sub-zero temperatures shows an increase in fatigue strength, it can be considered that the guidelines are very conservative, especially for thermal cut edges at sub-zero temperatures.

From the literature, it is known that the fatigue of ferritic structural steel undergoes a ductile to brittle transition at the FTT, which is related to a decrease in fatigue strength because the fatigue crack growth is accelerated. However, the results of the fatigue tests show that the fatigue strength of the tested thermal cut edges increased up to the lowest test temperature of −50 °C. This is probably due to the fact that the transition takes place at lower temperatures than those tested here. An indication of this could be found in the fracture surfaces of the tested specimens, so it can be observed that all tested specimens failed in the same ductile manner, independent of the test temperature.

The fatigue strength of thermal cut steel edges in the outer shell of ships, e.g., windows or balcony doors, does not solely depend on the cut edge quality and temperature. Rather, the shape of the cut outs, especially the fatigue critical opening corners, affect the fatigue strength. Thus, the increase in fatigue strength of thermal cut steel edges at sub-zero temperatures found in this study cannot be directly transferred to the cut outs. However, it can be assumed that the fatigue strength of those cut outs increased at sub-zero temperatures tested in this study, but maybe not as much as the test results suggest. More importantly, the test results of this study show that sub-zero temperatures do not pose an increased risk to thermal cut steel edges in terms of fatigue.

5. Conclusions

Thermal cut edge steel specimens of AH36 were tested at sub-zero temperatures and at room temperature. The best fit S-N curves were evaluated with the linear regression and the maximum likelihood method. Furthermore, it was found that the cracks did not always occur at the smallest cross section. Consequently, the local nominal stresses at the actual crack position were calculated and the stress ranges were corrected. With the modified data, the statistical evaluation was repeated. For the comparison of the fatigue strength of the different test series and for a comparison with the guidelines, an evaluation using the linear regression model was also performed with fixed slopes for all test series, both the modified and the not modified ones. The findings of this study are summarized in the following points:

- After analyzing the statistically evaluated best-fit S-N curves, it is clear that the fatigue strength of thermal cut steel edges of AH36 increase at sub-zero temperatures in comparison to room temperature. This is independent of the statistical method and corresponds to the experience gained with welds at sub-zero temperature [9,10].
- A fixed slope of $m = 4$ seems applicable for the evaluation with fixed slopes as the slopes that are found in the evaluation of the best fit S-N curves are close to that or greater. The modification of the stress range leads to steeper S-N curves, which are even closer to a slope of $m = 4$.
- Using fixed slopes shows the increase in fatigue strength in an even clearer manner. The highest increase in fatigue strength evaluated with a fixed slope of $m = 4$ is about 15.2% at a test temperature of −50 °C and in the case of the modified data, the increase is still about 13.5%.
- The comparison with the FAT-class values of the design curves of the common guidelines has shown that all test series surpass the FAT-class recommendations. This shows that the design curves are very conservative for thermal cut edges, especially at sub-zero temperatures.
- A difference in the crack location is not found. All cracks occur in the corner of the cut edges, independent to the test temperature.

Author Contributions: The study was conducted in accordance with the Conceptualization, methodology, software, validation, M.B. (Moritz Braun) and M.B. (Marten Beiler); formal analysis and investigation, M.B. (Marten Beiler); data curation, M.B. (Marten Beiler); writing—original draft preparation, M.B. (Marten Beiler); writing—review and editing, M.B. (Moritz Braun), F.v.B.u.P. and J.-H.G.; visualization, M.B. (Marten Beiler) and T.-N.B.; supervision, M.B. (Moritz Braun) and J.-H.G. All authors have read and agreed to the published version of the manuscript.

Funding: This research received no external funding.

Institutional Review Board Statement: Not applicable.

Informed Consent Statement: Not applicable.

Data Availability Statement: Not applicable.

Conflicts of Interest: The authors declare no conflict of interest.

Nomenclature

Symbols	**Unit**	**Description**
T_s	-	Scatter ratio in strength
m	-	Slope of S-N curve in linear regression
m_1	-	Slope before the knee point in maximum likelihood method
m_2	-	Slope behind the knee point in maximum likelihood method
N_f	Cycles	Cycles to failure
N_k	Cycles	Number of cycles at knee point
$\Delta\sigma$	MPa	Stress range
$\Delta\sigma_k$	MPa	Stress range at the knee point
R	-	Stress ratio
S_k	MPa	Stress range at knee point
$P_{tot,max}$	-	Highest probability of occurrence
P_i	-	Probability of occurrence for one test point
$P_{i=1,fail}$	-	Probability of occurrence for a failure
$P_{i=2,runout}$	-	Probability of occurrence for a runout
$stdv$	-	Standard deviation in the direction of the stress range

Abbreviations

FDBT	Fatigue ductile–brittle transition
FTT	Fatigue transition temperature
RT	Room temperature

References

1. von Bock und Polach, F.; Kahl, A.; Braun, M.; von Selle, H.; Ehlers, S. Analysis of governing parameters on the fatigue life of thermal cut edges. In Proceedings of the International Conference on Ships and Offshore Structures ICSOS, Orlando, FL, USA, 4–8 November 2019.
2. Lipiäinen, K.; Ahola, A.; Skriko, T.; Björk, T. Fatigue strength characterization of high and ultra-high-strength steel cut edges. *Eng. Struct.* **2021**, *228*, 111544. [CrossRef]
3. Diekhoff, P.; Hensel, J.; Nitschke-Pagel, T.; Dilger, K. *Fatigue Stregth of Thermal Cut Edges-Influence of ISO 9013 Quality Groups*; XIII-2687-17; International Institute of Welding: Liguria, Italy, 2017. [CrossRef]
4. Hobbacher, A.F. *Recommendations for Fatigue Design of Welded Joints and Component*, 2nd ed.; Springer International Publishing: Cham, Switzerland, 2016.
5. Walters, C.; Alvaro, A.; Johan, M. The effect of low temperatures on the fatigue crack growth of S460 structural steel. *Int. J. Fatigue* **2016**, *82*, 110–118. [CrossRef]
6. Von Bock und Polach, F.; Klein, M.; Kubiczek, J.; Kellner, L.; Braun, M.; Herrnring, H. State of the art and knowledge gaps on modelling structures in cold regions. In Proceedings of the ASME 2019 38th International Conference on Ocean, Offshore and Arctic Engineering, Glasgow, UK, 9–14 June 2019. [CrossRef]
7. Wallin, K. A Simple Theoretical Charpy-V–KIc correlation for irradiation embrittlement. *ASME-PVP* **1989**, *170*, 93–100.
8. Dzioba, I.; Lipiec, S. Fracture Mechanism of S355 Steel–Experimental Research, FEM Simulation and SEM Observation. *Materials* **2019**, *12*, 3959. [CrossRef] [PubMed]
9. Braun, M.; Scheffer, R.; Fricke, W.; Ehlers, S. Fatigue strength of fillet-welded joints at subzero temperatures. *Fatigue Fract. Eng. Mater. Struct.* **2020**, *43*, 403–416. [CrossRef]
10. Braun, M.; Milaković, A.; Ehlers, S.; Kahl, A.; Willems, T.; Seidel, M.; Fischer, C. Sub-zero temperatures fatigue strength of Butt-welded normal and high-strength steel joints for ships and offshore structures in arctic regions. In Proceedings of the ASME 2020 39th International Conference on Ocean, Offshore and Arctic Engeneering, Virtual, 3–7 August 2020. [CrossRef]
11. Sperle, J.-O. *Influence of Parent Metal Strength on the Fatigue Strength of Parent Material with Machined and Thermally Cut Edges*; IIW Document XIII-2174-07; International Institute of Welding: Paris, France, 2007.
12. Grimm, J.-H.; Castro, J.; von Selle, H.; Braun, M.; von Bock und Polach, F.; Ehlers, S.; Hensel, J.; Hesse, A.-C.; Dilger, K. The influence of edge treatment on fatigue behavior of thermal cut edges. In Proceedings of the ASME 2022 41st International Conference on Ocean, Offshore and Arctic Engineering, Hamburg, Germany, 5–11 June 2022.
13. *Deutsche Fassung EN ISO 9013:2017*; Thermisches Schneiden–Einteilung Thermischer Schnitte–Geometrische Produktspezifikation und Qualität (ISO 9013:2017). Deutsches Institut für Normung: Berlin, Germany, 2017.
14. *Merkblatt DVS 2403. Empfehlung für die Durchführung, Auswertung und Dokumentation von Schwingfestigkeitsversuchen an Schweißverbindungen Metallischer Werkstoffe*; Deutscher Verband für Schweißen und verwandte Verfahren: Düsseldorf, Germany, 2019.
15. Störzel, K.; Baumgartner, J. *Statistical Evaluation of Fatigue Tests Using Maximum Likelihood*; Walter de Gruyter GmbH: Berlin, Germany; Boston, MA, USA, 2021. [CrossRef]
16. von Bock und Polach, F.; Kahl, A.; Braun, M.; Sperle, J.-O.; von Selle, H.; Ehlers, S. Analysis of the scatter in fatigue life testing of thick thermal cut plate edges. *Ships Offshore Struct.* **2023**, *18*, 217–230. [CrossRef]
17. Diekhoff, P.; Hensel, J.; Nitschke-Pagel, T.; Dilger, K. *Investigation on Fatigue Strength of Cut Edges of High Strength Steels (S355M, S690Q)*; IIW Annual Assembly, Document XIII—2732-18; University of Braunschweig: Braunschweig, Germany, 2018.
18. Haibach, E. *Betriebsfestigkeit, Verfahren und Daten zur Bauteilberechnung*; Springer: Berlin/Heidelberg, Germany, 2006.
19. DNVGL. *Fatigue Assessment of Ship Structures*; DNVGL-CG-0129; Class Guideline, Edition October 2015; DNV GL AS: Oslo, Norway, 2015.

Journal of
Marine Science and Engineering

MDPI

Article

Fatigue Strength of Structural Steel-Welded Connections with Arc-Sprayed Aluminum Coatings and Corrosion Behavior of the Corresponding Coatings in Sea Water

Andreas Gericke [1,*], Michél Hauer [1], Benjamin Ripsch [1], Michael Irmer [2], Jonas Nehlsen [2] and Knuth-Michael Henkel [3]

[1] Thermal Joining Engineering, Fraunhofer Institute for Large Structures in Production Engineering IGP, 18059 Rostock, Germany
[2] Coating, Weathering and Corrosion Protection, Fraunhofer Institute for Large Structures in Production Engineering IGP, 18059 Rostock, Germany
[3] Faculty of Mechanical Engineering and Marine Technologies, Joining Technology, University of Rostock, 18059 Rostock, Germany
* Correspondence: andreas.gericke@igp.fraunhofer.de; Tel.: +49-381-49682-37

Abstract: The influence of thermally sprayed aluminum coatings (Al99%; arc spraying) on the fatigue strength of gas metal arc welded (GMAW) non-alloyed structural steel specimens with respect to foundations for offshore wind turbines was investigated. Additionally, the corrosion protection effect of such coatings for water conditions similar to the Baltic Sea was determined. Wöhler tests were carried out on test specimens with different weld details in the as-welded condition as well as in the thermal spray coat under the consideration of different kinds of surface preparation (blast cleaning with corundum and grit). Substrate and coating were characterized by scanning electron microscopy and the influence on the residual stress states was determined. Corrosion rate monitoring via LPR measurements was carried out as well as the monitoring of the galvanic current between coated and uncoated steel to characterize the coatings' sacrificial capability for minor defects. Fatigue strength was significantly increased through thermal spraying, especially for test specimens with welded transverse stiffeners ($\Delta\sigma_{c,var}$ = 127 MPa after coating compared to $\Delta\sigma_{c,var}$ = 89 MPa as welded). With a characteristic value of the stress range of $\Delta\sigma_{c,var}$ = 153 MPa, the welded butt joint specimens already exhibited a high fatigue strength in the as-welded condition. The corrosion studies demonstrated that thermally sprayed Al99% coatings have a high resistance to corrosion in seawater environments and are suitable as planar sacrificial anodes sufficiently polarizing bare steel below 0.8 V. The combination of fatigue strength improvement and corrosion protection makes the thermally sprayed Al coatings promising for design and operation of e.g., offshore structures.

Keywords: fatigue; corrosion; welding; thermal spraying; Al99; offshore; structural steel; blast cleaning; LPR; GMAW

Citation: Gericke, A.; Hauer, M.; Ripsch, B.; Irmer, M.; Nehlsen, J.; Henkel, K.-M. Fatigue Strength of Structural Steel-Welded Connections with Arc-Sprayed Aluminum Coatings and Corrosion Behavior of the Corresponding Coatings in Sea Water. *J. Mar. Sci. Eng.* **2022**, *10*, 1731. https://doi.org/10.3390/jmse10111731

Academic Editors: Sören Ehlers and Moritz Braun

Received: 19 October 2022
Accepted: 24 October 2022
Published: 11 November 2022

1. Introduction

For offshore wind energy plants, achieving the longest possible service life of foundation and tower structure is decisive for their economic efficiency. Due to challenging mechanical and corrosive conditions, structural steel foundations under water are at a high risk of degradation. Cyclic loading from waves, wind and plant operation can lead to 10^8 load cycles or more within the aspired 25 years of service. The design service life of towers and tower foundations is mainly determined by the fatigue strength of their welded connections. Apart from the apparent butt-welded connections, the attachment of support elements (non-load carrying welded attachments; detail category 80 acc. to EN 199319 [1]) limits the fatigue strength of the structure and leads to the use of plate thicknesses of more than 130 mm [2].

The fatigue strength of welded components is mainly determined by the geometrical and technological notch effect of the welds. The higher the quality level of the welds, the less distinct the notch effect and the better the fatigue strength of the welded detail. Other factors, such as the base material strength or welding process, are of minor importance for the fatigue strength of welded components [3]. Offshore wind turbine towers and tower foundations are manufactured at onshore facilities with high standards regarding the quality level of execution and quality assurance, so limitations in fatigue strength can only be raised by applying post-weld treatment methods.

Different kinds of post-weld treatment methods are frequently used for improving the fatigue strength of welded connections. Principally, post-weld treatment methods rely on reducing the notch effect of the weld by the reduction of stress concentration at the weld toe and on the introduction of beneficial compressive residual stresses [4]. In addition to the more commonly known and regulated burr grinding, TIG dressing, and HFMI processes, blast cleaning can lead to significant increases in fatigue strength of structural steel as well. Before coating, the welded components of wind turbine towers are blast cleaned for surface preparation, so the process can be used as an economic post-weld treatment method that does not require additional manufacturing steps [5–7].

Generally, post-weld treatment methods are also available for offshore foundations. However, because of the additional manufacturing effort and the comparatively small benefit for design due to regulation constraints, they are rarely implemented in production. Also, wind turbine foundations are rarely coated, so the corrosive effect of seawater must be expected to diminish the positive effect of potential post-weld treatment methods. Enhanced benefits from post-weld treatment methods in structural components located under water can be achieved by considering the coating of the corresponding areas.

Until recently, only active (ICCP) and passive (sacrificial anodes) corrosion protection systems have been used in the underwater area of wind turbine foundations. A new and increasingly used method for corrosion protection is the aluminum coating of the entire foundation structure by means of thermal spraying, see Figure 1. Here, the entire structure exposed to seawater is preserved by means of arc spraying with a 99% aluminum coating (Al 99%) as well as an organic sealer [8].

(**a**)

(**b**)

Figure 1. (**a**) Offshore wind turbines; (**b**) Facility for thermal spraying of offshore foundations as well as uncoated and coated monopiles.

The arc-spray process is a simple and cost-effective thermal spray technology, which is determined by a few key factors [9,10]. It is well established for the corrosion protection of large structures and on-site repairs [11,12].

The effect of thermal spraying on fatigue strength was investigated in different contexts. Depending on the substrate material, surface preparation, coating process, coating system, and a potential corrosive medium, thermal spraying can lead to the improvement as well as the deterioration of the substrates' fatigue strength characteristics. The introduction of tensile residual stresses as well as the increase of the component roughness were identified as significant variables influencing the fatigue strength of the coated component [13–18].

Arc voltage and current as well as the gas flow, pressure, and type mainly affect the resulting coating quality [10,19]. In addition, residual stresses inside the coating are heavily influenced by the aforementioned parameters and the spray kinematics [19,20]. In the literature, alternating quenching stresses (generating from particles hindered in shrinking by the substrate as soon as they hit the surface) are reported to dominate the composition of residual stresses in arc spraying, while thermal mismatch stresses (caused by different coefficients of thermal expansion for substrate and coating) also play a major role [19–21]. However, this might be altered by phase changes, e.g., due to heat treatment [21,22]. Concerning the superposition of residual stresses and external stresses or an excess of a certain coating thickness, further studies have revealed pronounced coating delamination and thus significantly reduced service lives [21]. Both findings were confirmed by our own investigations [20]. Furthermore, our investigations on cavitation erosion resistance of thermally sprayed coatings show correlations between the spray parameters, the kinematics, and the gases used as well as the residual stresses of the coatings [22]. Coatings made of fatigue-resistant materials were found to be particularly durable in regard to cavitation erosion. However, the substrate used was pure bulk material instead of welded detail [22]. Consequently, it can be assumed that the longevity and residual stress state of the coating has a significant influence on the fatigue strength of the welded and subsequently blast-cleaned and thermally sprayed connections.

The influence of thermally sprayed Al—coatings on welded steel structures subjected to cyclic loading has not been investigated to date and could be beneficial in terms of service life and manufacturing costs of offshore foundations.

The sacrificial behavior concerning the cathodic protection of exposed steel as well as the self-corrosion rate of arc-sprayed aluminum coatings have been investigated thoroughly [23,24]. However, investigations on the corrosion protection performance as well as on the sacrificial behavior of thermal-spray aluminum coatings in sea water with low salinity via quantitative methods are absent from the literature. Galvanic testing of a combination of thermally coated and uncoated steel can provide insight into the sacrificial behavior of thermal-spray aluminum coatings [25]. Regarding large subsea structures this particular characteristic of thermal-spray aluminum coatings needs to be studied as not only their barrier properties, but also the polarizing effect on mild steel, are key factors to their corrosion protection effect.

Hence, first-time investigations on the fatigue resistance of basic weld details coated with thermally sprayed aluminum under consideration of different surface preparation methods are summarized in this paper. Moreover, investigations regarding the corrosion protection performance and sacrificial behavior of corresponding coatings are presented.

2. Materials and Methods

Based on welded connections typical for steel towers and tower foundations, arc-welded butt joint specimens, as well as specimens with transverse stiffeners, were manufactured from plates of non-alloy structural steel EN 10025-2-S355J2+N [26] with a sheet thickness of t = 8 mm. A S5 SpeedPulse XT welding unit (Lorch Schweißtechnik GmbH) was used for welding. A welding tractor was used for guiding the welding torch to ensure reproducible welding results, a constant heat input, and the constant quality of the welds. After welding, the plates were cut into individual specimens on a band saw. Sharp edges were removed by breaking them manually using a file. Specimen geometries as well as the concomitant welding symbols for the GMAW welds are shown in Figure 2.

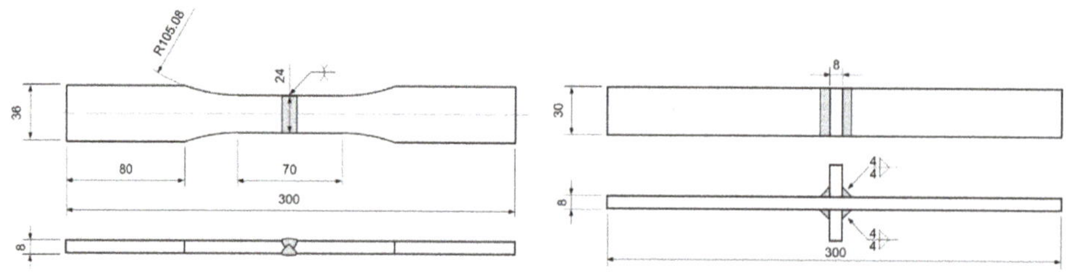

Figure 2. Specimen geometry and welding symbols acc. to ISO 2553 [27] (**a**) Butt joint specimen; (**b**) Specimen with transverse stiffeners.

Prior to thermal spraying, the surfaces of the specimens were blast cleaned by using corundum and grit, respectively. Blast cleaning with corundum was performed at the facilities of Linde AG. Grit blasting was carried out at a manufacturer of steel pipes for wind energy turbine towers using blasting parameters typical for surface preparation of the tower segments before coating. To determine the blasting intensity of the blasting processes, Almen intensity tests according to SAE J433 [28] were carried out at each blasting site using a sentenso Almen gage model TSP-3 and Almen test strips Type A (Grade 2) according to SAE J442 [29]. Parameters, consumables, and quality requirements for welding and blasting are shown in Table 1 (heat input calculated with a coefficient of thermal efficiency of k = 0.8 for GMAW processes according to [30,31]).

Table 1. Parameters, consumables and quality requirements for welding and blasting.

Welding Parameter	Value	Blasting Parameters	Value
Arc Voltage Welding Current	27 V 300 A	Almen Intensity Steel Grit Almen Intensity Corundum	628 µmA (T = 2.5 s) 467 µmA (T = 18.2 s)
Wire feed rate	10 m/min	**Blasting Consumables**	**Type & Particle Size**
Welding speed Heat input Shielding gas flow rate Electrode stickout	35 cm/min 1.1 kJ/mm 12 L/min 18–22 mm	Steel grit Corundum	ISO 11124 M/HCS/G100 [32] nominal particle size 1.0 mm F24 (FEPA 42-1 [33]) 600–850 µm
Welding wire diameter	1.2 mm	**Blasting Quality Criteria**	**Requirement**
Torch angle Travel direction	10° Push welding	Surface preparation grade Coverage	Sa3 (ISO 8501-01 [34]) 100% (SAE J2277 [35])
Welding Consumables	**Designation**	**Welding Quality Criteria**	**Requirement**
Welding wire Shielding gas	ISO 17632-A—T 46 6 M M21 1 [36] ISO 14175—M21—ArC—18 [38]	General Quality Level	B (ISO 5817 [37])

Thermal spraying of the specimens was carried out at the facilities of Linde AG utilizing the arc spray process. For the spray experiments, a power source Sparc 400 equipped with a Shark 400 RE gun on a robot was used (both GTV Verschleißschutz GmbH, Luckenbach, Germany). The spray parameters for the experiments were kept constant and can be found in Table 2. A meander-shaped type of spray pattern was employed. The specimens were coated on all sides. The two wires corresponded to type Al 99% (Metco Aluminium, Oerlikon Metco GmbH, Kelsterbach, Germany; Ø 1.6 mm).

Table 2. Thermal spray parameters.

Gas	Flow Rate in m^3/h	p_{Gas} in Bar	Robot Speed in m/min	Robot Offset in mm	Stand-Off Distance in mm	Number of Passes	Voltage in V	Current in A	Wire Feed Rate in m/min
Air	93.0	3.5	40	9	100	2	28	107	6.2

Micro-sections of the specimens were prepared by cold mounting (two-phase system: liquid hardener and powder resin), gradual grinding, and subsequent polishing up to 3 µm polishing suspension, which was finished by OP-S polishing. Moreover, Nital etching (3%) was applied for enhancing the contrast.

Micrographs of the micro-sections and fracture surfaces were taken using an optical microscope (OM) Leica DM6000M (Leica Microsystems GmbH, Wetzlar, Germany) and the software DHS tool (dhs GmbH, Greifenstein-Beilstein, Germany). Moreover, representative analyses regarding the morphology were carried out using a scanning electron microscope (SEM) JEOL JSM-IT100 (JEOL Germany GmbH, Freising, Germany; with an acceleration voltage 10 kV, a backscatter detector, and a low vacuum mode). Further investigations were carried out on the microstructure of the substrate materials regarding the impact of blast cleaning and thermal spraying. Additionally, the fracture surfaces were analyzed particularly focusing on the origin of the fatigue cracks. Both were investigated using the same equipment as described earlier.

The coating thickness was determined quantitatively in the OM micrographs, recording 3 times the 7 measured values, while eliminating maxima and minima values for each. In addition, the number of coating defects (i.e., porosity, oxidation, and cracks) was quantified at 3 random areas of the SEM images. For that purpose, the software ImageJ (National Institutes of Health, USA; in the region of interest using Despeckle filter, normalization, and finally the Trainable Weka Segmentation tool) was applied.

Residual stress measurements were carried out for specimen characterization prior to fatigue testing. The measurements were performed by the hole drilling method combined with electronic speckle pattern interferometry (ESPI) using the PRISM system and the PrismS software (both Stresstech GmbH, Rennerod, Germany) with three measurements for each specimen. The coated specimens and the reference material (without grit blasting or thermal spraying) were tested after the application of a developer spray (typically used in the non-destructive penetration test) to reduce surface reflections and, in the case of the coated samples, in the as-sprayed state. The holes were processed in varying steps using high-speed steel end mills (2 fluted, TiN coated, and Ø 0.8 mm) up to a final profile depth of 400 µm. The inner integration radius was twice the hole diameter, while the outer integration radius corresponded to four times the hole diameter. The Poisson ratio used for calculation was 0.334, while the Young's modulus was assumed to be 71 GPA. Both values correspond to bulk aluminum, which should be kept in mind regarding the evaluation. A Tikhonov regularization factor of 0.01 was applied in stress calculation.

Fatigue tests were performed under axial loading and generally according to the ISO/TR 14345 [39]. A resonance testing machine, the Power Swing 100 kN from SincoTec Test Systems GmbH, was used for carrying out the tests. The testing frequency was f = 60 Hz for the butt-welded specimens and f = 110 Hz for the specimens with transverse attachments. Runouts were defined for 5×10^6 cycles. Specimen failures aimed for an even distribution over the target live range of $5 \times 10^4 < N < 5 \times 10^6$ cycles. A complete crack of the specimens or a visually detectable crack close to being a complete crack were defined as failure criteria. Specimens were loaded by constant amplitude loading with a stress ratio of R = 0.5. The limit to the maximum stress applied was the base material yield strength determined via the tensile tests, which led to a maximum stress range of $\Delta\sigma_{max}$ = 200 MPa. Statistical evaluation of the fatigue test results was carried out according to the background document [40] to the Eurocode 3 Part 1–9 [1]. The runouts as well as specimens with cracks in the base metal area were excluded from the statistical evaluation.

The corrosion protection capability of the Al99% coating was tested in a close-to-real environment. For this purpose, a test stand consisting of 5 Intermediate Bulk Containers (IBC) was set up indoors at a constant temperature of 21 ± 1 °C. Each of the IBCs holds 600 L water, while the one where the testing occurred was also partly filled with sediment. The IBCs are interconnected and a circulative flow of 600 L/h is generated by a submersible pump. Water and sediment were recovered from the Baltic sea at an area 35 km northeast of the island Rügen at 40 m depth. Table 3 provides an overview of the water composition analyzed by ion chromatography and the additional water properties.

Table 3. Seawater composition and properties.

Component	Sodium	Potassium	Calcium	Magnesium	Chloride	Bromide	Sulphate
Concentration in mg/L	2906.9	99.6	227.4	480.6	6518.7	47.8	800.1
Property	**Salinity**	**pH**	**O_2**	**Temperature**			
Value	0.81 g/L	7.5 ± 0.1	7 ± 0.3 mg/L	21 ± 1 °C			

The coated test specimens are made of S355J2+N Steel with a size of 175 mm × 50 mm × 5 mm. Prior to arc spraying the steel was blast cleaned to Sa 3 to a roughness between 60 µm and 100 µm according to ISO 8501-1 [34]. The coating thickness ranged between 400 µm and 600 µm due to manual application. The bare steel coupons that simulate damage to the coating were sized at 17 mm × 13 mm × 10 mm. All test specimens were provided with a threaded bore with a diameter of 4 mm on the topside for electrical connection. The electrical connection as shown in Figure 3 was established by the standard banana plug measurement cables cut to length and tightened to the coupons. Insulation from the surrounding electrolyte was guaranteed by a silan-modified polymer sealant; with the sealant covering the whole topside of the bare steel coupons and their remaining surface area representing 5% of the surface area of the coated specimens.

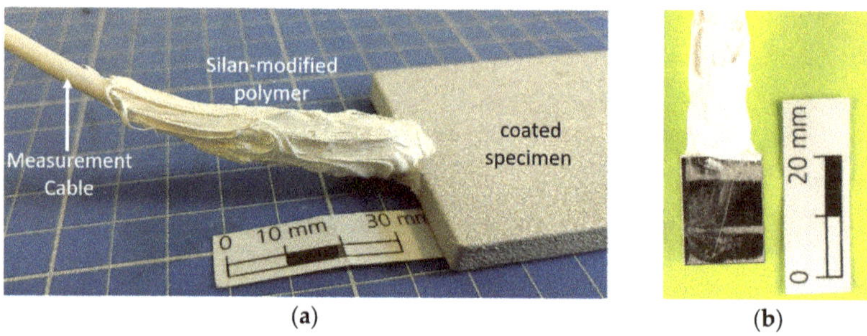

Figure 3. (**a**) The electrical connection of a thermal spray-coated specimen with insulation by silan-modified polymer sealant; (**b**) A bare steel specimen with the electrical connection.

Three different specimen configurations were tested simultaneously. Four thermal spray-coated S355J2+N specimens as well as four bare steel specimens were exposed as stand-alone for reference purposes. As a third configuration the thermal spray-coated specimens were electrically coupled to bare steel specimens through a specifically developed Arduino-based connection box in between the measurements and through the potentionstat while measuring. This way an electrical connection between the coated specimens and the bare steel was always provided with the protection current being quantifiable on a regular basis.

Prior to immersion, all specimens were degreased using isopropanol and deionized water. As shown in Figure 4 all configurations were positioned in one IBC around a single Ag/AgCl reference electrode (RE) which was replaced daily. The counter electrode (CE) made from graphite with the dimensions 400 mm × 400 mm × 30 mm was placed on the side of the IBC. The specimens were hung by their cables with a fixed distance of 5 cm (minimum) between the galvanic couples.

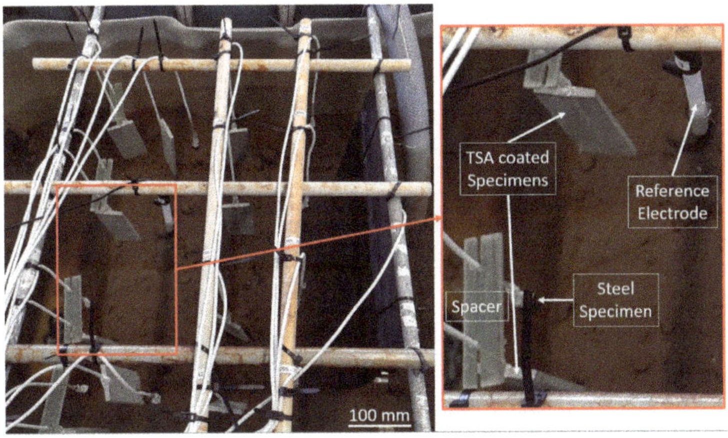

Figure 4. The test setup in IBC with the graphite counter electrode to the right, the reference electrode in the middle, and the specimens arranged around it. The galvanic couples are on the left side. The thermal spray-coated specimens are in the middle of the picture and the bare steel specimens are placed to their right.

The potentiostat used was a Metrohm Autolab PGSTAT 302 N equipped with 3 multiplexers for the serial measurement of 12 test specimens or specimen couples. The test duration for the corrosion test was 30 days. At the beginning and end of the test the potentiodynamic polarization curves were recorded using additional specimens. These specimens were polarized down to −1.2 V vs. RE-potential before scanning up to −0.5 V vs. RE-potential using a scan rate of 0.5 mV/s.

The open circuit potential (OCP) of all specimen configurations and the current between the galvanic couples were recorded every 4 h. At the beginning of each measurement the potentiostat established an electrical measurement circuit parallel to the connection box. After 10 s the wiring through the Arduino-based box was cut for the recording duration (see Figure 2). The values of every channel were tracked for 120 s before reconnecting through the arduino-box and disconnecting the potentiostat.

Once per week and twice during the first week LPR measurements were performed. After 120 s of OCP scanning time the specimens and specimen couples were polarized at ±10 mV around OCP with a scan rate of 0.1 mV/s. The results from the LPR measurements and the Tafel constants ba and bc gathered from the polarization curves were used to calculate the corrosion rates using the Stern–Geary [41] equation. For the calculation of the corrosion rates of the galvanic couples the equivalent weight of aluminum was used due to the surface area ratio between the coated specimens and the bare steel.

After the exposure period the images in a wet and dried state, the SEM images (same equipment as above), and the energy dispersive x-ray spectrometry (EDX) maps (JEOL Dry SD25 detector (JEOL Germany GmbH, Freising, Germany; acceleration voltage 10 kV)) of selected specimens were acquired. The deposits formed on the coating were analyzed by IR-spectroscopy (FTIR-Spectroscope ALPHA (Bruker Optik GmbH, Leipzig, Germany).

For the base material characterization, the base material yield strength was determined via tensile tests according to ISO 6892-1 [42] and the chemical composition of the base material was verified by optical emission spectrometry (OES) using a Spectromaxx (Spectro Analytical Instruments GmbH, Kleve, Germany).

3. Results

3.1. Materials Characterization

3.1.1. Base Metal Properties

Tensile tests according to ISO 6892-1 [42] resulted in a medium upper yield strength of R_{eH} = 403 MPa (s = 7 MPa; n = 4). The OES analysis results and content limits according to EN 10025-2 [13] are shown in Table 4.

Table 4. The OES analysis results and content limits according to EN 10025-2 [13].

Steel	Content in %	C	Si	Mn	P	S	N	Cu	Ni	Cr	Mo
S355J2+N acc. to [43]	Max. content acc. to [13]	0.20	0.55	1.60	0.025	0.025	-	0.55	0.42	0.29	0.11
	$\overline{x}$ ($n = 5$)	0.159	0.197	0.72	0.0085	0.0042	0.022	0.021	0.042	0.056	0.012
	s_x	0.0083	0.0038	0.0040	0.0002	0.0001	0.0025	0.0001	0.0021	0.0009	0.0002

3.1.2. Weld Seam Quality

Figure 5 shows the macro sections of a butt-welded specimen and a specimen with transverse stiffeners (both blast cleaned and thermally sprayed). Quality level B according to ISO 5817 [37] was met for the imperfections of the welded joints in the specimens and verified via visual testing.

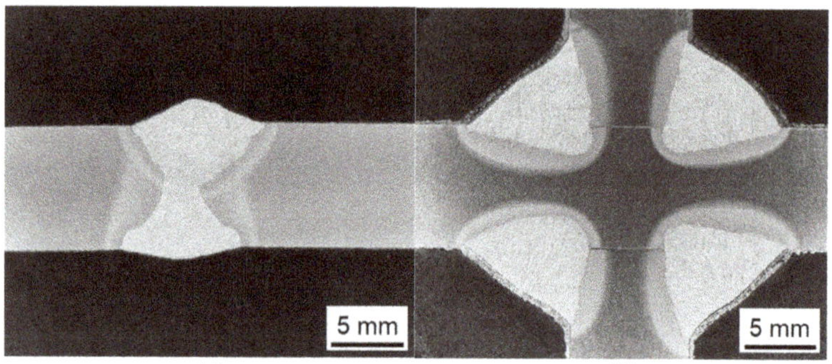

Figure 5. Macro-sections of thermally sprayed specimens, etched with Nital (5%), (**a**) butt-weld; (**b**) fillet weld.

3.1.3. Surface and Coating Properties

Regarding surface preparation, grit blasting shows a stronger impact on the near surface microstructure of the steel substrate compared to blasting with corundum. A more severe plastic deformation and a visible compression of the microstructure of the surface material layer can be seen in the micro-section of the grit-blasted specimen compared to the specimen blasted with corundum, see Figure 6.

This stronger deformation is also reflected in the subsequent coating analyses. First, it is obvious that the grit-blasted samples exhibit higher coating thicknesses, which are also constant over a larger area in the section, see Figure 7. Apart from this, a typical lamellar morphology with good bonding to the substrate can be observed. Both types of surface preparation share the non-uniform distribution of cluster-like coating defects, which are marked in Figure 7. Yet, these variations are apparently more frequent for the specimens blasted with corundum. In contrast, layered oxides typical for arc spraying are hardly visible. However, coating breakouts can be detected in part for both surface preparations as a result of the polishing process, although they are more pronounced for the specimens blasted with corundum.

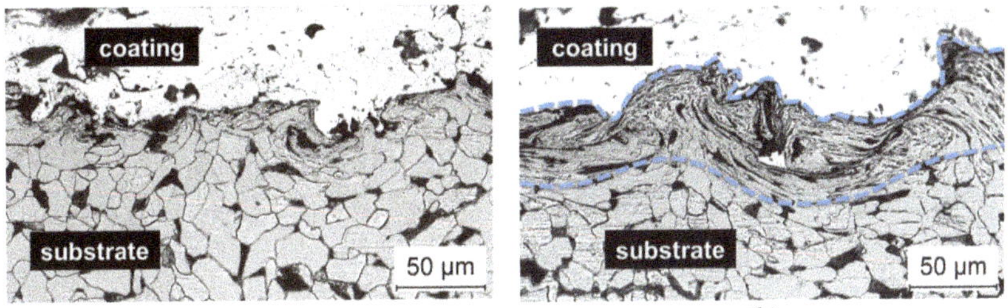

(a) (b)

Figure 6. The impact of different surface preparation methods on the base metal microstructure; micro-sections etched with Nital (3%), (**a**) substrate blasted with corundum; (**b**) substrate blasted with steel grit (deformed microstructure bordered by the dashed line).

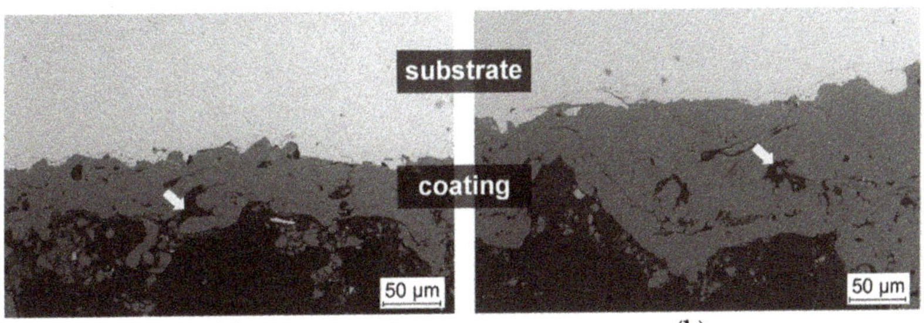

(a) (b)

Figure 7. Representative SEM images showing the substrate on top and the coating below (specimen bottom) for the samples blasted with (**a**) corundum; (**b**) grit.

The observations on the coating microstructure can further be confirmed quantitatively, see Table 5. For example, the grit-blasted specimens had higher coating thickness than the corundum-blasted specimens, with similar percent standard deviations for both surface preparations.

Table 5. The quantitative coating analysis showing thicknesses and defects.

Blasting Procedure	Coating Thickness in μm	Amount of Defects (Porosity, Cracks, Oxides) in %
Corundum	78 ± 20	4.7 ± 2.3
Grit	97 ± 26	6.2 ± 0.5

Likewise, according to the representative analyses, a uniform coating structure is recognizable for specimens blasted with grit. Although the absolute number of defects is somewhat higher, the higher standard deviation of the corundum blasted samples indicates an increased inhomogeneity.

3.1.4. Residual Stress Measurements

The residual stress measurements show a high amount of compressive residual stresses in the blast-cleaned steel substrate (Figure 8). In the as-welded condition, the untreated specimen shows significant tensile residual stresses in the area of the weld transition. The measurements were performed at a distance of 5 mm from the weld toe in the base material. The compressive stresses in the grit-blasted specimens reach deeper underneath the specimen surface and gain a higher compressive stress value of up to −200 MPa compared to the specimens blasted with corundum, with a compressive stress maximum of about −100 MPa, see Figure 8. At a drilling depth within the range of a coating thickness of approximately 0.05 to 0.1 µm, a transition from tensile stresses to compressive stresses can be seen in the residual stress depth profiles obtained. Tensile residual stresses in the coating of the specimens are higher in the specimens blasted with steel grit and reach levels of up to 200 MPa, compared to less than a 100 MPa tensile stress in the coating of the specimen blasted with corundum. At a depth of 0.35 mm, nearly no compressive stresses could be measured in the specimen blasted with corundum, whereas the grit-blasted specimen still showed compressive stresses of more than −50 MPa at the same drilling depth.

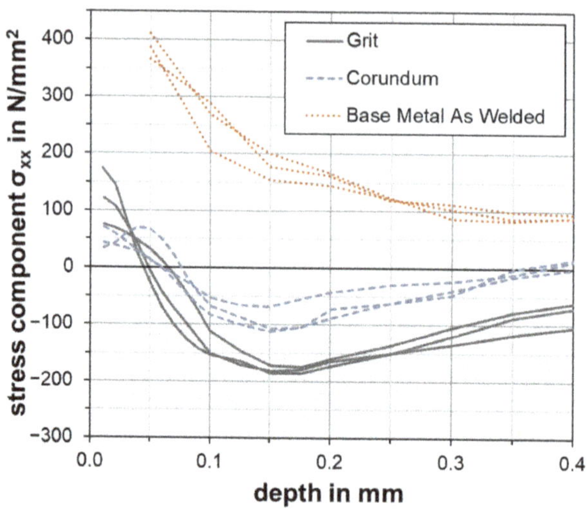

Figure 8. The residual stress measurement results of thermally sprayed specimens previously blasted with grit or corundum and non-treated reference specimens in the as-welded state at the weld toe.

3.2. Fatigue Tests

Fatigue strength evaluation by statistical analysis with a variable slope according to [40] leads to a characteristic value of the stress range of $\Delta\sigma_{c,var}$ = 153 MPa with a slope of m_{var} = 9.9 for the butt-welded reference specimens (n = 5). The specimens with transverse attachments show a characteristic value of $\Delta\sigma_{c,var}$ = 89 MPa with a slope of m_{var} = 4.0 (n = 9). Blast-cleaned and subsequently thermally sprayed butt weld specimens show a nearly identical characteristic value of $\Delta\sigma_{c,var}$ = 152 MPa (m_{var} = 14.7; n = 8) compared to the non-blasted butt weld specimens. Fractures generally occurred at the weld toe of the specimens. However, with the blast-cleaned and thermally sprayed butt joint specimens, fractures also occurred in the base material area of several specimens (n = 11). Regarding the non-blasted, non-coated specimens, fractures in the base material area did not occur. Blast-cleaned and subsequently thermally sprayed specimens with transverse attachments show a characteristic value of $\Delta\sigma_{c,var}$ = 127 MPa (m_{var} = 6.6; n = 18). For both types of specimens, the kind of surface preparation (grit blasting and blasting with corundum) does not seem to make a difference in the fatigue test results. The corresponding S–N diagrams including the testing parameters and characteristic values of the stress range can be seen in Figure 9.

The different types of crack locations are shown in Figure 10. As shown, for some of the specimens with transverse stiffeners, cracks appeared at both sites of the stiffeners simultaneously.

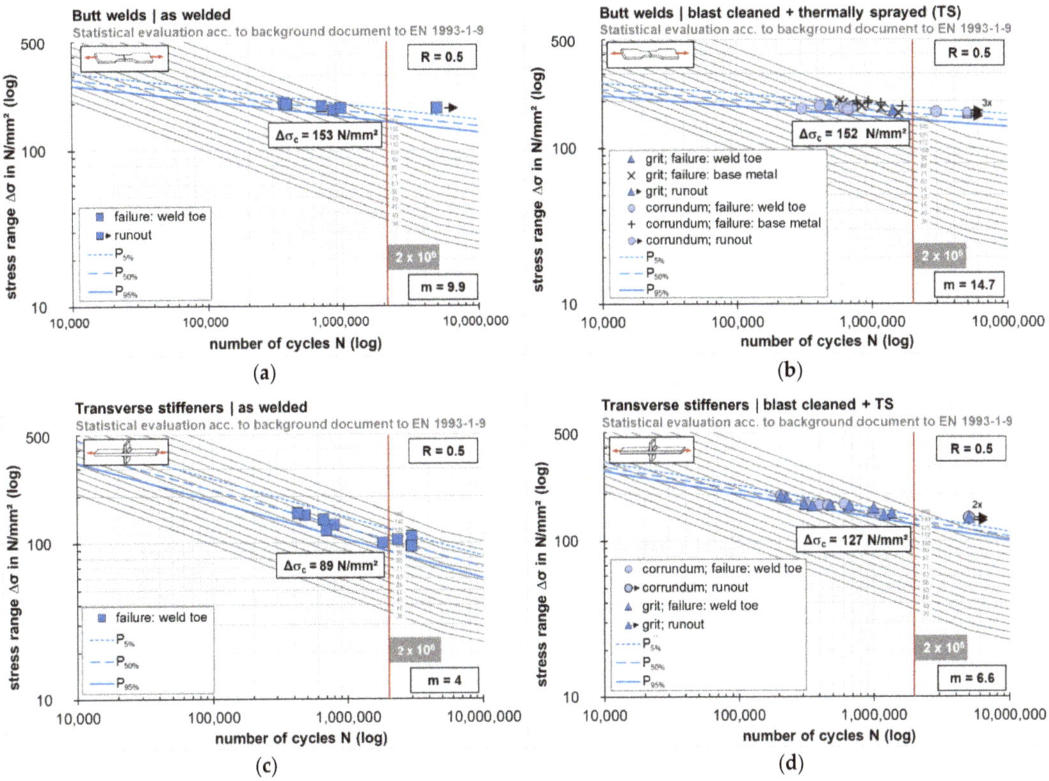

Figure 9. S–N Diagrams. (**a**) butt-welded specimens, as welded; (**b**) butt-welded specimens, blast cleaned and thermal spray coated; (**c**) specimens with transverse stiffeners, as welded; (**d**) specimens with transverse stiffeners, blast cleaned, and thermal spray coated.

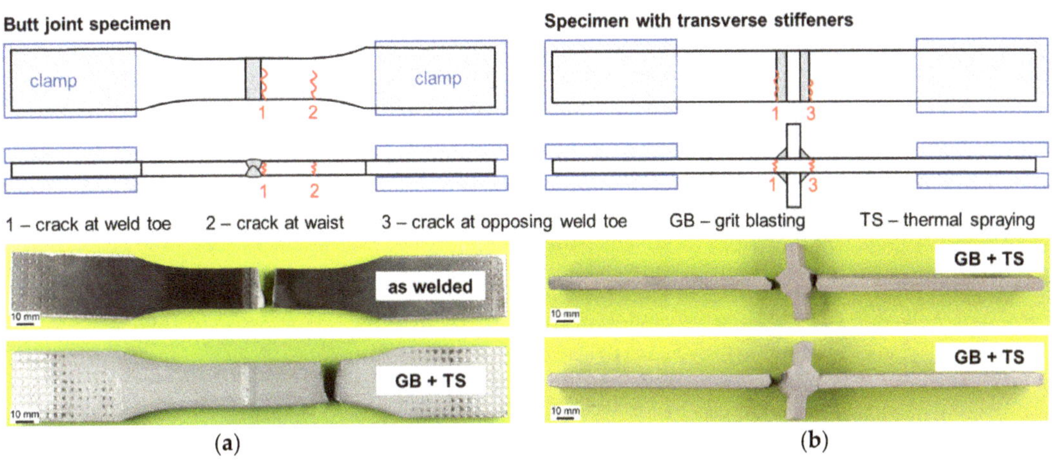

Figure 10. Failure locations. (**a**) Butt joint specimens; (**b**) Specimens with transverse stiffeners.

Concerning the fracture surfaces shown in Figure 11, the fatigue fracture zone with a plain surface orthogonal to load direction and the fast fracture zone with distinct plastic deformation of the material (ductile fracture) can be clearly distinguished. The fatigue fracture zone of substrate and coating is characterized by different types of cracks. While orthogonal cracks with a large opening angle are predominantly observed for the substrate, horizontal cracks are occasionally found in the coating. These mostly occur at the interface or nearby, thus indicating lower adhesion in these areas. Yet, in general the coatings still adhere well to the substrate over most of the observed fracture surfaces. In contrast, orthogonal cracks could not be observed in the coating during the analyses before fatigue testing. Moreover, small particles from the substrate could partly be detected in the coatings (not shown). A single origin of the fatigue fracture could not be found for the specimen shown. Instead, multiple crack fronts originating at multiple points across the weld toe seem to blend together over the width of the specimen. With some of the grit-blasted and thermally sprayed specimens, fatigue cracks occurred at both sides simultaneously or, on the contrary, only at one side of the weld toe. No evidence was found to indicate whether the cracks initiated from the coating or the substrate material. These observations apply to both kinds of surface preparation equally.

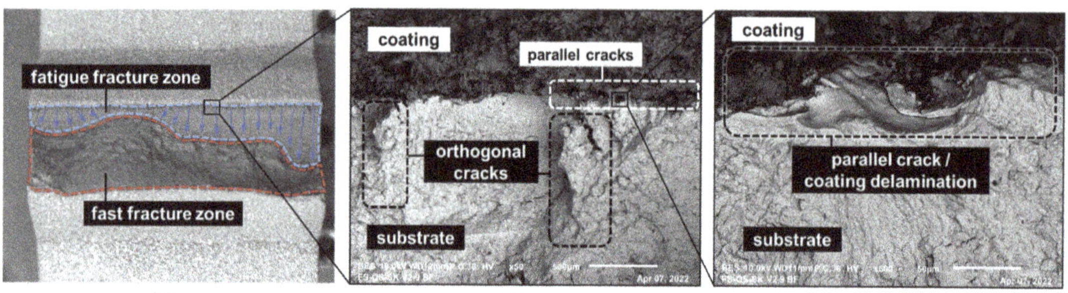

(a) (b) (c)

Figure 11. The fracture surface of a grit-blasted specimen with transverse stiffeners. (**a**) Overview image showing the fatigue fracture zone and the fast fracture zone; arrows indicate the origin and direction of fatigue crack propagation. (**b**) The BSE image of the interface area showing different orientations of secondary cracks in the fatigue fracture zone. (**c**) The BSE image of the parallel crack at the interface of the coating.

3.3. *Corrosion Testing*

Figure 12a shows the OCP values during the immersion. The steel's OCP started at around −0.55 V and stabilized after 3 days of immersion between −0.65 V and −0.67 V, respectively. During the whole immersion time no further changes in the steel's OCP were observed, except for a decreasing difference between measurements. From the initial values between −0.80 V and −0.86 V the coating's solo potential declined within 3 to 10 days to less than −1.0 V. Subsequently, it increased and seemed to stabilize at around −0.95 V by the end of the test duration. The values from the galvanic couples were observed to be in-between these values. The OCP values of the couples started at around −0.65 V and declined within 2 to 3 days to values close to −1.0 V. At the end of the test duration the combined OCP of the galvanic couples ranged slightly higher than those from the stand-alone coating and were in the region of −0.9 V.

The galvanic current density over time is shown in Figure 12b. This was calculated using the surface area of the bare steel specimens. The values of all the couples were initially between 130 mA/m^2 and 180 mA/m^2. From the outset of the experiment, the values started to rise reaching their maxima after 3 to 5 days. All maxima lay between 400 mA/m^2 and 500 mA/m^2. After the maximum was reached, the current density declined with the curve being shaped asymptotically. After 20 days of immersion the values of all the couples were slightly above 100 mA/m^2, but still decreasing. At the end of the test all the values were lower than 100 mA/m^2.

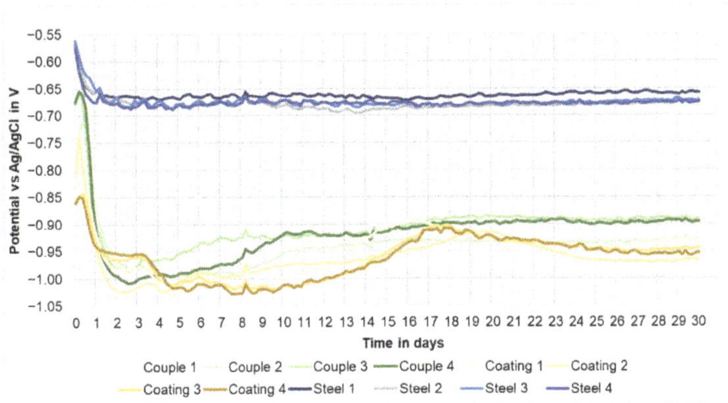

(a)

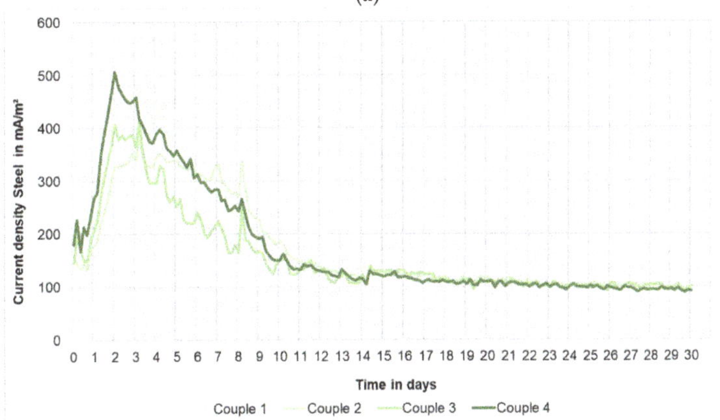

(b)

Figure 12. (a) OCP vs. Ag/AgCl of the single specimens and specimen couples over the test duration; (b) Galvanic current density of the thermal spray coated and bare steel specimen couples over exposure time.

Figure 13a presents the difference between a steel specimen that was electrically coupled to a coated specimen on the left side and to stand-alone steel on the right side. The coating cathodically protected the bare steel sufficiently during the test. No signs of corrosion were visible on the surface of any bare steel specimens that were coupled to coated specimens in the wet state directly after being retrieved from exposure. Conversely, a thin layer of corrosion products formed on the earlier-protected bare steel coupled in the dried state. There was no observed effect of the orientation of the steel specimen on the coated plates.

Figure 13b shows the coated specimen that was electrically coupled to the bare steel specimen shown in Figure 13a on the left side. IR Spectroscopy identified the white deposits as aluminum oxide. No optical difference between stand-alone coated specimens and the coated specimens from the galvanic couples after exposure was detected.

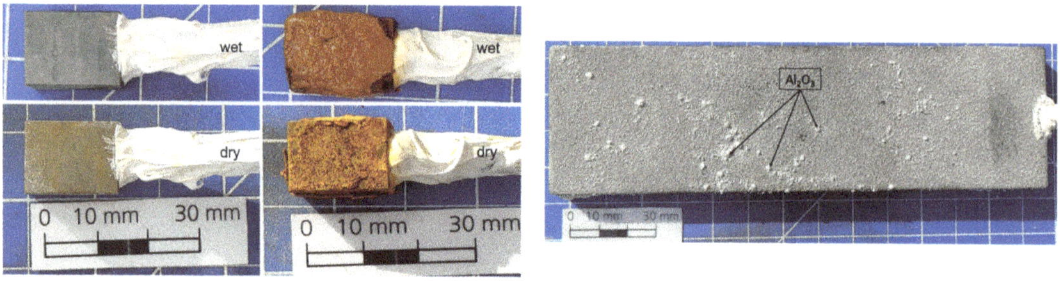

(**a**) (**b**)

Figure 13. (**a**) Bare steel specimens after 30 days of immersion in Baltic sea water; (**b**) a thermal spray coated specimen that was electrically coupled to a bare steel specimen.

The SEM analysis of the steel coupon in the dried state primarily showed iron oxides on the surface. This corroded layer was also optically visible and established during drying after the exposure. In addition, sulfur, magnesium, and sodium were also detected. The SEM images and EDX maps shown in Figure 14 demonstrate an even distribution of the corresponding elements.

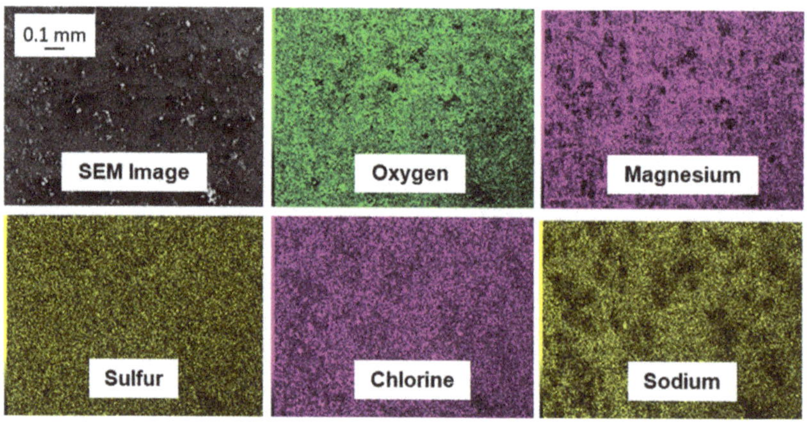

Figure 14. The SEM images and EDX maps of a bare steel coupon that was continually galvanically coupled to a coated specimen.

Figure 15a shows the corrosion rates of coated specimens and Figure 15b shows the galvanic couples' corrosion rates during exposure. Initially, the values of the stand-alone coating were lower than those for the galvanic couples. They ranged from 30 μm/a to 50 μm/a. Over the exposure period the corrosion rates decreased to as low as 20 μm/a at the experimental completion. The corrosion rates of the galvanic couples—one could also call them a coating with a 5% defect—were initially higher, ranging from 55 μm/a to 70 μm/a. Instead of directly decreasing after the start of the test, the values rose to a maximum of 80 μm/a. During the 14 days it took the galvanic current density to decrease to 100 mA/m^2, the corrosion rates also decreased to around 30 μm/a. At the end of the exposure time the corrosion rates of the galvanic couples reached the same levels as the stand-alone thermal spray coating with the slope still pointing to even lower values.

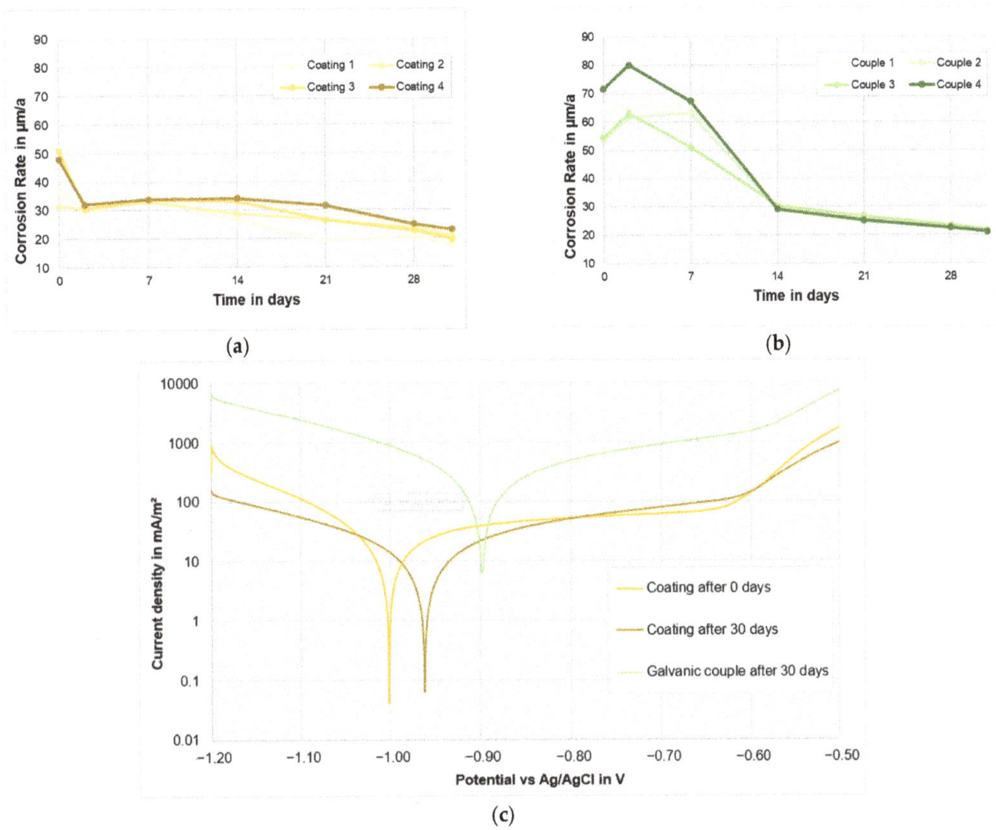

Figure 15. (**a**) The corrosion rate of the thermal spray coating; (**b**) The corrosion rate of the galvanic couples; (**c**) Potentiodynamic polarization curves of the coating at the start of the test, and of the coating and a galvanic couple after 30 days.

Potentiodynamic polarization curves of the coating at the start of the test and of the coating and a galvanic couple after 30 days are displayed in Figure 15c.

4. Discussion

The fatigue strength of the butt-welded non-blasted specimens is close to base material fatigue strength of $\Delta\sigma_c = 160$ MPa (m = 3) according to EN 1993-1-9 [1]. This surpasses the general recommendation of detail category 90 for butt-welded connections according to the standard by far. This result can be explained by the good execution of the welds with low levels of imperfections, especially regarding angular and axial misalignment, and a general benefit in fatigue strength known for thin-walled specimens [44,45]. The steeper slope and the lower fatigue strength of the as-welded specimens with transversal attachments were expected and are congruent with the regulatory values (detail category 80 for plates with transversal attachments according to [1]). The process of blast cleaning only can lead to different results regarding fatigue strength of the blasted components, depending on the abrasive and blasting parameters used: in prior investigations, blasting with corundum led to a reduction in fatigue strength due to an increased surface roughness and only minor effects on the near surface microstructure, whereas grit blasting led to a significant fatigue strength improvement due to the increased compression of the near surface microstructure, the cold hardening effects, and an increased introduction of compressive stresses [7].

Given that the thermally sprayed specimens show the same fatigue strength characteristics regardless of the blasting process used for the surface preparation, the thermal spraying itself seems to lead to an increase in fatigue strength as well: for the specimens blasted with corundum the thermal spraying seems to compensate for the issues that are normally associated with this kind of surface preparation [7].

Looking at the micrographs, the generally negative effect of increased surface roughness on the fatigue strength of the blast-cleaned specimens might be reduced by the coating material filling out the dents on the substrate surface, which might hinder fatigue crack initiation and growth. This seems to be confirmed by the relative oxide-free coatings, which are instead dominated by clustered porosity. The microstructural analyses and evaluations of the fracture surfaces generally showed a good bonding of the coatings and, in particular, a denser and more uniform coating structure for grit. However, this difference does not seem to have a direct influence on the fatigue strength. Further research is necessary to explain the effect sufficiently. The considerable number of post-weld treated butt-weld specimens with fractures occurring in the base material area shows that the maximum increase in fatigue strength may have been reached for this type of specimen. For these, the reduction of the notch effect of the welds led to other specimen characteristics being more critical, e.g., the tapering of the specimens, which resulted in cracks occurring in the base metal area rather than at the weld toe. For most of the grit-blasted and thermally sprayed specimens, the missing single point of origin of the fatigue cracks indicates a uniform notch effect across the weld toe.

Whilst the Almen intensities of the different blasting processes proved to be on a comparable level, the results of the residual stress measurements confirm the earlier findings already mentioned, which state that blasting structural steel with corundum has more of an abrasive effect and that in contrast grit blasting, with particles larger and more dense than corundum, induces a higher number of compressive stresses into the material and leads to a more distinctive cold work hardening of the near surface areas blasted [5,7]. The more distinct microstructural plastic deformation of the steel substrate surface layer of the grit-blasted specimens compared to the specimens blasted with corundum indicates a stronger cold work hardening effect in the surface area of the grit-blasted specimens and goes along with the residual stress measurement results. The tensile stresses prevailing in the thermal spray coating do not seem to transfer into the blast-cleaned base material nor do they seem to have a negative impact on the fatigue strength of the coated specimens, as the fatigue strength characteristics determined are the same regardless of the kind of blasting process used for surface preparation, or the height of tensile stresses in the specimen coating, respectively. Instead, it might also be assumed, that a certain stress compensation can take place beyond the coating–substrate interface and thus favorable compressive residual stresses could prevail, which is generally in line with the investigation. However, the applied elastic parameters of the residual stress measurements solely apply to one single material and usually cannot include transitions. In addition, parameters for the bulk material were used, although it is known that the Young's moduli of thermally sprayed coatings are in part significantly lower [20,46].

The results of the corrosion test indicate sufficient galvanic protection of areas with defects in Al99% coatings, where bare steel is exposed to sea water with a low salinity. No corrosion products were formed on exposed steel surfaces if they were electrically connected to Al99% coated specimens. During the whole exposure period the OCP values of the stand-alone coating, as well as of the galvanic couples, were below the protection level for bare steel [46].

5. Summary

The investigations show that Al99% coatings applied by thermal spraying to welded structural steel specimens of type S355 J2+N lead to a significant improvement in the fatigue strength of the basic structure and that the arc-sprayed coatings show sufficient adhesion even under a fatigue load. For blast-cleaned and coated test specimens with welded transverse stiffeners, the characteristic value of the stress range was determined with $\Delta\sigma_{c,var}$ = 127 MPa at a variable slope of m_{var} = 6.6, which is 40 % higher than characteristic value determined for the as-welded reference specimens ($\Delta\sigma_{c,var}$ = 89 MPa).

Test specimens with welded butt joints exhibited a very high fatigue strength in the as-welded condition already, which can be attributed to the low level of geometric imperfections of the weld seams and specimens (angular and axial misalignment). Subsequent coating by thermal spraying resulted in a fatigue strength comparable to that in the reference specimens, with $\Delta\sigma_{c,var}$ = 152 N/mm^2. Crack initiation in the thermally sprayed specimens occurred more in the base material area and less at the weld toe, indicating a reduction of the notch effect of the weld.

The increase in fatigue strength is independent of the blasting intensity and the abrasive used for the surface preparation. However, compressive residual stresses introduced by grit blasting are much higher and extend deeper below the surface compared to blasting with corundum. Also, the cold work hardening effects in the near surface area, which lead to a delay in crack initiation, are more pronounced after blast cleaning with steel grit. Since these differences did not lead to a difference in fatigue strength of the specimens, the thermally sprayed coating itself seems to contribute to fatigue strength improvement as well, e.g., by compensating the generally negative effect of increased surface roughness as well as the lack of sufficient cold work hardening effects associated with the specimens that were blast cleaned with corundum only. This may be attributed to the closure of dents on the substrate surface acting similarly to a reduction in roughness of the specimen surface and/or the stress compensation effects beyond the substrate coating interface and these will be investigated further in future studies.

Concerning the corrosion protection properties, it can be concluded that Al99% coating sufficiently works as a sacrificial anode for exposed steel in aerobic Baltic sea water with a salinity of 0.8% at room temperature. Further tests can be conducted to determine functionality at lower temperatures, in seabed sediment, and for larger defective areas.

Overall, the studies indicate that, in combination, surface preparation by blast cleaning and the application Al99% thermal spray coatings, can sufficiently protect steel structures from corrosion and simultaneously lead to an increase in fatigue strength of welded components. This is especially interesting for, e.g., offshore structures exposed to cyclic loading. The effects examined should be validated in future investigations, for instance in studies that also consider a combination of cyclic loading and corrosive wear.

Author Contributions: Conceptualization, K.-M.H. and A.G.; fatigue test specimen manufacturing (supervision of welding and surface preparation), A.G., B.R. and M.H.; thermal spraying (supervision and writing), M.H.; residual stress measurements, base metal surface analysis, and coating analysis (supervision, investigation, writing, and visualization), M.H. and B.R.; base metal properties and fatigue tests (supervision, formal analysis, writing, and visualization), B.R.; fracture surface examination (investigation, writing, and visualization), M.H. and B.R.; corrosion specimen manufacturing and corrosion testing (supervision, investigation, writing, and visualization), J.N. and M.I.; overall supervision, A.G. All authors have read and agreed to the published version of the manuscript.

Funding: This research received no external funding.

Institutional Review Board Statement: Not applicable.

Informed Consent Statement: Not applicable.

Data Availability Statement: All data that supports the findings in this study are available to researchers upon reasonable request.

Acknowledgments: The authors would like to thank all co-workers involved in the study, namely listed in alphabetical order: R. Arndt, K. Hasche, A. Herhaus, J. Hilbert, F. Knöchelmann, W. Krömmer, and S. Schneider.

Conflicts of Interest: The authors declare no conflict of interest.

References

1. *DIN EN 1993-1-9:2005+AC:2009*; Eurocode 3: Design of Steel Structures—Part 1–9: Fatigue. DIN Deutsches Institut für Normung: Berlin, Germany, 2009. [CrossRef]
2. Nikitas, G.; Arany, L.; Aingaran, S.; Vimalan, J.; Bhattacharya, S. Predicting long term performance of Offshore Wind Turbines using Cyclic Simple Shear apparatus. *Soil Dyn. Earthq. Eng.* **2017**, *92*, 678–683. [CrossRef]
3. Hobbacher, A.F. *Recommendations for Fatigue Design of Welded Joints and Components—IIW Document IIW-2259-15*; Springer International Publishing: Cham, Switzerland, 2016. [CrossRef]
4. Haagensen, P.J.; Maddox, S.J. *IIW Recommendations on Methods for Improving the Fatigue Strength of Welded Joints*; Woodhead Publishing Limited: Cambridge, UK, 2013; pp. 4–12. [CrossRef]
5. Hensel, J.; Eslami, H.; Nitschke-Pagel, T.; Dilger, K. Fatigue Strength Enhancement of Butt Welds by Means of Shot Peening and Clean Blasting. *Metals* **2019**, *9*, 744. [CrossRef]
6. Glienke, R.; Wegener, F.; Gericke, A.; Hobbacher, A.; Henkel, K.; Marten, F.; Günther, H. Experimentelle und numerische Untersuchungen zur Schwingfestigkeit an Schweißdetails in Stahlrohrtürmen für Windenergieanlagen. *Stahlbau* **2019**, *88*, 968–987. [CrossRef]
7. Gericke, A.; Glienke, R.; Brandenburg, K.; Walter, C.; Ripsch, B.; Henkel, K.; Frithjof, M. *Untersuchung des Reinigungsstrahlens als Nahtnachbehandlungsmethode zur Verbesserung der Schwingfestigkeit Geschweißter Strukturen*; DVS Berichte 2018—Band 344; DVS Media GmbH: Düsseldorf, Germany, 2018; pp. 350–357.
8. Price, S.J.; Figueira, R.B. Corrosion Protection Systems and Fatigue Corrosion in Offshore Wind Structures: Current Status and Future Perspectives. *Coatings* **2017**, *7*, 25. [CrossRef]
9. Kim, J.-H.; Lee, M.-H. A Study on Cavitation Erosion and Corrosion Behavior of Al-, Zn-, Cu-, and Fe-Based Coatings Prepared by Arc Spraying. *J. Therm. Spray Technol.* **2010**, *19*, 1224–1230. [CrossRef]
10. Lang, F.; Krömmer, W. Economic and ecological benefits of using gas mixtures for arc spraying. In Proceedings of the 10th Colloquium High Velocity Oxy-Fuel Flame Spraying, Erding, Germany, 29–30 October 2015; pp. 79–88.
11. GTV Verschleißschutz GmbH, Krebs Korrosionsschutz GmbH Rostock, EEW SPC GmbH. Available online: https://www.youtube.com/watch?v=O5v73kPugm0 (accessed on 19 July 2022).
12. Barth, A. Application areas of arc spraying. In Proceedings of the Seminar "Modern Coating Technologies", Dortmund, Germany, 5 November 2014.
13. Voorwald, H.J.C.; Souza, R.C.; Pigatin, W.L.; Cioffi, C.O.H. Evaluation of WC–17Co and WC–10Co–4Cr thermal spray coatings by HVOF on the fatigue and corrosion strength of AISI 4340 steel. *Surf. Coat. Technol.* **2005**, *190*, 155–164. [CrossRef]
14. Puchi-Cabrera, E.S.; Staia, M.H.; Ortiz-Mancilla, M.J.; La Barbera-Sosa, J.G.; Ochoa Pérez, O.C.; Villalobos-Gutiérrez, C.; Bellayer, S.; Traisnel, M.; Chicot, D.; Lesage, J. Fatigue behavior of a SAE 1045 steel coated with Colmonoy 88 alloy deposited by HVOF thermal spray. *Surf. Coat. Technol.* **2010**, *205*, 1119–1126. [CrossRef]
15. McGrann, R.T.R.; Greving, D.J.; Shadley, J.R.; Rybicki, E.F.; Kruecke, T.L.; Bodger, B.E. The effect of coating residual stress on the fatigue life of thermal spray-coated steel and aluminum. *Surf. Coat. Technol.* **1998**, *108–109*, 59–64. [CrossRef]
16. Kovářík, O.; Haušild, P.; Medřický, J.; Tomek, L.; Siegl, J.; Mušálek, R.; Curry, N.; Björklund, S. Fatigue Crack Growth in Bodies with Thermally Sprayed Coating. *J. Therm. Spray Technol.* **2016**, *25*, 311–320. [CrossRef]
17. Yamaguchi, N.; Shiozaki, T.; Tamai, Y.; Okuda, K.; Ichikawa, Y.; Ogawa, K. Effect of cold spray deposition on fatigue strength of arc-welded ultra-high strength steel sheet. *Int. J. Fatigue* **2022**, *161*, 2411–2502. [CrossRef]
18. Al-Fadhli, H.Y.; Stokes, J.; Hashmi, M.S.J.; Yilbas, B.S. HVOF coating of welded surfaces: Fatigue and corrosion behaviour of stainless steel coated with Inconel-625 alloy. *Surf. Coat. Technol.* **2006**, *200*, 4904–4908. [CrossRef]
19. Chi, J.; Zhang, A.; Xie, S.; Jin, C. Process Optimization and Residual Stress Measurement for Arc Spraying Rapid Tooling. *Acad. J. Xi'an Jiaotong Univ.* **2014**, *48*, 126–130.
20. Hauer, M.; Henkel, K.; Krebs, S.; Krömmer, W. Study of Traverse Speed Effects on Residual Stress State and Cavitation Erosion Behavior of Arc-Sprayed Aluminum Bronze Coatings. *J. Therm. Spray Technol.* **2017**, *26*, 217–228. [CrossRef]
21. Sampath, S.; Jiang, X.Y.; Matejicek, J.; Prchlik, L.; Kulkarni, A.; Vaidya, A. Role of thermal spray processing method on the microstructure, residual stress and properties of coatings: An integrated study for Ni–5 wt.% Al bond coats. *Mater. Sci. Eng.* **2004**, *364*, 216–231. [CrossRef]
22. Chen, Y.X.; Liang, X.B.; Liu, Y.; Wei, S.C.; Xu, B.S. Effect of heat treatment on microstructure and residual stress of wire arc sprayed high carbon steel coating. *Surf. Eng.* **2010**, *26*, 407–412. [CrossRef]
23. Syrek-Gerstenkorn, B.; Paul, S.; Davenport, A.J. Sacrificial Thermally Sprayed Aluminium Coatings for Marine Environments: A Review. *Coatings* **2020**, *10*, 267. [CrossRef]
24. Ryen, A.; Johnsen, R.; Iannuzzi, M.; Årtun, L. Cathodic protection by distributed sacrificial anodes—Performance at elevated Temperature and in mud. In Proceedings of the Corrosion Conference and Expo 2018, Phoenix, AZ, USA, 15–19 April 2018; pp. 2934–2948.
25. Grinon-Echaniz, R.; Refait, P.; Jeannin, M.; Sabot, R.; Paul, S.; Thornton, R. Study of cathodic reactions in defects of thermal spray aluminium coatings on steel in artificial seawater. *Corros. Sci.* **2021**, *187*, 109514. [CrossRef]
26. *DIN EN 10025-2:2019-10*; Hot Rolled Products of Structural Steels—Part 2: Technical Delivery Conditions for Non-Alloy Structural Steels. European Committee for Standardization: Brussels, Belgium, 2019. [CrossRef]

27. *DIN EN ISO 2553:2019-03*; Welding and Allied Processes—Symbolic Representation on Drawings—Welded Joints. DIN Deutsches Institut für Normung: Berlin, Germany, 2019. [CrossRef]
28. *SAE J443:2017-08*; Procedures for Using Standard Shot Peening Almen Test Strip. Society of Automotive Engineers: Warrendale, PA, USA, 2017. [CrossRef]
29. *SAE J442:2022-05*; Test Strip, Holder, and Gage for Shot Peening. Society of Automotive Engineers: Warrendale, PA, USA, 2022. [CrossRef]
30. *ISO/TR 18491:2015-05*; Welding and Allied Processes—Guidelines for Measurement of Welding Energies. International Organization for Standardization: Geneva, Switzerland, 2015.
31. *ISO/TR 17671-1:2002-02*; Welding—Recommendations for Welding of Metallic Materials—Part 1: General Guidance for Arc Welding. International Organization for Standardization: Vernier, Switzerland, 2012.
32. *DIN EN ISO 11124-1:2018-12*; Preparation of Steel Substrates before Application of Paints and Related Products—Specifications for Metallic Blast-Cleaning Abrasives—Part 1: General Introduction and Classification. DIN Deutsches Institut für Normung: Berlin, Germany, 2018. [CrossRef]
33. *FEPA-Standard 42-1:2006*; Grains of Fused Aluminium Oxide, Silicon Carbide and Other Abrasive Materials for Bonded Abrasives and for General Applications Macrogrits F 4 to F 220. Federation of European Producers of Abrasives: Paris, France, 2006.
34. *DIN EN ISO 8501-1:2007-12*; Preparation of Steel Substrates before Application of Paints and Related Products—Visual Assessment of Surface Cleanliness—Part 1: Rust Grades and Preparation Grades of Uncoated Steel Substrates and of Steel Substrates after Overall Removal of Previous Coatings. International Organization for Standardization: Geneva, Switzerland, 2007. [CrossRef]
35. *SAE J2277:2013-04*; Shot Peening Coverage Determination. Society of Automotive Engineers: Warrendale, PA, USA, 2013. [CrossRef]
36. *DIN EN ISO 17632:2016-05*; Welding Consumables—Tubular Cored Electrodes for Gas Shielded and Non-Gas Shielded Metal arc Welding of Non-Alloy and Fine Grain Steels—Classification. DIN Deutsches Institut für Normung: Berlin, Germany, 2016. [CrossRef]
37. *DIN EN ISO 5817:2014-06*; Welding—Fusion-Welded Joints in Steel, Nickel, Titanium and Their Alloys (Beam Welding Excluded)—Quality Levels for Imperfections. DIN Deutsches Institut für Normung: Berlin, Germany, 2014. [CrossRef]
38. *DIN EN ISO 14175:2008-06*; Welding Consumables—Gases and Gas Mixtures for Fusion Welding and Allied Processes. DIN Deutsches Institut für Normung: Berlin, Germany, 2008. [CrossRef]
39. *ISO/TR 14345:2012-06*; Fatigue—Fatigue Testing of Welded Components—Guidance. International Organization for Standardization: Vernier, Switzerland, 2012.
40. ECCS Technical Committee 6. *Background Information on Fatigue Design Rules—Statistical Evaluation*, 2nd ed.; European Convention for Constructional Steelwork: Brussels, Belgium, 2018.
41. Stern, M.; Geary, A.L. Electrochemical Polarization: I. A Theoretical Analysis of the Shape of Polarization Curves. *J. Electrochem. Soc.* **1956**, *104*, 56–63. [CrossRef]
42. *DIN EN ISO 6892-1:2019-11*; Metallic Materials—Tensile Testing—Part 1: Method of Test at Room Temperature. DIN Deutsches Institut für Normung: Berlin, Germany, 2019. [CrossRef]
43. *DIN EN 10027-1:2017-01*; Designation Systems for Steels—Part 1: Steel Names. DIN Deutsches Institut für Normung: Berlin, Germany, 2017. [CrossRef]
44. Leitner, M.; Stoschka, M.; Eichsleder, W. Fatigue Enhancement of Thin-Walled, High-Strength Steel Joints by High Frequency Mechanical Impact Treatment. *Weld. World* **2014**, *58*, 29–39. [CrossRef]
45. Hauer, M.; Gärtner, F.; Krebs, S.; Klassen, T.; Watanabe, M.; Kuroda, S.; Kroemmer, W.; Henkel, K.-M. Process Selection for the Fabrication of Cavitation Erosion-Resistant Bronze Coatings by Thermal and Kinetic Spraying in Maritime Applications. *J. Therm. Spray Technol.* **2021**, *30*, 1310–1328. [CrossRef]
46. *DNV-RP-B401*; Cathodic Protection Design. Det Norske Veritas: Bærum, Norway, 2010.

Journal of
Marine Science and Engineering

MDPI

Article

Study on Fatigue Spectrum Analysis and Reliability Analysis of Multilayer Flexible Riser

Jianxing Yu [1], Fucheng Wang [1,2,*], Yang Yu [1,2], Haoda Li [1,2], Xin Liu [1,2] and Ruoke Sun [1,2]

1 State Key Laboratory of Hydraulic Engineering Simulation and Safety, Tianjin University, Tianjin 300072, China
2 Tianjin Key Laboratory of Port and Ocean Engineering, Tianjin University, Tianjin 300072, China
* Correspondence: tjuwfc1993@163.com

Abstract: Multilayer composite flexible risers have been widely used in engineering. However, this type of structure is complex, as there are influences between layers. Moreover, a range of uncertain factors need to be considered in fatigue analysis. Therefore, it is difficult to perform the fatigue analysis research of multilayer flexible risers. In this paper, the fatigue spectrum analysis and reliability analysis method of a nine-layer flexible riser structure are proposed, and a complete fatigue and reliability analysis process for multilayer structures is developed. The theoretical basis of the fatigue spectrum analysis method is introduced, and the calculation program is described. Finite element software is used to analyze the stress of the multilayer flexible riser under the influence of the upper platform structure movement and the ocean current. Moreover, the stress response of the riser structure of each layer is obtained. According to this, the irregular wave load is simulated by the method of random number simulation, and the stress response spectrum is formed. Then, the appropriate S-N curve is selected to calculate the fatigue damage degree of each layer, and the fatigue damage nephogram is displayed, so as to analyze the structural fatigue damage. Finally, the uncertainty in the process of fatigue damage calculation is analyzed. According to the results, the methods of multilayer riser analysis are summarized and the future research directions are put forward.

Keywords: multilayer flexible riser; uncertainty; spectrum analysis; fatigue damage; reliability

Citation: Yu, J.; Wang, F.; Yu, Y.; Li, H.; Liu, X.; Sun, R. Study on Fatigue Spectrum Analysis and Reliability Analysis of Multilayer Flexible Riser. *J. Mar. Sci. Eng.* **2022**, *10*, 1561. https://doi.org/10.3390/jmse10101561

Academic Editors: Sören Ehlers and Moritz Braun

Received: 16 August 2022
Accepted: 17 October 2022
Published: 21 October 2022

1. Introduction

With the structural complexity of offshore deep-water and ultra-deep-water platforms, the difficulty of structural safety assessment increases significantly [1]. Flexible risers have been used in marine structures since 1970. In the process of deep-water oil and gas drilling, flexible risers connect the seabed and the platform structure, and transport the oil and gas resources extracted from the seabed. This structure plays a key role, and any damage will carry serious consequences [2]. At present, most of the flexible risers are multilayer structures, and they are very complex [3]. Interactions exists between each layer. This makes it difficult to study the fatigue strength of composite flexible risers. In the multilayer flexible composite riser, the structural load is mainly borne by the metal layer structure, including the tensile armor layer, compressive armor layer and so on, while other connection structures are polymeric materials and do not participate in load bearing [4]. Therefore, these layers are generally not considered in the study of structural strength. In current research, the skeleton layer structure and the compressive/tensile layer structure are regarded as mutually orthogonal cylindrical structures. The structure simulation method is effective to obtain the circumferential pressure load, and the stress response process of the flexible riser under a circumferential load can be proposed [5,6].

Due to the existence of repeated loading in the deep-sea environment, the fatigue damage of the riser structure is also a focus of research. The fatigue damage analysis method of common riser structures has been studied so far. In the composite flexible riser, the spiral structure is the most prone to damage. Therefore, the research focus is usually on the helical structure. SKEIE [7] focused on the stress and fatigue life of spiral components under the action of friction caused by stick and slip in the bending process, and he focused on the influence of friction stress on fatigue life. As for the effect of friction force, Sævik [8] analyzed the nonlinear moment curvature relationship and sandwich beam theory; two theoretical models were proposed and verified via experiments. However, in the current research on riser fatigue damage, most of the research objects are the overall structures, and few are local structures. In his study, De Sousa [9] proposed that, on the basis of calculating the overall stress response of the riser, the global response can be transferred to the local structures so as to calculate the stress of the local helix structure. In a further study, he considered the interlayer interaction of the flexible riser and proposed a theoretical calculation method for the overall fatigue life of the flexible riser. Kunpeng Wang et al. [10] proposed a global analysis model of the flexible riser, conducted a global analysis of the structure, focused on the fatigue damage of the spiral steel wire near the touchdown point, and considered the interlayer friction.

However, in the process of fatigue life calculation, many variables are uncertain. This will affect the accuracy of the calculation. Therefore, these uncertainties also need to be taken into account in the research. In this paper, based on the analysis of multilayer flexible risers and the established model, the fatigue calculation method of marine structures is integrated into the existing system, the uncertainty in the fatigue analysis process is fully considered, and the solutions to various uncertain factors are given.

In this paper, the solution of load uncertainty is proposed through the spectral analysis method. The basic flow of spectrum fatigue analysis is introduced, the program of fatigue spectrum analysis is designed to calculate the fatigue damage according to the time history stress data, and the result is displayed in the form of a cloud diagram. Then, a nine-layer composite flexible pipe is taken as an example for calculation, and load-bearing structures are selected to analyze the distribution of fatigue damage. Then, considering uncertain factors, with the exception of load, a reliability analysis method is proposed. The results of fatigue spectrum analysis are combined with the analysis of other uncertain factors through reliability analysis, and the uncertain factors in fatigue research are taken into account in a more comprehensive and concrete manner. Finally, the future research directions of this work are proposed.

2. Fatigue Spectrum Analysis

Most previous fatigue studies adopted deterministic methods. In a given stress range, the cycle times of structural fracture were calculated via the S-N curve. However, we usually take approximate preliminary estimations by probabilistic method. There will be some uncertainty in the approximation algorithm. This uncertainty mainly arises from whether the S-N curve adopted is consistent with the material and welding form of the actual structure, and whether the cyclic load selected is consistent with the load form of the actual conditions [11]. Due to the huge amount of time history load data, the calculation time will be greatly increased if the uncertainty of load is taken into account. In recent years, due to the rapid development of computing software, scholars have begun to take the accuracy of fatigue assessment into account [12].

Fatigue spectrum analysis fully takes the randomness and uncertainty of loads into account [13] and provides more reasonable statistics of the load cycle process on the structure by establishing a load response spectrum or stress response spectrum. This makes spectral analysis more reliable in the fatigue damage assessment of marine structures [14]. Accordingly, spectral analysis also requires more detailed statistics of the load and fully considers its irregularity [15].

This section introduces the whole process of fatigue spectrum analysis. Firstly, the appropriate wave load power spectrum is selected according to the environmental conditions of marine structures, and then the stress response spectrum is calculated according to the peak distribution of stress in an alternating process in a short-term sea state. Finally, the cumulative fatigue damage results are calculated through the Palmgren–Miner method.

2.1. Stress Response Spectrum

There are a large number of standardized spectra in marine structure design standards. The most common are the Pierson–Moskowitz spectrum (referred to as P-M spectrum) [16] and JONSWAP spectrum [17]. The power spectral density selected in this paper is the two-parameter P-M spectrum recommended by the International Ship Structure Conference (ISSC). The P-M spectrum can be written in the form of different parameters. If the two parameters of meaningful wave height H_s and mean span zero period T_Z are used to express the wave spectrum, the expression is

$$G_{\eta\eta}(\omega) = \frac{H_s^2}{4\pi}\left(\frac{2\pi}{T_z}\right)^4 \omega^{-5} \exp\left(-\frac{1}{\pi}\left(\frac{2\pi}{T_z}\right)^4 \omega^{-4}\right) \tag{1}$$

In the analysis, the actual response frequency should be the encounter frequency ω_e, which is related to the wave frequency ω as follows.

$$\omega_e = \omega(1 + \frac{2\omega U}{g}\cos\theta) \tag{2}$$

where θ is the wave direction angle (rad); U is the sailing speed (Knot).

The energy of the corresponding wave frequency element is unchanged; the input wave energy spectrum should be transformed and expressed in the form of the encounter frequency and wave angle. The transformed expression is as follows

$$G_{\eta\eta}(\omega)d\omega = G_{\eta\eta}(\omega_e, \theta)d\omega_e \tag{3}$$

According to Equations (2) and (3), the wave energy spectrum expressed by the encounter frequency can be obtained as follows:

$$G_{\eta\eta}(\omega_e, \theta) = \frac{G_{\eta\eta}(\omega)}{1 + \frac{2\omega U}{g}\cos\theta} \tag{4}$$

For the working condition with wave angle θ, the response spectrum of stress can be obtained from Equation (4), which is expressed as

$$G_{XX}(\omega_e, \theta) = |H_\sigma(\omega_e, \theta)|^2 \cdot G_{\eta\eta}(\omega_e, \theta) \tag{5}$$

where $H_\sigma(\omega_e, \theta)$ is the stress response transfer function.

2.2. Short-Term Distribution of Stress Range

The stress peak of the stress alternating process in the short-term sea state follows the Rayleigh distribution. The probability density function is

$$f_\sigma(\sigma) = \frac{\sigma}{m_0}\exp\left(-\frac{\sigma^2}{2m_0}\right),\ 0 \le \sigma < +\infty \tag{6}$$

where σ is peak stress; m_0 is the zero-order moment of $G_{XX}(\omega_e)$, the power spectral density of the alternating stress process.

Then, the probability density function and distribution function of the corresponding stress range can be obtained, which can be expressed as

$$f_{\Delta\sigma}(\Delta\sigma) = \frac{\Delta\sigma}{4m_0}\exp\left(-\frac{\Delta\sigma^2}{8m_0}\right),\ 0 \le \Delta\sigma < +\infty \tag{7}$$

$$F_{\Delta\sigma}(\Delta\sigma) = 1 - \exp\left(-\frac{\Delta\sigma^2}{8m_0}\right),\ 0 \leq \Delta\sigma < +\infty \tag{8}$$

where $\Delta\sigma$ is the stress range.

When computing M, care should be taken to replace the frequency ω with the encounter frequency ω_e. When calculating the average cross-zero rate, the second moment of the response spectrum is also required, which can be expressed as

$$\begin{aligned} m_n &= \int_0^{+\infty} \omega_e^n \cdot G_{XX}(\omega_e, \overline{\theta}) d\omega_e \\ &= \int_0^{+\infty} \omega_e^n \cdot \int_{-\pi/2}^{\pi/2} f(\beta)\left[H(\omega_e, \overline{\theta} - \beta)\right]^2 \cdot G_{\eta\eta}(\omega_e, \overline{\theta} - \beta) d\beta d\omega_e \end{aligned} \quad (n = 0,\ 2) \tag{9}$$

In order to obtain the number of stress cycles within a given time, it is necessary to give the average cross-zero rate ν of the alternating stress process, i.e., the average number of times that we cross the zero mean with a positive slope in unit time. Its expression is as follows:

$$\nu = \frac{1}{2\pi}\sqrt{\frac{m_2}{m_0}} \tag{10}$$

2.3. Stress Response Spectrum

For the local structure of risers, the stress amplitude and the average stress required in the fatigue analysis are scalars. However, the stress calculated by time history is a tensor, so it is necessary to calculate a scalar value based on the stress tensor for fatigue stress analysis. The calculated scalar value is fatigue stress, which is calculated via different methods according to different materials and structures. In the fatigue check in this study, the Von Mises stress of each position was calculated, and it was applied for fatigue analysis. Von Mises stress was calculated by the following formula:

$$\sigma_{\rm vm} = \sqrt{\frac{1}{2}((\sigma_x - \sigma_y)^2 + (\sigma_y - \sigma_z)^2 + (\sigma_z - \sigma_x)^2 + 6(\tau_{xy}^2 + \tau_{yz}^2 + \tau_{xz}^2))} \tag{11}$$

Specifically, for the cylindrical riser structure with an isotropic structure, the stress–strain relationship can be expressed as follows [2]:

$$\begin{bmatrix} \sigma_x \\ \sigma_y \\ \tau_{xy} \end{bmatrix} = \begin{bmatrix} \frac{E_x}{1-\upsilon_{xy}\upsilon_{yx}} & \frac{\upsilon_{yx}E_x}{1-\upsilon_{xy}\upsilon_{yx}} & 0 \\ \frac{E_y\upsilon_{yx}}{1-\upsilon_{xy}\upsilon_{yx}} & \frac{E_x}{1-\upsilon_{xy}\upsilon_{yx}} & 0 \\ 0 & 0 & G_{xy} \end{bmatrix} \begin{bmatrix} \varepsilon_x \\ \varepsilon_y \\ \gamma_{xy} \end{bmatrix} \tag{12}$$

where E is the Young's modulus, υ is the Poisson's ratio, and G is the shear modulus.

In this study, the local riser structure is mainly subjected to three types of loads: tension and compression, bending, and torsion. The stress response of the riser structure under the action of these three types of loads is studied separately. Finally, through the combined action of the three types of loads, the structural stress response of the riser structure under the overall load is studied.

The fatigue of structures is caused by cyclic dynamic loads, but a structure may be subjected to multiple complex loads at the same time during service. In this paper, the superposition method is used for load calculation. In the superposition method, it can be divided into several independent load conditions according to the load characteristics. If a structure is subjected to n load conditions at the same time, based on the maximum load $F_{\max}$ in the load conditions, then the time history load can be expressed as a time history process $F_i(t)$ with the maximum load as a coefficient. The superposition method is used to calculate the time history stress of the structure. We apply the maximum load in the working condition to the structure model, calculate the stress distribution σ_i^1 of the structure, and the time history stress of the structure under all loading conditions can be obtained by superposition:

$$\sigma(t) = \sum \sigma_i^1 F_i(t) \tag{13}$$

2.4. Fatigue Calculation of Structure by Rain-Flow Counting Method

The rain-flow counting method is generally applied to deal with the time-calendar stress variation of structures. The method of transforming an irregular, random time history load into a series of cycles is called "cycle counting". The simplified rain-flow counting method is suitable for a repeating course based on a typical load spectrum. Since the load is the repetition of a typical segment, taking the start and end of the maximum peak and valley as the typical segment will not cause a loss of generality, as shown in Figure 1.

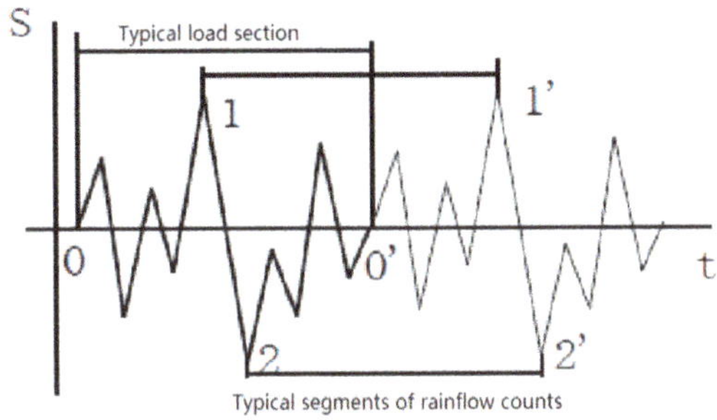

Figure 1. The selection of a typical section of rain-flow counting.

Based on the stress process statistics given by the rain-flow counting method, fatigue analysis can be carried out further. Specifically, the load cycle times n_i can be obtained from the variable amplitude load spectrum S_i. Then, the damage of the component D_i under the action cycle n_i under the stress level S_i can be calculated as $D_i = \frac{n_i}{N_i}$. Under the action of stress levels S_i and each cycle n_i, the total damage can be defined as follows:

$$D_i = \sum_i^k D_i = \sum_i^k \frac{n_i}{N_i} (i = 1, 2, \ldots k,) \tag{14}$$

The destruction criteria are

$$D = \sum \frac{n_i}{N_i} = 1 \tag{15}$$

where n_i is the cycle number under the stress S_i, which is given by the load spectrum; N_i is the life cycle to destruction under the stress S_i, determined by the S-N curve.

3. Example of Fatigue Spectrum Analysis of Multilayer Riser Structure

The procedure of fatigue strength analysis for complex flexible riser structures at local locations is shown in Figure 2.

Risers are subjected not only to external loading but also to internal loading. Much research has been done on the effect of internal pressure in multilayered risers and pipelines [18–22]. The internal pressure usually does not have a significant effect on fatigue during stable oil and gas collection. Therefore, it is assumed in this paper that the influence of internal loading is ignored. The riser selected in this paper has a total of 9 layers. From the interior to exterior, they are the skeleton layer, internal sheath layer, compressive armor layer, internal anti-friction layer, internal tensile armor layer, external anti-friction layer, external tensile armor layer, four-fold anti-friction layer, and external sheath layer. Among them, the four-fold anti-friction layer is a four-fold structure, which is treated as one layer in the analysis. The solid structure of the riser is shown in Figure 3.

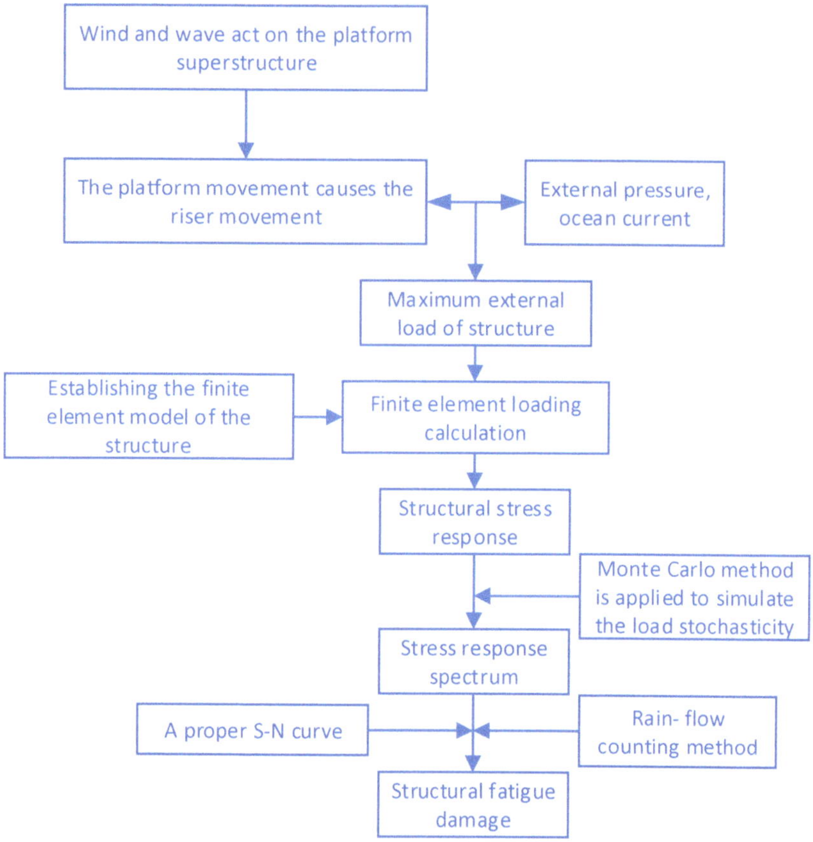

Figure 2. The fatigue spectrum analysis process.

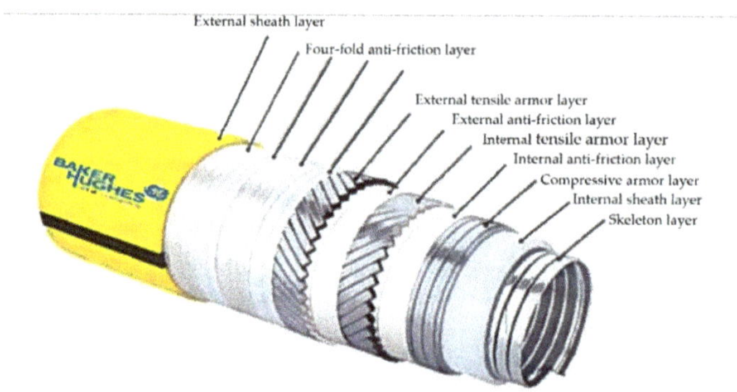

Figure 3. A schematic diagram of the multilayer flexible riser structure.

The basic parameters of the riser are as Table 1:

Table 1. The material properties of multilayer risers.

Layer	Material	ρ (g/cm^3)	E	ν	O.D. (Outer Diameter) [mm]	Thick [mm]
Skeleton layer	Duplex 2205	8.0	200 GPa	0.3	248.58	6.84
Internal sheath layer	PA 12 Natural	1.01	8700 MPa	0.3	262.58	7.00
Compressive armor layer	Steel 110 ksi UTS	8.0	210 Gpa	0.3	278.58	8.00
Internal anti-friction layer	Tape PA 11 P20 30mil	1.04	4000 MPa	0.3	281362	1.52
Internal tensile armor layer	Steel 190ksi UTS	8.0	200 Gpa	0.3	287.62	3.00
External anti-friction layer	Tape PA 11 P20 30mil	1.04	4000 MPa	0.3	290.66	1.52
External tensile armor layer	Steel 190ksi UTS	8.0	200 Gpa	0.3	296.65	3.00
Four-fold anti-friction layer	Polypropylene High Density Glass Filament Polypropylene Tape Polyester Fabric	8.0	200 Gpa	0.3	297.24 298.52 299.11 299.93	0.30 0.64 0.30 0.41
External sheath layer	PE100 Grade GP100OR	1.0	900 MPa	0.3	317.93	9.00

Figures 4 and 5 show the local segment model of the multilayer riser established in finite element analysis software ABAQUS.

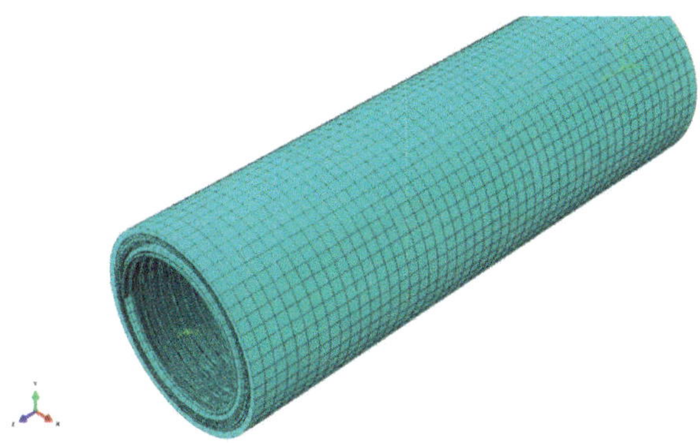

Figure 4. A finite element model of a multilayer flexible riser.

Set geometric reference point RP1 at (0,0,0) and RP2 at (0,0,1000). Defined RP1 as the bottom and RP2 as the top of the riser respectively. All nodes of the bottom section of each layer were coupled to RP1, and all nodes of the top section of each layer were coupled to RP2. The coupling method was kinematic coupling.

RP1 boundary conditions were rigid and fixed, and RP2 is set according to different working conditions, shown as Figure 6.

Figure 5. The sectional view of a multi-storey structure.

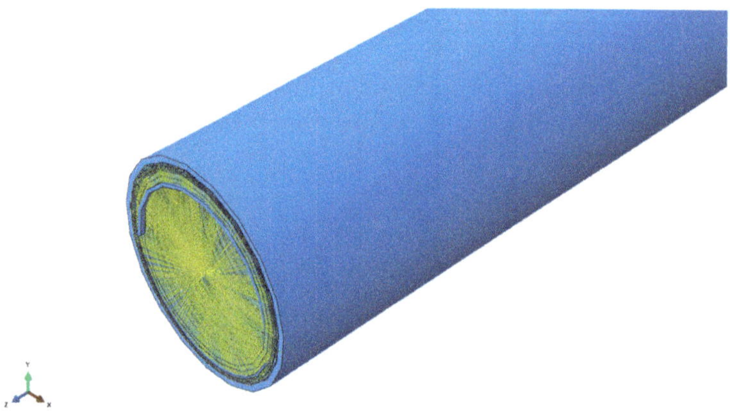

Figure 6. The boundary conditions of a multi-story structure.

The loads studied include axial compression, bending moment, and torsion. The stress response of the structure under external pressure is also calculated. However, the external pressure load is constant and will not cause fatigue damage, so it is not considered in this paper. The load is obtained from the maximum force under conventional load conditions in the corresponding sea area. The loading situation is shown as Table 2.

Table 2. The boundary conditions of a multi-story structure.

Loading Condition	Load	Boundary Condition of RP1	Boundary Condition of RP2
Axial compression	F = 150 kN	Rigid fixed	Free
Bending moment	M = 9 kN·m	Rigid fixed	Free
Torsion	T = 4000 N·m	Rigid fixed	Axial restraint

The structural stress of each metal layer under axial pressure is as Figure 7.

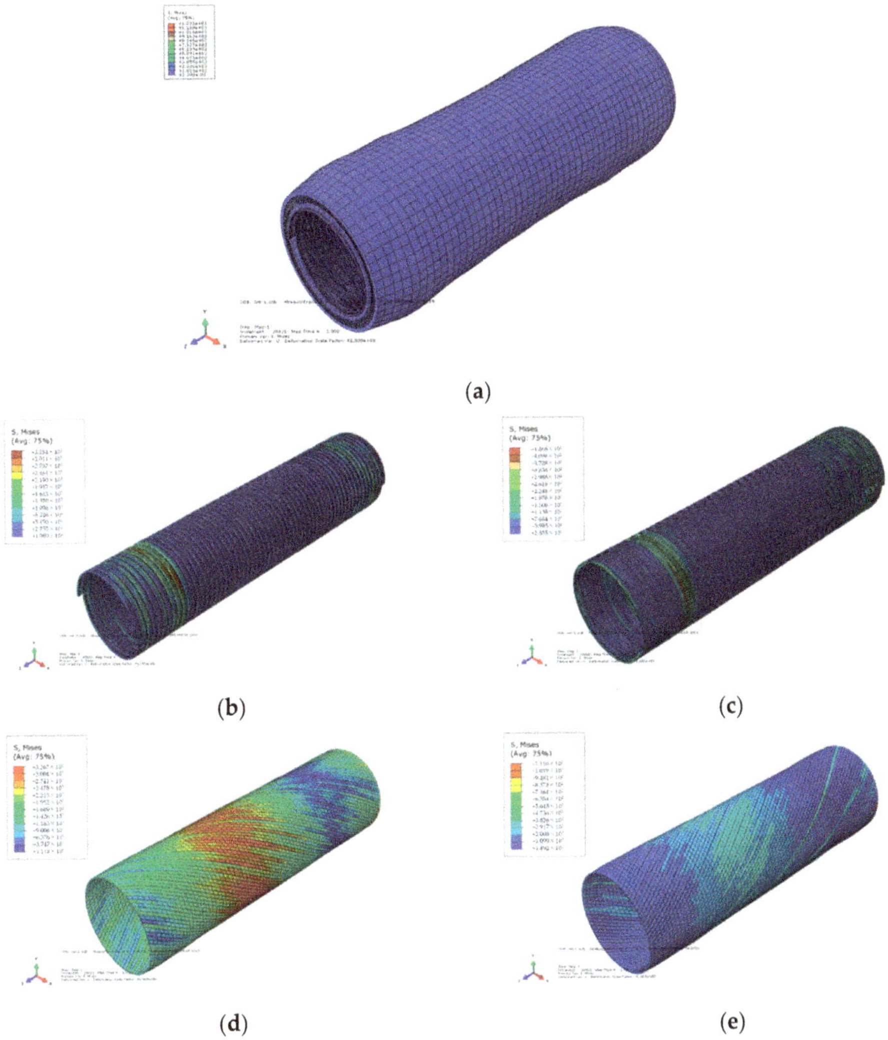

Figure 7. A stress nephogram under axial pressure. (**a**) The riser; (**b**) Skeleton layer; (**c**) Compressive armor layer; (**d**) Internal tensile armor layer; (**e**) External tensile armor layer.

The structural stress of each metal layer under the bending moment is as Figure 8.

The structural stress of each metal layer under torsion is as Figure 9.

Then, the fatigue reliability of the structure is analyzed on the premise of ensuring that the structure meets the ultimate strength requirements. The load of random wave action is assumed to be normally distributed. The Monte Carlo method is applied to form the time history load spectrum and simulate the time history load acting on the structure under the action of random waves during the design life of the structure.

In this paper, the stress response of each layer in the multilayer riser under maximum load is calculated by the finite element method. According to the fatigue spectrum analysis and calculation method described above, the cumulative fatigue damage degree of each layer is calculated by using a self-designed program of fatigue spectrum analysis. As mentioned above, the metal layers mainly play the bearing role in the multilayer riser, so the metal layers are mainly analyzed in this section. The stress calculated above is the maximum stress value under each load condition. Therefore, the fatigue stress response spectrum in this paper is the ratio ranges from 0 to 1 based on the maximum stress response. The stress response spectrum is shown in Figure 10. Load transformation was carried out every 0.1 s, and 4000 s was set as a group of cycle. The stress response spectrum in the figure cycles 25 times.

Figure 8. A stress nephogram under bending moment. (**a**) The riser; (**b**) Skeleton layer; (**c**) Compressive armor layer; (**d**) Internal tensile armor layer; (**e**) External tensile armor layer.

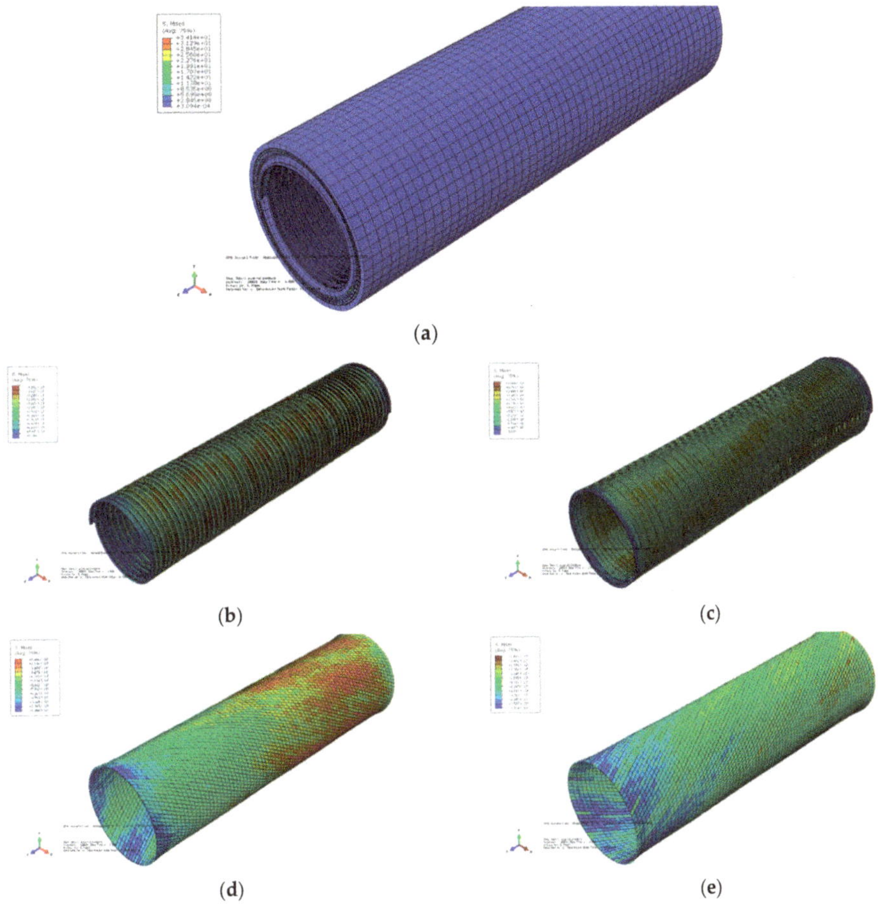

Figure 9. A stress nephogram under torsion. (**a**) The riser; (**b**) Skeleton layer; (**c**) Compressive armor layer; (**d**) Internal tensile armor layer; (**e**) External tensile armor layer.

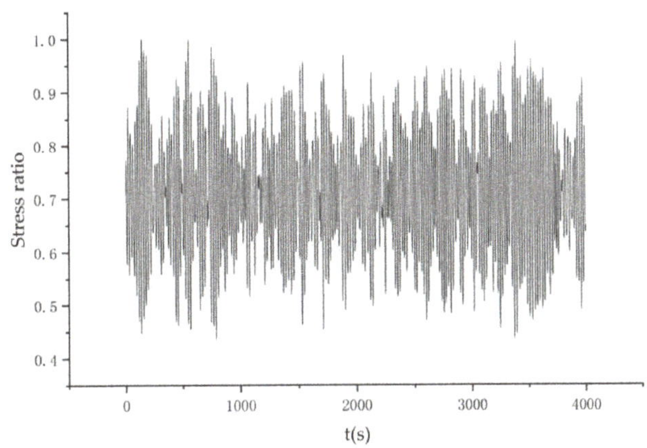

Figure 10. The stress response spectrum for the riser.

In the multilayer riser structure, the high strength steel mainly bears the load of the structure. The S-N curve of the high strength steel structure is given in the specification. The material properties of steel structures in this paper are basically the same. In order to facilitate calculation, the same curve can be approximately used for calculation. According to the high-strength stress steel formula given by the DNV-RP-C203 specification [23], the fatigue S-N curve equation should be selected for the complex flexible riser multilayer structure as follows:

$$\log N = 17.446 - 4.70 \log\Delta\sigma \tag{16}$$

The fatigue damage degree of each node is calculated. The results are visually expressed in the form of a cloud image in the self-programmed display software. The stress and fatigue damage cloud diagram of each metal layer is given and the results are analyzed.

Under the action of tension and compression load, the fatigue damage nephogram of the skeleton layer and cumulative layer are as follows.

The fatigue damage nephograms of the skeleton layer under the three loads are as Figure 11:

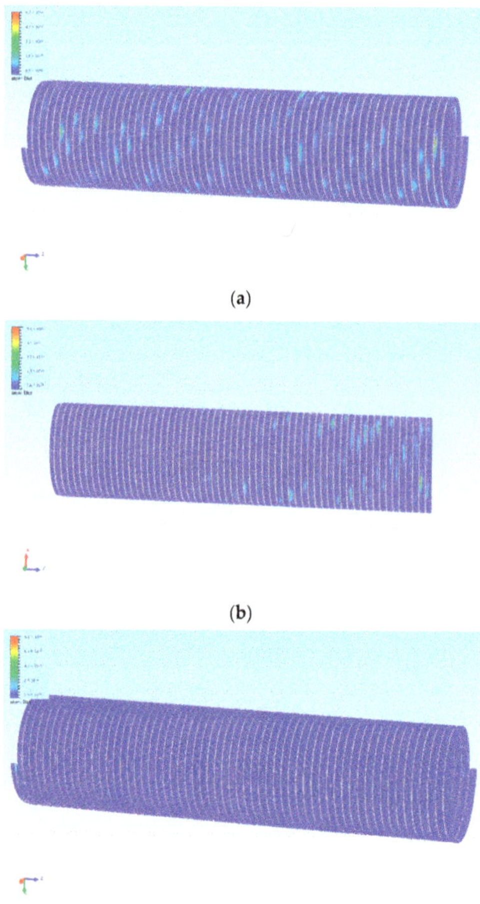

Figure 11. Fatigue damage of the skeleton layer. (**a**) The fatigue damage nephogram under axial compressive load; (**b**) The fatigue damage nephogram under bending moment; (**c**) The fatigue damage nephogram under torsion.

The fatigue damage nephograms of the compressive armor layer under the three loads are as Figure 12:

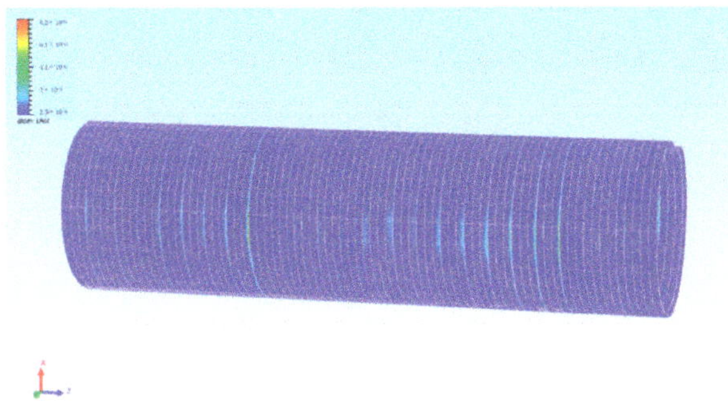

(**a**)

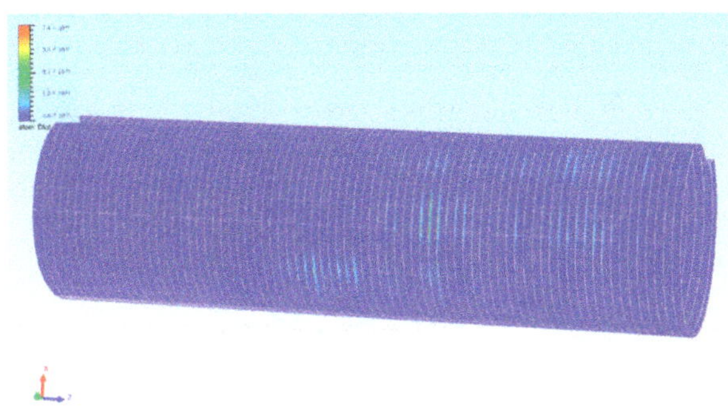

(**b**)

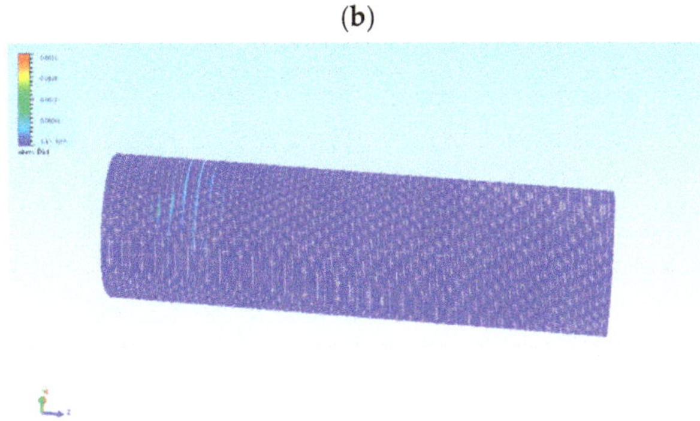

(**c**)

Figure 12. Fatigue damage of a compressive armor layer. (**a**) The fatigue damage nephogram under axial compressive load; (**b**) The fatigue damage nephogram under bending moment; (**c**) The fatigue damage nephogram under torsion.

The fatigue damage nephograms of the internal tensile armor layer under the three loads are as Figure 13:

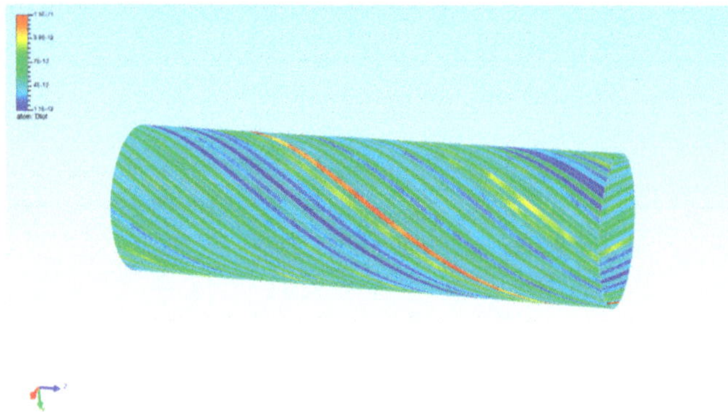

(**a**)

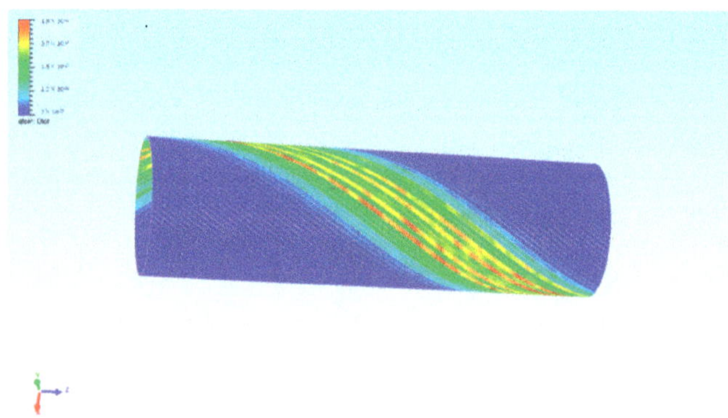

(**b**)

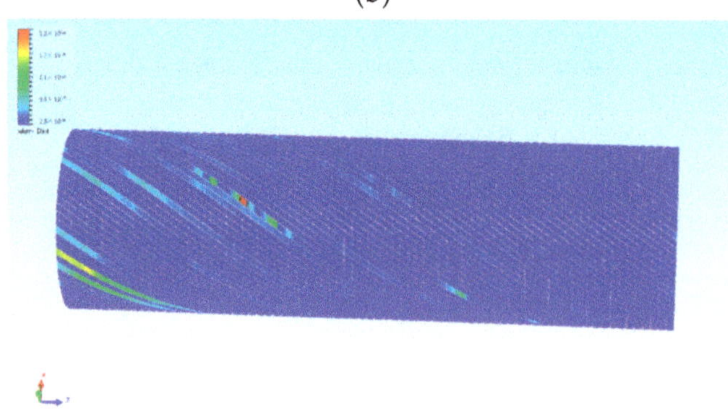

(**c**)

Figure 13. Fatigue damage of an internal tensile armor layer. (**a**) The fatigue damage nephogram under axial compressive load; (**b**) The fatigue damage nephogram under bending moment; (**c**) The fatigue damage nephogram under torsion.

The fatigue damage nephograms of the external tensile armor layer under the three loads are as Figure 14:

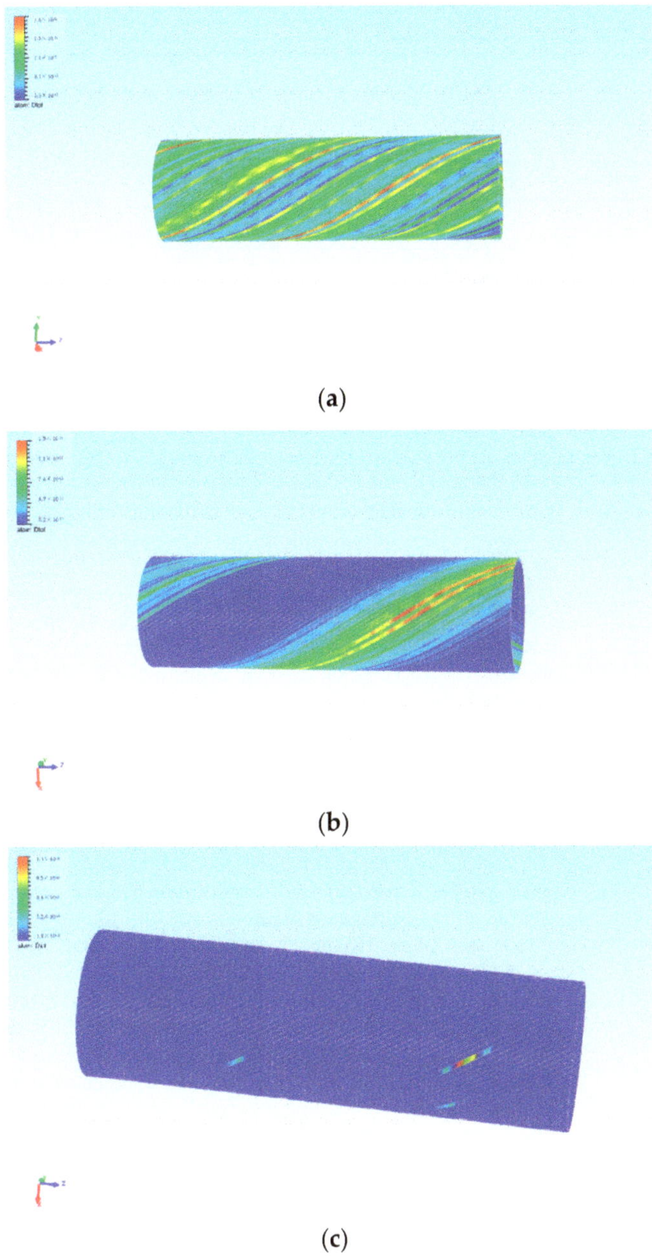

Figure 14. Fatigue damage of an external tensile armor layer. (**a**) The fatigue damage nephogram under axial compressive load; (**b**) The fatigue damage nephogram under bending moment; (**c**) The fatigue damage nephogram under torsion.

The maximum fatigue damage under the three loads is shown in Table 3.

Table 3. The maximum fatigue damage under the three loads.

Layer	Fatigue Damage		
	Axial Compression	Bending Moment	Torsion
Skeleton layer	6.3×10^{-13}	5.4×10^{-13}	8.1×10^{-16}
Compressive armor layer	8.2×10^{-12}	7.4×10^{-13}	0.0024
Internal tensile armor layer	1.3×10^{-11}	4.9×10^{-12}	2.2×10^{-16}
External tensile armor layer	1.5×10^{-11}	1.5×10^{-11}	1.3×10^{-14}

In the calculation results, there is a concentrated area of fatigue damage under torsion. This is mainly caused by stress concentration under torsion.

The stress concentration of the riser occurs in the middle position of the riser, and there will be greater fatigue damage at the stress concentration. Even in layers without helical form, such as the internal tensile armor and external tensile armor, the cumulative fatigue damage degree will also appear as a spiral form. In the stress cloud, no such spiral result exists. Especially for the tensile armor layer, its structure is completely symmetric, and, since the load applied in this section is also symmetric, both its stress response and fatigue cumulative damage should appear in a relatively symmetric form. However, the results presented in this section indicate that an interlayer interaction obviously exists and cannot be ignored in the analysis of multilayer riser structures. The structural form of the adjacent layers will cause the interlayer load, which will affect the fatigue damage of the structure.

When simulated by ABAQUS, it is observed from the stress cloud diagram of some structures that there is stress concentration in some positions of the structure. Especially in the case of a symmetrical load applied to the symmetrical structure, the stress concentration is not symmetric, which can also indicate the interaction between the layers. In Figures 10 and 11, for example, the stress response is asymmetrical. However, in the stress response of some structures, this situation is not obvious. For example, in the stress cloud diagram obtained under the action of tension and compression load in FIG. 6, the stress situation is in a symmetric form on the whole, and there is no obvious stress concentration node. However, in the fatigue damage cloud image, it is obvious that the damage degree has a local concentration. In the stress response analysis, the structure is only loaded once. Although there is stress concentration in some locations, it is not obvious. This concentration may be reflected in numerical values, but it is difficult to determine in the stress distribution graph. However, after a large number of load cycles, the stress concentration effect is obviously enlarged, and finally the cumulative fatigue damage is concentrated.

In previous fatigue analyses, typical welding joints were mostly selected as calculation nodes [24]. According to the specification, the selected nodes are usually located in the structural corners, large openings of coamings, and other positions where stress concentration can easily occur. On this basis, combined with the finite element analysis results of the whole structure, the stress concentration location was selected for fatigue analysis. Through the findings presented in this section, it can be observed that in the selection of fatigue nodes, this method cannot select all fatigue danger points. Some nodes without obvious stress concentration also have the possibility of developing concentrated fatigue damage, which affects the accuracy of fatigue analysis. In this section, the fatigue damage of all nodes is obvious in the form of the fatigue damage nephogram, which makes the fatigue analysis result more comprehensive and objective.

4. Time-Variant Fatigue Reliability Analysis of Multilayer Flexible Risers

The sources of uncertainty in the process of fatigue damage assessment have been discussed in many studies [25,26]. As for structure, comparing risers in practical engineering with finite element models, there may be some detail deviation in the materials and structures. Moreover, welding defects and residual stresses in the actual engineering processing cannot be predicted in numerical simulation analysis, which will cause the numerical simulation analysis results to be different from the structural stress in practical engineering, thus affecting the prediction of fatigue damage. As for loads, there is some randomness in the loads of practical structures. Regarding fatigue calculation methods, the widely used Palmgren–Miner linear cumulative damage theory holds that the structure is considered to be damaged as long as the linear sum of fatigue damage in all sea states reaches 1. However, this is an approximate method, and the cumulative damage degree is not always exactly 1 when the structure is damaged under practical circumstances. There are some studies on this shortcoming, and researchers are looking for alternative [27]. This rule is linear and does not take into account the loading sequence effects. Therefore, the fatigue calculation above is based on the assumption of linear damage accumulation. Recently, some nonlinear fatigue models have been established [28,29]. However, the Palmgren–Miner criterion is still widely adopted in the evaluation of ships and marine structures due to its simplicity.

As for the load uncertainty, the Monte Carlo method is applied in the fatigue load spectrum simulation above, i.e., random numbers are used to simulate the random sea state. Therefore, the randomness of the load is not considered again in this section. In this section, considering the other two factors, the reliability method is applied to reduce the impact of uncertainty.

4.1. Establishment of Reliability Analysis Model

For the multilayer flexible riser structure considered in this paper, in order to ensure its safety, it is necessary that each layer structure is not damaged. As mentioned above, only metal layers are considered for fatigue assessment. Since the failure of each layer is independent, the multilayer riser can be regarded as a serial system. If there is fatigue failure in any layer, the entire multilayer flexible riser is considered to have failed. Thus, the structural reliability can be defined as

$$R = R_s \cdot R_{ca} \cdot R_{ita} \cdot R_{eta} \tag{17}$$

where $R_s, R_{ca}, R_{ita}, R_{eta}$ are the reliability of the skeleton layer, compressive armor layer, internal tensile armor layer, and external tensile armor layer.

When the fatigue life is shorter than the design life, fatigue failure occurs. The limit state equation is expressed as follows:

$$g(z) = T - T_A \tag{18}$$

where Z is the vector related to Δ, B, and D. Structural fatigue reliability is

$$R_f = R(T > T_A) = R[g(z) > 0] \tag{19}$$

The failure probability is

$$P_f = 1 - R_f \tag{20}$$

Due to the influence of uncertain factors in each layer of the structure, the fatigue life of each layer is dispersed and superimposed, resulting in great dispersion of the overall structure. Therefore, two uncertain factors in each layer should be quantified. At present, most research adopts the probability method to simulate this, i.e., quantifying the uncertainty through a random process and random variable.

Considering the uncertain factors, the fatigue life of the structure is

$$T = \frac{\Delta T_A}{BD} \tag{21}$$

where T_A is the design life, D is the fatigue reliability, Δ is the random variable considering that the cumulative damage degree is not always 1 when fatigue damage occurs to the structure, and B is the random variable considering the influence of errors between modeling and practical structures. The reliability indicator is

$$\beta = -\phi^{-1}(P_f) \tag{22}$$

4.2. Reliability Calculation

In the fatigue life prediction of marine structures, the fatigue reliability calculation process is as follows:

(1) Conduct random sampling and select samples according to the statistical characteristics of random variables;
(2) The random number obtained by random sampling is used as the initial parameter to calculate the fatigue life of the structure;
(3) According to the fatigue life of each considered detail, combined with the limit state equation, the reliability of the system is finally calculated.

Two types of uncertain factors are analyzed. Considering the uncertainty of the deviation between the model and the practical structures, it affects the fatigue damage calculation of the structure by affecting the stress results calculated. Therefore, its distribution type can be considered as the load distribution. Since the influence of load has been considered in the previous section, the mean coefficient of load here can be regarded as 1. The mean coefficient and variation coefficient of other parameters can be selected according to the study of Wirsching [23]. The values are described in Table 4.

Table 4. The median and variation coefficients of random variables.

Variable	Distribution	Mean Value	Coefficient of Variation
B	lognormal distribution	1	0.2
Δ	lognormal distribution	1	0.3

According to the selection structure of this paper, and referring to the calculation results in Section 2, we select nodes with the largest degree of fatigue damage in each calculation layer (numbered 1–4 from interior to exterior). Substituting the values of parameters from the preceding table into the formula, the reliability calculation results are as shown in Table 5.

Table 5. Reliability calculation results of the four nodes.

Layer	Reliability Specifications
Skeleton layer	3.916
Compressive armor layer	2.585
Internal tensile armor layer	3.097
External tensile armor layer	4.435

Based on the structural attenuation model of ships and marine structures summarized in previous studies [30], and introducing the fatigue cumulative damage increasing function, the power function model was used to represent the annual increasing coefficient of cumulative fatigue damage. The expression is as follows:

$$D(n) = D(0)\alpha(n) \tag{23}$$

$$\alpha(n) = (1 + k\frac{n}{T_D})^2 \tag{24}$$

where $D(n)$ is the cumulative damage after n years, $D(0)$ is the initial cumulative damage, $\alpha(n)$ is the increasing coefficient of cumulative damage, T_D is the design life for structural fatigue. The design life of this structure is 20 years. The reliability index of the structure in the design life changes with the years of use, as shown in the following table. The reliability index of the structure in the design service life changes with time, as shown in the following table. The reliability index β of the structure in the design service life changes with the year, as shown in Table 6. The higher the reliability index, the lower the probability of fatigue failure.

Table 6. Reliability indicators for different years.

Layer	Reliability Specifications				
	0 Years	**5 Years**	**10 Years**	**15 Years**	**20 Years**
Skeleton layer	5.916	5.847	5.781	5.716	5.654
Compressive armor layer	2.585	2.516	2.450	2.385	2.323
Internal tensile armor layer	5.097	5.028	4.962	4.899	4.836
External tensile armor layer	4.435	4.365	4.298	4.233	4.170

With the increase of service life, the structural reliability index decreases gradually. Figure 15 shows the changes in the reliability indicators in different years.

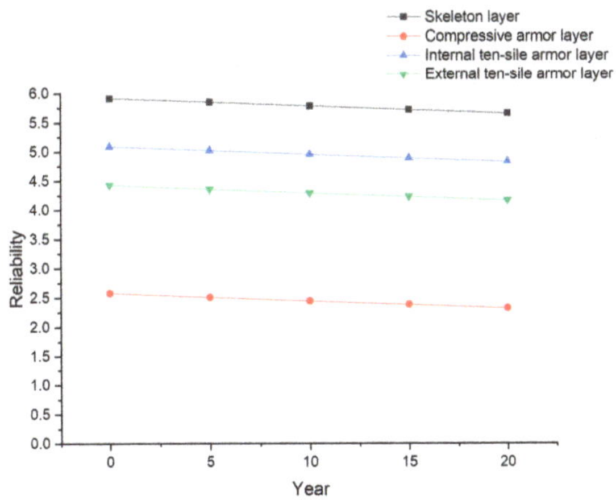

Figure 15. The reliability of metal layer changes with time.

5. Conclusions

In this paper, the fatigue life and reliability of local composite multilayer flexible riser structures are studied. Firstly, the paper analyzed many uncertain factors in fatigue problems, and the solutions for different types of uncertainties were given. For the uncertainty of load, the paper adopted the spectral analysis method, established the random wave load spectrum, and deduced the stress response spectrum; the fatigue life could then be calculated. Then, taking a multilayer riser structure as an example, the cumulative fatigue damage of each layer was calculated and displayed. The influence of the interlayer interaction on the fatigue life was analyzed via the fatigue damage through the nephogram. Finally, the reliability calculation method of multilayer structures was put forward as a means to deal with uncertain factors besides the load, and the calculation example was analyzed. The highlights of the article are as follows.

(1) Unlike previous reliability studies, this paper isolates the load factor, and the influence of load uncertainty is directly taken into account in fatigue calculation by spectrum analysis. This approach reduces the number of uncertainties that need to be considered

in reliability analysis. Moreover, compared with the reliability method in which random variables are directly applied to the fatigue damage degree results, the results obtained by considering randomness in the calculation process are more accurate.

(2) In the programming of cumulative fatigue damage calculation, the author added the module of the fatigue damage degree nephogram. It is easy to see from the analysis results presented in this paper that the fatigue damage data obtained by this method are more intuitive, and, compared with the previous algorithm of selecting typical nodes, this method is more comprehensive for the evaluation of cumulative fatigue damage.

At the same time, the research in this paper also has certain limitations, and further improvements can be made in the following two aspects.

(1) The structure selected in this paper is a local structure, which has fewer nodes than the overall structure, which means less computation time. If the fatigue life of all the nodes of the whole structure is studied and calculated, the time cost of stress analysis and fatigue analysis will increase greatly. At present, the method of typical node analysis is still the first choice for the fatigue life assessment of ships and marine structures. In subsequent research, researchers should consider developing some methods to improve the computational efficiency, so as to make the structural fatigue analysis more comprehensive and efficient.

(2) In the fatigue analysis of multilayer structures, the interaction between layers is analyzed from the cumulative fatigue damage nephogram. In fact, for the multilayer structure, the interlayer interaction already exists under the action of load, so it should be reflected in the stress cloud diagram. However, in the stress nephogram presented in this paper, the effect of interlayer interaction is not obvious. In the future, the interlayer influence of multilayer flexible riser structures under load should be further studied.

Author Contributions: Conceptualization, J.Y.; methodology, F.W.; software, Y.Y.; validation, F.W., H.L. and X.L.; formal analysis, H.L.; investigation, Y.Y.; resources, X.L.; data curation, F.W.; writing—original draft preparation, F.W.; writing—review and editing, F.W.; visualization, X.L.; supervision, R.S.; project administration, Y.Y.; funding acquisition, J.Y. All authors have read and agreed to the published version of the manuscript.

Funding: This research was funded by The National Natural Science Foundation of China (Grant No. 51879189) and National Natural Science Foundation of China (Grant No. 52071234).

Institutional Review Board Statement: Not applicable.

Informed Consent Statement: Not applicable.

Data Availability Statement: The data presented in this study are available on request from the corresponding author.

Conflicts of Interest: The authors declare no conflict of interest.

References

1. Drumond, G.P.; Pasqualino, I.P.; Pinheiro, B.C.; Estefen, S.F. Pipelines, Risers and Umbilicals Failures: A Literature Review. *Ocean Eng.* **2018**, *148*, 412–425. [CrossRef]
2. Song, X.; Wang, S. A Novel Spectral Moments Equivalence Based Lumping Block Method for Efficient Estimation of Offshore Structural Fatigue Damage. *Int. J. Fatigue* **2019**, *118*, 162–175. [CrossRef]
3. Moore, F. Materials for Flexible Riser Systems: Problems and Solutions. *Eng. Struct.* **1989**, *11*, 208–216. [CrossRef]
4. Api, R.P. Recommended Practice for Flexible Pipe. 2008, 4. Available online: https://images.prod-techstreet.com/direct/tocs/API/API_RP_17B_e4_toc.pdf (accessed on 15 August 2022).
5. de Sousa, J.R.M.; Magluta, C.; Roitman, N.; Ellwanger, G.B.; Lima, E.C.P.; Papaleo, A. On the Response of Flexible Risers to Loads Imposed by Hydraulic Collars. *Appl. Ocean Res.* **2009**, *31*, 157–170. [CrossRef]
6. de Sousa, J.R.M.; Magluta, C.; Roitman, N.; Londoño, T.V.; Campello, G.C. A Study on the Response of a Flexible Pipe to Combined Axisymmetric Loads. In *International Conference on Offshore Mechanics and Arctic Engineering*; American Society of Mechanical Engineers: New York, NY, USA, 2013; Volume 55379, p. V04BT04A040.

7. Skeie, G.; Sødahl, N.; Steinkjer, O. Efficient Fatigue Analysis of Helix Elements in Umbilicals and Flexible Risers: Theory and Applications. *J. Appl. Math.* **2012**, *2012*. [CrossRef]
8. Sævik, S. Theoretical and Experimental Studies of Stresses in Flexible Pipes. *Comput. Struct.* **2011**, *89*, 2273–2291. [CrossRef]
9. de Sousa, J.R.M.; de Sousa, F.J.M.; de Siqueira, M.Q.; Sagrilo, L.V.S.; de Lemos, C.A.D. A Theoretical Approach to Predict the Fatigue Life of Flexible Pipes. *J. Appl. Math.* **2012**, *2012*. [CrossRef]
10. Wang, K.; Ji, C.; Xue, H.; Tang, W. Fatigue Damage Study of Helical Wires in Catenary Unbonded Flexible Riser Near Touchdown Point. *J. Offshore Mech. Arct. Eng.* **2017**, *139*. [CrossRef]
11. Lotsberg, I. Development of Fatigue Design Standards for Marine Structures. *J. Offshore Mech. Arct. Eng.* **2019**, *141*. [CrossRef]
12. Yue, J.; Yang, K.; Peng, L.; Guo, Y. A Frequency-Time Domain Method for Ship Fatigue Damage Assessment. *Ocean Eng.* **2021**, *220*, 108154. [CrossRef]
13. American Bureau of Shipping. Guide for the Fatigue Assessment of Offshore Structures. 2003. Available online: https://safety4sea.com/wp-content/uploads/2016/06/Fatigue_Guide_e-Mar18.pdf (accessed on 15 August 2022).
14. Thompson, I. Validation of Naval Vessel Spectral Fatigue Analysis Using Full-Scale Measurements. *Mar. Struct.* **2016**, *49*, 256–268. [CrossRef]
15. Rörup, J.; Garbatov, Y.; Dong, Y.; Uzunoglu, E.; Parmentier, G.; Andoniu, A.; Quéméner, Y.; Chen, K.C.; Vhanmane, S.; Negi, A. Round Robin Study on Spectral Fatigue Assessment of Butt-Welded Joints. *Marit. Transp. Harvest. Sea Resour.* **2017**, *5*, 663–671.
16. DNV GL. Class Guideline DNVGL CG 0130: Wave Loads. 2018. Available online: https://rules.dnv.com/docs/pdf/DNV/CG/2018-01/DNVGL-CG-0130.pdf (accessed on 15 August 2022).
17. Hasselmann, K.; Barnett, T.P.; Bouws, E.; Carlson, H.; Cartwright, D.E.; Enke, K.; Ewing, J.A.; Gienapp, A.; Hasselmann, D.E.; Kruseman, P. Measurements of Wind-Wave Growth and Swell Decay during the Joint North Sea Wave Project (JONSWAP). 1973. Available online: https://pure.mpg.de/pubman/faces/ViewItemOverviewPage.jsp?itemId=item_3262854 (accessed on 15 August 2022).
18. Pavlou, D.G. Flow-Induced Instability of Multi-Layered Anisotropic Pipelines. *Int. J. Comput. Methods Exp. Meas.* **2016**, *4*, 543–553. [CrossRef]
19. Pavlou, D. Inner Flow-Induced Buckling of FRP Catenary Risers. *J. Offshore Mech. Arct. Eng.* **2020**, *142*, 1–28. [CrossRef]
20. Pavlou, D.G. Longitudinal–Flexural–Torsional Dynamic Behavior of Liquid-Filled Pipelines: An Analytic Solution. *J. Offshore Mech. Arct. Eng.* **2020**, *142*, 011701. [CrossRef]
21. Pavlou, D.G.; Correia, J.A. Dynamic Response of Pipelines under Impact and Harmonic Loading. In *Proceedings of the Institution of Civil Engineers-Maritime Engineering*; Thomas Telford Ltd: London, UK, 2019; Volume 172, pp. 15–22.
22. Pavlou, D.G. Dynamic Response of a Multi-Layered FRP Cylindrical Shell under Unsteady Loading Conditions. *Eng. Struct.* **2016**, *112*, 256–264. [CrossRef]
23. Veritas, D.N. Fatigue Design of Offshore Steel Structures. 2010. Available online: https://cdm.ing.unimo.it/dokuwiki/_media/wikitelaio2017/rp-c203.pdf (accessed on 15 August 2022).
24. Li, L.; Jia, Q.; Wan, Z.; Zhang, Y.; Qian, P.; Li, J. Experimental and Numerical Investigation of Effects of Residual Stress and Its Release on Fatigue Strength of Typical FPSO-Unit Welded Joint. *Ocean Eng.* **2020**, *196*, 106858. [CrossRef]
25. Soares, C.G.; Garbatov, Y.; von Selle, H. Fatigue Damage Assessment of Ship Structures Based on the Long-Term Distribution of Local Stresses. *Int. Shipbuild. Prog.* **2003**, *50*, 35–55.
26. Garbatov, Y.; Guedes Soares, C. Uncertainty Assessment of Fatigue Damage of Welded Ship Structural Joints. *Eng. Struct.* **2012**, *44*, 322–333. [CrossRef]
27. Santecchia, E.; Hamouda, A.M.S.; Musharavati, F.; Zalnezhad, E.; Cabibbo, M.; el Mehtedi, M.; Spigarelli, S. A Review on Fatigue Life Prediction Methods for Metals. *Adv. Mater. Sci. Eng.* **2016**, *2016*, 9573524. [CrossRef]
28. Pavlou, D.G. The Theory of the SN Fatigue Damage Envelope: Generalization of Linear, Double-Linear, and Non-Linear Fatigue Damage Models. *Int. J. Fatigue* **2018**, *110*, 204–214. [CrossRef]
29. Pavlou, D. A Deterministic Algorithm for Nonlinear, Fatigue-based Structural Health Monitoring. *Comput. Aided Civ. Infrastruct. Eng.* **2022**, *37*, 809–831. [CrossRef]
30. Torng, T.Y.; Wirsching, P.H. Fatigue and Fracture Reliability and Maintainability Process. *J. Struct. Eng.* **1991**, *117*, 3804–3822. [CrossRef]

Journal of
Marine Science and Engineering

MDPI

Article

Experimental and Numerical Study of Cyclic Stress–Strain Response and Fatigue Crack Initiation Life of Mid-Carbon Steel under Constant and Multi-Step Amplitude Loading

Kasumi Morita [1,2,*], Masashi Mouri [2], Riccardo Fincato [1] and Seiichiro Tsutsumi [1]

[1] Department of Civil Engineering, Osaka University, Osaka 5650871, Japan
[2] IHI Corporation, Yokohama 2358501, Japan
* Correspondence: morita0101@ihi-g.com

Abstract: This paper investigates the fatigue cyclic deformation behavior of mid-carbon steel. Uniaxial tensile loading tests and fatigue tests under constant and multi-step amplitude loading steps are performed to characterize the influence of loading history. The material is shown to exhibit different uniaxial ratcheting behavior depending on loading history. A smooth and gradual increase in cyclic softening is observed under smaller stress/strain conditions. Based on experimental characterization, numerical investigations are carried out to reproduce the cyclic stress–strain behavior under different variable amplitude load ranges. The nonlinear material behavior is reproduced by means of an elastoplasticity model called the Fatigue SS Model (hereafter, FSS model). The main feature of the FSS model is the ability to describe the cyclic softening behavior within a macroscopically elastic stress state. The good agreement between experimental and numerical results proves the reliability of the model to catch a realistic material response in fatigue problems. Furthermore, the present study introduces a method for the prediction of fatigue crack initiation life under variable loading conditions based on cumulative plastic work.

Keywords: fatigue crack initiation life; cyclic plasticity; variable amplitude load

Citation: Morita, K.; Mouri, M.; Fincato, R.; Tsutsumi, S. Experimental and Numerical Study of Cyclic Stress–Strain Response and Fatigue Crack Initiation Life of Mid-Carbon Steel under Constant and Multi-Step Amplitude Loading. *J. Mar. Sci. Eng.* **2022**, *10*, 1535. https://doi.org/10.3390/jmse10101535

Academic Editors: Sören Ehlers and Moritz Braun

Received: 28 September 2022
Accepted: 10 October 2022
Published: 19 October 2022

1. Introduction

Steel bridges are subjected to cyclic loading conditions during their service life and, to fulfil safety criteria, they are required to perform within a local stress or strain elastic regime, below the pertinent fatigue limit [1]. However, it is known that during fatigue tests, performed under a stress regime lower than the macroscopic yield stress, the initial elastic cyclic response smoothly transforms into an elastoplastic response with a significant increase in the strain amplitude due to the generation of irreversible deformations. The phenomenon is known as cyclic softening behavior and is typical not only of carbon steels, but has been experimentally observed in various structural materials over the past decades [2–8].

Conventional plasticity models [9] are unsuitable for the description of fatigue problems characterized by stress states lower than the macroscopic yield stress, as they assume the pure elastic response of materials. On the other hand, several unconventional plastic models have been proposed [10–14]; however, those models were originally designed to simulate inelastic deformation under low-cycle fatigue problems, where the applied cyclic stress is relatively large and exceeds the initial yield stress.

In recent years, an increasing number of models have been developed to describe the damage process during high-cycle fatigue loading. Based on a fatigue continuum damage model framework [15,16], Barbu et al. [17] proposed a stepwise load-advancing strategy for the prediction of high cycle fatigue performance. Furthermore, Van Do [18] developed continuum a damage mechanics model for multiaxial high cycle fatigue to describe the welding residual stresses of welded joints. In addition, Zhu et al. [19] proposed a damage accumulation model based on Miner's rule to consider the coupled damage due to HCF–LCF interaction by introducing new load parameters. One common aspect of these models is that the damage becomes a function of the number of cycles, adapting a phenomenological description of the continuum damage variable to the high-cycle fatigue theory. However, the cyclic softening process, depending on the loading history, is generally neglected with these constitutive models.

Cumulative fatigue damage analysis plays a key role in life prediction of components and structures subjected to field load histories [20]. The increasing interest in the cumulative fatigue damage concept led to the formulation of life prediction models. In order to consider the accumulation of plastic deformation with cyclic loading conditions lower than the macroscopic yield stress, cyclic stress–strain curves (i.e., CSSCs) have been taken into fatigue design [21,22] and several methods have been proposed to describe CSSC in previous works [23–26].

In order to describe the softening response observable in steels under cyclic loading conditions lower than the macroscopic yield stress, the authors have proposed an elastoplastic model [27,28], the Fatigue SS model (hereafter, the FSS model). The FSS model is based on an unconventional plasticity theory [10], further enriched by including the elastic boundary concept [29] and cyclic damage concept [27], for the description of strain-softening behavior within a macroscopically elastic stress state [27,28]. However, previous study of the FSS model focused only on constant amplitude loading, whereas the applicability on more general loading conditions, such as multi-step loading, remains unclear.

In order to contribute to the reasonable and safe design of bridges, the first goal of this work is to prove the reliability of the FSS model to provide a realistic material description under variable fatigue loading conditions. Both experimental and numerical campaigns were carried out to clarify the evolution of cyclic stress–strain responses of mid-carbon steel under relatively low-stress regimes. Firstly, the static mechanical properties of the material were experimentally characterized by performing monotonic tensile tests at room conditions. Subsequently, fatigue experiments were conducted to investigate the uniaxial ratcheting behavior under constant loading, two-step amplitude loading, and repeated increasing/decreasing amplitude loading conditions. Based on the experimental campaign, the numerical analyses were carried out by means of the FSS model. The FSS model material constants were calibrated by minimizing the differences between the experimental and numerical cyclic stress–strain curves to give a realistic description of the ratcheting and softening behaviors. The second aim of this paper is to propose a fatigue crack initiation life prediction method that can consider the loading history based on the FSS model. In this study, two approaches, using the total strain range and the cumulative plastic strain damage, are compared and the criterion of cumulative plastic strain damage for fatigue crack initiation life prediction is proposed based on the fatigue tests result.

2. Experiment

2.1. Uniaxial Tensile Tests

A series of uniaxial tensile tests was performed to characterize the mechanical and stress–strain behavior of mid-carbon steel. The specimens were machined from a 16 mm thick mid-carbon steel plate named SM490A, as specified by the JIS G 3106 2008 [30]. The loading axis was assumed parallel to the rolling direction of the plate. Figure 1 displays the microstructure of a specimen.

As can be seen, the microstructure of the material clearly shows microstructural bands parallel to the rolling direction. The present work does not investigate material anisotropy; future works will consider this aspect by analyzing the different material response depending on the loading direction and by adopting an anisotropic yield criterion for the numerical analyses (e.g., Hill48 [31], Yld-2004-18p [32], CB2001 [33], among others). The details of the chemical composition and the geometric dimensions of the specimen are reported in Figure 2 and Table 1, respectively.

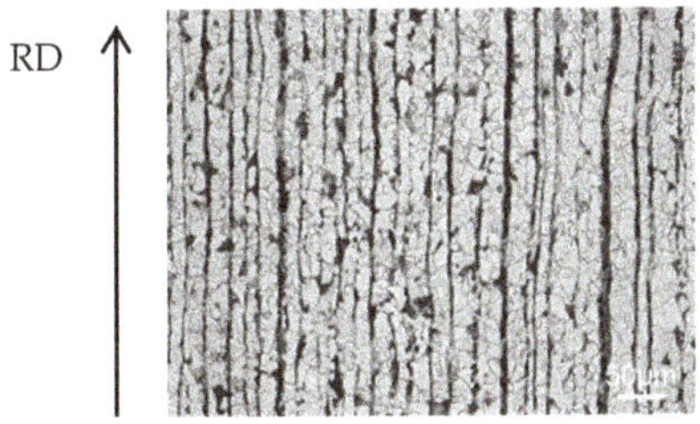

Figure 1. Specimen microstructure.

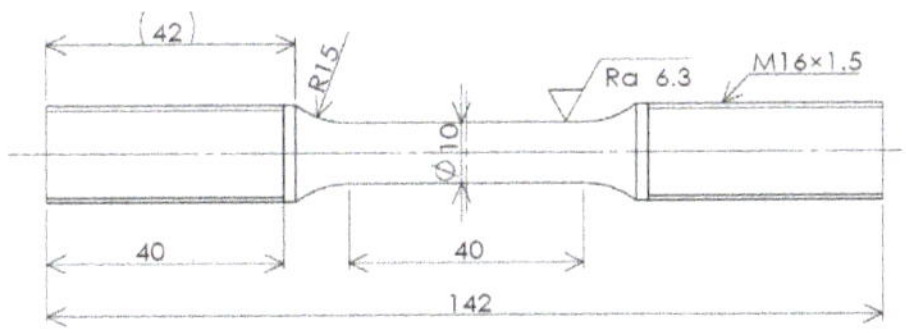

Figure 2. Specimen geometry; all measures are in mm.

Table 1. Chemical composition of the mid-carbon steel.

Chemical Composition /%				
C	Si	Mn	P	S
0.16	0.36	1.42	0.015	0.004

The value of the stress concentration factor K_t of the specimen is 1.11. After machining, an annealing stress relief process was conducted at 600 °C for 1 h, with a subsequent cooling process in air at room temperature. To monitor the diameter of the gauge section during tests, a displacement meter was attached to the specimen, as shown in Figure 3.

Figure 3. Detail of the equipment for measuring the strain and the specimen diameter.

Following the standard method proposed by JIS 2241:2011 2009 [34], uniaxial tensile tests were carried out by a servo-hydraulic testing machine. All tensile tests were conducted at a displacement rate of 0.06 mm/min before specimen yield, and then at a crosshead rate of 0.9 mm/min after the yield behavior was recorded. Four specimens were tested at room temperature and the mean values of the static mechanical properties were computed. Figure 4 presents one of the uniaxial tensile stress–strain curves, whereas the averaged static mechanical constants are shown in Table 2.

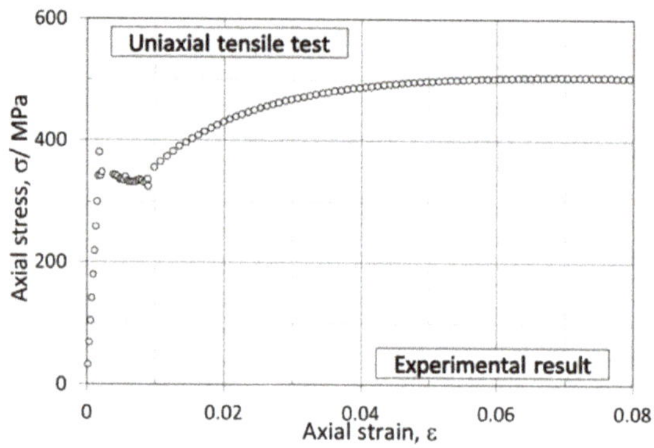

Figure 4. Experimental monotonic tensile stress–strain curve.

Table 2. Mechanical properties of averaged material obtained by monotonic tensile tests.

Upper yield stress/MPa	402
Lower yield stress/MPa	336
Young's modulus/GPa	208

2.2. Fatigue Tests

Uniaxial fatigue tests under various loading conditions were performed to characterize the fatigue behavior of the mid-carbon steel. The geometry and heat treatment conditions of the specimens tested in this section were the same as those used in uniaxial tensile tests, as shown in Figure 2. Before the fatigue test, the gauge section was polished along the loading axis with emery papers from a grit size of #220 to #1500. All the tests were conducted with fully reversed loading conditions (i.e., stress ratio $R = -1$) at room temperature and the fatigue failure was defined when the displacement of the load cell exceeded 2 mm. Table 3 shows the test conditions applied in experiments using a servo-hydraulic testing machine.

Table 3. Fatigue test conditions.

Name	Control	Stress Ratio ($R = \sigma_{min}/\sigma_{max}$)	Cyclic Amplitude
Constant amplitude loading test	Strain Load	−1 −1	ε_a = 0.0014/0.005/0.008 σ_a = 210/230/250/270/290 MPa
Two-step amplitude loading test	Load	−1	σ_{a1} = 240 (2000 cycles) →σ_{a2} = 280 MPa (2000 cycles)
Repeated increasing and decreasing amplitude loading test	Load	−1	σ_a = 160→200→240→280→240→200→160→200 MPa (Every 100 cycles, changing the amplitude)

To monitor the diameter of the gauge section and control the strain range during tests, a displacement meter and a clip gauge were attached to the specimen, as shown in Figure 3. All specimens were assumed as run-outs at 2,000,000 cycles.

Uniaxial cyclic loading tests were carried out with constant strain amplitude ranges from ε_a = 0.0014 to 0.008 to investigate the strain range dependence. The stress–strain hysteresis loops for each amplitude are presented in Figure 5a. Comparing the stabilized loop sizes under different strain amplitudes, it is possible to notice that the stable loops do not overlap due to the increase in the elastic domain induced by material hardening. The evolution of the stress response versus the number of cycles is shown in Figure 5b. Cyclic hardening is observed in the initial cycles with a strain amplitude $\varepsilon_a = 0.008$; however, initial cyclic hardening is not observed under a lower strain amplitude $\varepsilon_a = 0.0014$. The stress range decreases during the first 1000 cycles (i.e., cyclic softening) and it subsequently saturates in the ongoing of the test (i.e., cyclic plastic shakedown). The rate of cyclic softening versus the number of cycles shows different tendencies under different strain range loading conditions. Under a low strain range regime, it shows a smooth and gradual increase with the number of cycles, whereas the increase is faster for larger strain range loadings.

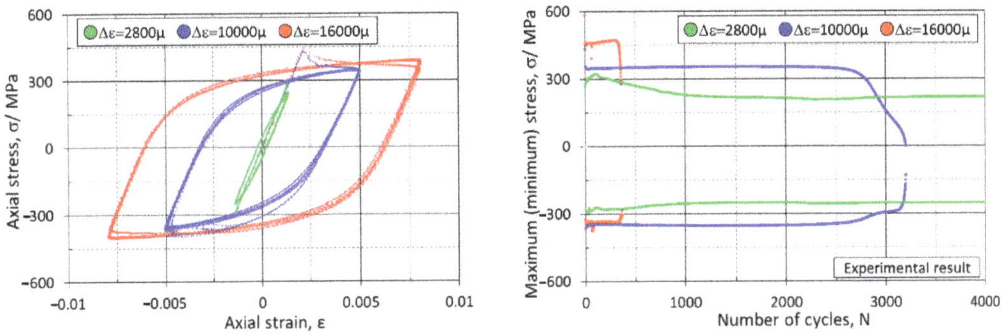

(a) (b)

Figure 5. (**a**) Experimental results, axial stress versus axial strain curves under constant strain amplitude loading; (**b**) experimental relationships between σ_{max}-σ_{min} and the number of cycles under constant strain amplitude loading.

Two-step amplitude loading tests were conducted by changing the stress amplitude, σ_{a1} = 240 MPa, and σ_{a2} = 280 MPa for 2000 cycles, to investigate the material response under variable loading conditions. Figure 6a presents the stress–strain hysteresis loops, whereas the evolutions of the stress response versus the number of cycles are displayed in Figure 6b. As can be seen in the two graphs, the plastic deformation is almost negligible during the first step (σ_{a1} = 240 MPa). However, as soon as the second step (σ_{a2} = 280 MPa) starts, cyclic softening is detected with the subsequent saturation of the stress range.

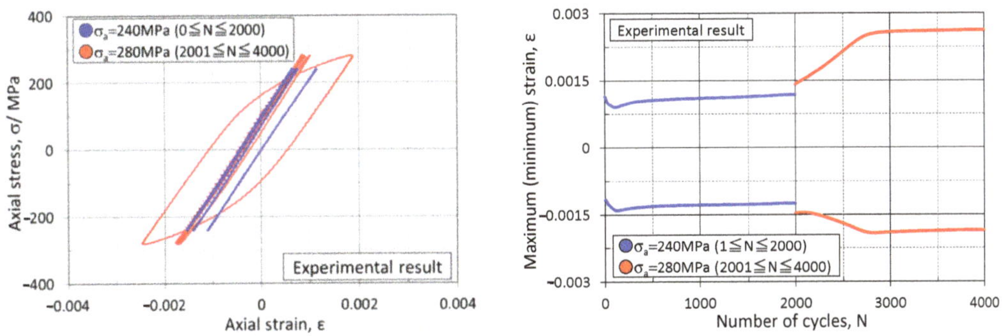

(a) (b)

Figure 6. (**a**) Experimental stress and strain curves under two-step amplitude loading test; (**b**) experimental relationships between ε_{max}-ε_{min} and the number of cycles under two-step amplitude loading test.

A repeated increasing and decreasing amplitude loading test was also performed to study the influence of the loading history. In this test, the stress amplitude σ_a was increased by 40 MPa every 100 cycles, starting from a lower value of 160 MPa up to an upper limit of 280 MPa. After reaching the upper value of 280 MPa, the stress was decreased in the same manner. The aforementioned loading history was performed until fatigue failure was detected. Figure 7 shows the evolution of the strain response versus the number of cycles. Almost elastic behavior is detected within the first 3000 cycles, and subsequently, an increase in the strain ranges is recorded between 3000 and 6000 cycles. After 6000 cycles, the saturation of the strain ranges is observed.

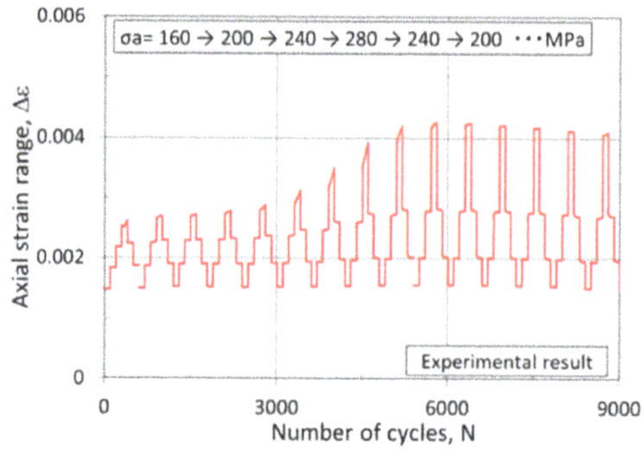

Figure 7. Experimental relationships between $\Delta\varepsilon$ and the number of cycles under increasing and decreasing amplitude loading test.

3. Numerical Simulation of the Cyclic Stress–Strain Response

3.1. The FSS Model

The FSS model was used to investigate the inelastic response of the material under cyclic loading conditions. The FSS model [27,28] describes the generation of plastic strain within the yield surface, which can be obtained through a similarity transformation from the conventional yield surface [10,35,36]. Classical theories distinguish the elastic and plastic regions, allowing an irreversible stretch only in the plastic region whenever the stress increment satisfies the loading criterion. In contrast, the FSS model abolishes the separation into domains, stating that a plastic response can be realized for every change in the stress state that satisfies the loading criterion. Furthermore, the use of a mobile similarity center, which is a function of the plastic strain, makes this theory particularly suitable for studying cyclic mobility problems. A detailed explanation of the model is beyond the scope of this paper; the reader can refer to references [27–29,37–46] for a more complete discussion.

3.2. Calibration of the FSS Model

The material parameters for the FSS model were calibrated against the experimental cyclic stress–strain curves obtained in the fatigue tests. Numerical analyses were performed by implementing the constitutive equations of the FSS model via a single-quadrature point in in-house numerical code. The simulations were carried out under the same loading conditions in experiments (see in Table 3) and by adopting the elastoplastic parameters of the FSS model listed in Table 4.

Table 4. Set of material constants obtained from the calibration of the numerical results against the experimental data.

Poisson's Ratio	Isotropic Hardening Parameter		Kinematic Hardening Parameter			High-Cycle Fatigue Parameter for the Computation of Damage		
μ	h_1	h_2	a_1	a_2	a_3	d_1	d_2	d_3
0.3	0.1	4.0	1.0	740	1.0	0.0004	0.0096	2.0

The description of material parameters can be found in references [28,29]. Figures 8a and 9a display the stress–strain hysteresis loops with constant strain amplitudes and two-step amplitude loading. As can be seen from Figures 8a and 9a, the size of the loops is realistically described by the model, even though the experimental hollow markers show a slight shift, approximately 7% difference, toward the compressive side in Figure 9a. The possible reason for this behavior is initial imperfections of the specimen and it was neglected in the calibration. The evolution of the stress response versus the number of cycles is shown in Figures 8b and 9b. Hollow markers report the experimental results discussed in Section 2.2. The solid continuous lines represent the numerical solutions. In Figures 8b and 9b, the variations in strain and stress against the number of cycles show good agreement with the experiments, including the plastic deformations.

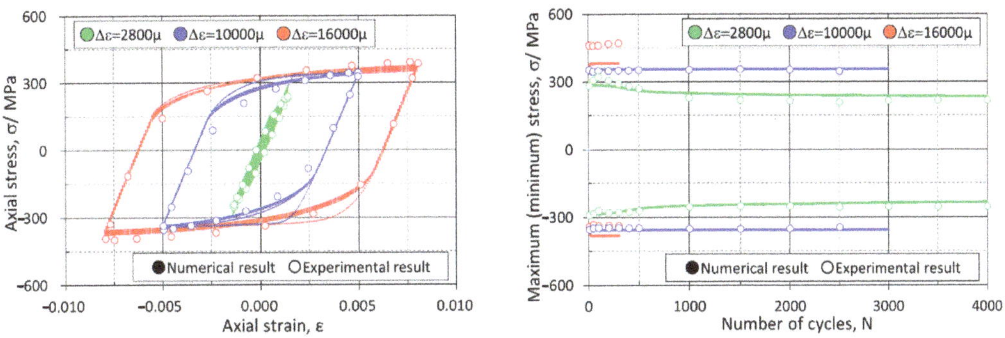

(a) (b)

Figure 8. (**a**) Numerical stress and strain curves under constant amplitude loading; (**b**) numerical relationships between σ_{max}-σ_{min} and the number of cycles under constant amplitude loading.

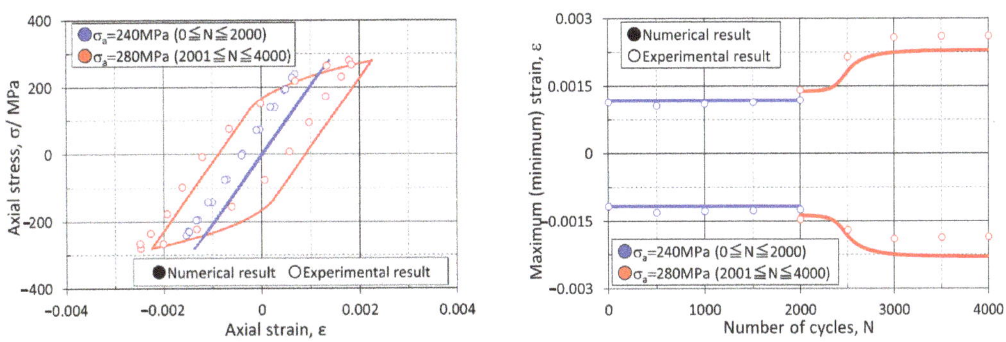

(a) (b)

Figure 9. (**a**) Numerical stress and strain curves under two-step amplitude loading; (**b**) numerical relationships between ε_{max}-ε_{min} and the number of cycles under two-step amplitude loading.

Figure 10 compares the numerical and experimental results under repeated increasing and decreasing amplitude loading. The strain amplitude behavior and the evolution of plastic deformation are similar to the experimental behavior. Overall, the trends in the numerical simulations are in good agreement with the experimental data under several loading conditions.

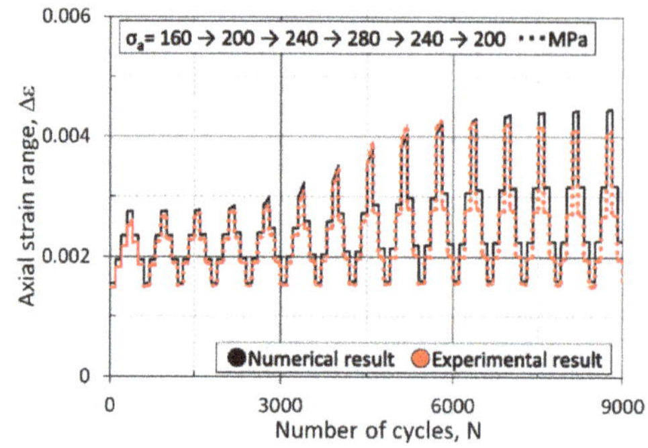

Figure 10. Numerical relationships between $\Delta\varepsilon$ and the number of cycles under the repeated increasing and decreasing amplitude loading condition.

4. Prediction of Crack Initiation Life

The prediction of the fatigue crack initiation life of the material was carried out adopting the two methods discussed in this section, namely strain range-based and cumulative plastic damage-based criteria. The first approach was investigated utilizing Iida's equation (see in Equation (1), [47]), which formulates a total strain range-based crack initiation life prediction. Figure 11 graphically reports the results of the fatigue tests, which are converted from fatigue failure lives into fatigue crack initiation lives [48], and the crack initiation life is predicted by Equation (1).

$$\frac{\Delta\varepsilon_t}{2} = 0.415N_c^{-0.606} + 0.00412N_c^{-0.115} \tag{1}$$

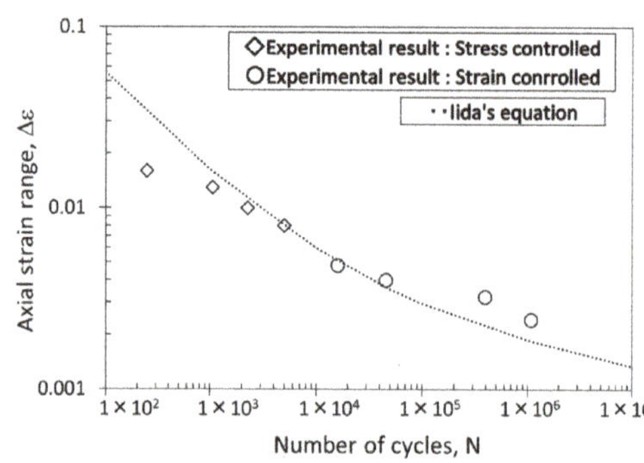

Figure 11. Relationships between strain range and fatigue crack initiation lives.

Rhomboidal hollow markers indicate strain-controlled tests, whereas circular hollow markers indicate load-controlled tests. As can be seen from Figure 11, Iida's prediction shows a good agreement with the experimental results obtained in this study. In general, it is well-known that strain range-based methods predict the fatigue (crack initiation) life well under constant amplitude loading conditions, as reported in previous works [49–51]. On the other hand, it is complicated to consider the dependency on the loading history when it comes to variable and multi-axial loading problems.

The second approach was investigated using the FSS model. The effect of the accumulation of plastic strain on damage evolution can be measured by the following functions.

$$D(H_d) = (1 - d_2)\left[1 + \left(\frac{d_1}{H_d}\right)^{d_3}\right]^{-1};\ H_d = \int \sqrt{\frac{2}{3}}|\boldsymbol{D^p}|Ddt \tag{2}$$

$$D\left(R\right) = (1 - k_2)\left[1 + \left(\frac{k_1}{R}\right)^{k_3}\right]^{-1} \quad (3)$$

The damage D in Equation (2) is representative of the damage accumulation in the material model [28,29] contributing to the opening of the hysteresis loops. However, it does not affect the elastic response of the material. In detail, d_1, d_2, and d_3 are material constants to be calibrated against experimental tests. The scalar plastic internal variable H_d represents the amount of accumulated damage in the form of accumulated plastic work, with $|\boldsymbol{D}^p|$ being the norm of the plastic strain rate tensor. The variable D helps to regulate the amount of damage induced for stress states below the macroscopic yield stress. The variable R in Equation (3) is introduced as a measure of the magnitude of the stress state within the conventional elastic domain. It assumes a null value for a null stress state and it is at a maximum when the stress reaches the macroscopic yield stress. As can be seen from Equation (3), for stress states close to the plastic potential, D tends to its maximum value $D = (1 - k_2)$, whereas D tends to be zero for lower stress regimes. An in-depth discussion of the role of the material constants d_1, d_2, and d_3 in Equation (2) and k_1, k_2, and k_3 in Equation (3) can be found in [29]. In this study, the cumulative plastic work, H_d, is used to investigate the crack initiation life. The relation between cumulative plastic work (i.e., H_d) versus the number of cycles under constant amplitude loading is illustrated in Figure 12. H_d increases with the number of cycles, and larger loading conditions return higher H_d rates.

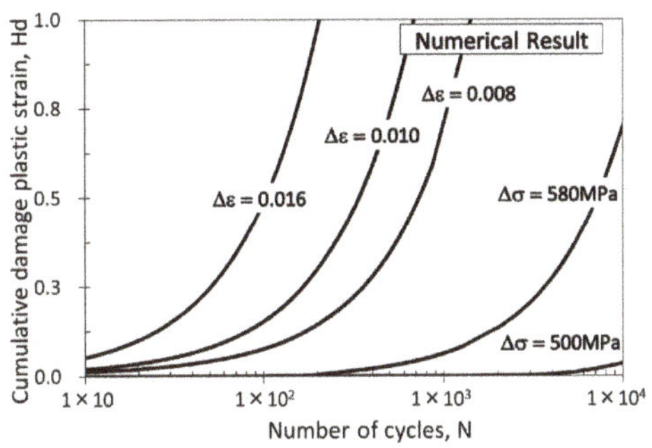

Figure 12. Relationship between cumulative damage plastic strain, H_d, and number of cycles under constant amplitude loading.

Figure 13 displays the relationships of strain amplitudes against the number of cycles, assuming the number of cycles when the variable H_d reaches a value of 2.0 as the criterion for the crack initiation life.

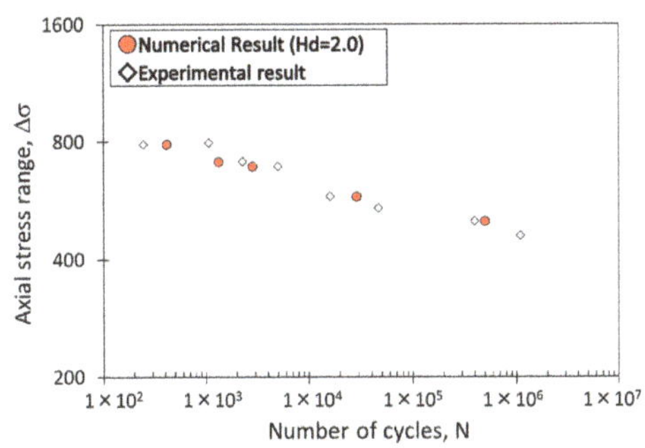

Figure 13. Relationship between cumulative damage plastic strain, H_d, and number of cycles under constant amplitude loading.

As indicated in the graph, the proposed crack initiation life criterion based on the cumulative plastic work agrees well with the experimental data. Finally, Figure 14 compares the experimental and numerical predicted number of cycles to failure under same loading conditions.

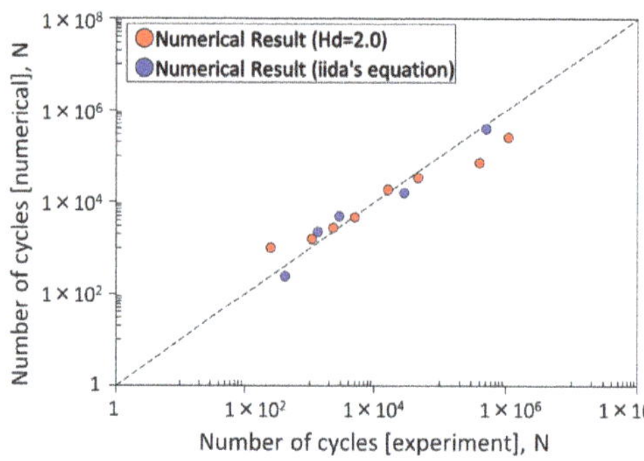

Figure 14. The experimental versus the numerical predicted number of cycles to crack initiation under the same loading conditions.

The difference between the number of cycles to crack formation obtained in the experiments and the one predicted by the two aforementioned criteria is equivalent adopting the proposed criterion (i.e., based on the H_d variable) compared against the strain range-based method. The applicability of the FSS model and cumulative plastic damage method to variable and multiaxial loading conditions will be investigated in future work.

5. Conclusions

The paper investigated the applicability of the FSS model to mid-carbon steel (SM490A) under variable uniaxial loading conditions and proposed a novel criterion for the evaluation of the fatigue crack life. The main aspects of the present work can be summarized as follows:

1. Uniaxial tensile tests were carried out to obtain the mechanical properties of SM490A.
2. Fatigue tests with various loading conditions were carried out to investigate the cyclic response of SM490A. Cyclic softening/hardening and cyclic shakedown were observed under different loading conditions.
3. The calibration of the material constants using the FSS model was performed under constant and two-step amplitude loading and repeated increasing and decreasing conditions. The numerical results in terms of stress and strain responses showed good agreement with the experimental results.
4. Fatigue crack initiation life prediction using Iida's equation, based on the total strain range, agreed quite well with the experiment under constant amplitude loading. The criterion based on the cumulative plastic damage strain calculated by the FSS model predicted a realistic fatigue crack initiation life as well as Iida's equation.

Author Contributions: Conceptualization, S.T. and K.M.; methodology, S.T. and R.F.; validation, K.M.; formal analysis, R.F.; investigation, K.M.; resources, M.M.; data curation, R.F.; writing—original draft preparation, K.M.; writing—review, S.T.; editing, K.M.; visualization, K.M.; supervision, S.T.; project administration, S.T. and M.M.; funding acquisition, M.M. All authors have read and agreed to the published version of the manuscript.

Funding: This research received no external funding.

Institutional Review Board Statement: Not applicable.

Informed Consent Statement: Not applicable.

Data Availability Statement: The data presented in this study are available upon request from the corresponding author.

Conflicts of Interest: The authors declare no conflict of interest.

References

1. Stephens, A.; Fatemi, R.R.; Stephens, H.F. *Metal Fatigue in Engineering [Hardcover]*; John Wiley & Sons: Hoboken, NJ, USA, 2000; ISBN 0471510599.
2. Plumtree, A. Cyclic Stress–Strain Response and Substructure. *Int. J. Fatigue* **2001**, *23*, 799–805. [CrossRef]
3. Jones, A. Cyclic Stress-Strain Curves Generated from Random Cyclic Strain Amplitude Tests. *Int. J. Fatigue* **1999**, *21*, 521–530. [CrossRef]
4. Paul, S.K.; Sivaprasad, S.; Dhar, S.; Tarafder, S. Cyclic Plastic Deformation and Cyclic Hardening/Softening Behavior in 304LN Stainless Steel. *Theor. Appl. Fract. Mech.* **2010**, *54*, 63–70. [CrossRef]
5. Gorash, Y.; MacKenzie, D. On Cyclic Yield Strength in Definition of Limits for Characterisation of Fatigue and Creep Behaviour. *Open Eng.* **2017**, *7*, 126–140. [CrossRef]
6. Hilditch, T.B.; Timokhina, I.B.; Robertson, L.T.; Pereloma, E.V.; Hodgson, P.D. Cyclic Deformation of Advanced High-Strength Steels: Mechanical Behavior and Microstructural Analysis. *Metall. Mater. Trans. A* **2009**, *40*, 342–353. [CrossRef]
7. Zhang, J.; Jiang, Y. An Experimental Study of Inhomogeneous Cyclic Plastic Deformation of 1045 Steel under Multiaxial Cyclic Loading. *Int. J. Plast.* **2005**, *21*, 2174–2190. [CrossRef]
8. Belattar, A.; Taleb, L.; Hauet, A.; Taheri, S. Dependence of the Cyclic Stress–Strain Curve on Loading History and Its Interaction with Fatigue of 304L Stainless Steel. *Mater. Sci. Eng. A* **2012**, *536*, 170–180. [CrossRef]
9. Drucker, D.C. Conventional and Unconventional Plastic Response and Representation. *Appl. Mech. Rev.* **1988**, *41*, 151–167. [CrossRef]
10. Hashiguchi, K. Subloading Surface Model in Unconventional Plasticity. *Int. J. Solids Struct.* **1989**, *25*, 917–945. [CrossRef]
11. Iwan, W.D. On a Class of Models for the Yielding Behavior of Continuous and Composite Systems. *J. Appl. Mech.* **1967**, *34*, 612–617. [CrossRef]
12. Mróz, Z. On the Description of Anisotropic Workhardening. *J. Mech. Phys. Solids* **1967**, *15*, 163–175. [CrossRef]
13. Dafalias, Y.F.; Popov, E.P. A Model of Nonlinearly Hardening Materials for Complex Loading. *Acta Mech.* **1975**, *21*, 173–192. [CrossRef]
14. Ohno, N.; Wang, J.-D. Kinematic Hardening Rules with Critical State of Dynamic Recovery, Part I: Formulation and Basic Features for Ratchetting Behavior. *Int. J. Plast.* **1993**, *9*, 375–390. [CrossRef]
15. Chaboche, J.L.; Lesne, P.M. A Non-Linear continuous fatigue damage model. *Fatigue Fract. Eng. Mater. Struct.* **1988**, *11*, 1–17. [CrossRef]
16. Oller, S.; Salomón, O.; Oñate, E. A Continuum Mechanics Model for Mechanical Fatigue Analysis. *Comput. Mater. Sci.* **2005**, *32*, 175–195. [CrossRef]
17. Barbu, L.G.; Oller, S.; Martinez, X.; Barbat, A. High Cycle Fatigue Simulation: A New Stepwise Load-Advancing Strategy. *Eng. Struct.* **2015**, *97*, 118–129. [CrossRef]
18. Van Do, V.N. The Numerical High Cycle Fatigue Damage Model of Fillet Weld Joint under Weld-Induced Residual Stresses. *IOP Conf. Ser. Earth Environ. Sci.* **2018**, *143*, 012037. [CrossRef]
19. Zhu, S.-P.; Yue, P.; Yu, Z.-Y.; Wang, Q. A Combined High and Low Cycle Fatigue Model for Life Prediction of Turbine Blades. *Materials* **2017**, *10*, 698. [CrossRef]
20. Fatemi, A.; Yang, L. Cumulative Fatigue Damage and Life Prediction Theories: A Survey of the State of the Art for Homogeneous Materials. *Int. J. Fatigue* **1998**, *20*, 9–34. [CrossRef]
21. Socie, D.F. Fatigue-Life Prediction Using Local Stress-Strain Concepts. *Exp. Mech.* **1977**, *17*, 50–56. [CrossRef]
22. Conle, F.A.; Chu, C.-C. Fatigue Analysis and the Local Stress–Strain Approach in Complex Vehicular Structures. *Int. J. Fatigue* **1997**, *19*, 317–323. [CrossRef]
23. Maier, H.; Christ, H.-J. Modeling of Cyclic Stress–Strain Behavior and Damage Mechanisms under Thermomechanical Fatigue Conditions. *Int. J. Fatigue* **1997**, *19*, 267–274. [CrossRef]
24. Li, B.; Reis, L.; Defreitas, M. Simulation of Cyclic Stress/Strain Evolutions for Multiaxial Fatigue Life Prediction. *Int. J. Fatigue* **2006**, *28*, 451–458. [CrossRef]
25. Manson, S.S. Fatigue: A Complex Subject–Some Simple Approximation. *Exp. Mech.* **1965**, *5*, 193–226. [CrossRef]
26. Wang, C.; Fan, J.; Xu, L.; Nie, X. Cyclic Hardening and Softening Behavior of the Low Yield Point Steel: Implementation and Validation. *Eng. Struct.* **2020**, *210*, 110220. [CrossRef]
27. Tsutsumi, S.; Murakami, K.; Gotoh, K.; Toyosada, M. Cyclic Stress-Strain Relation under High Cycle Fatigue Process. *J. Japan Soc. Nav. Archit. Ocean Eng.* **2008**, *7*, 243–250. [CrossRef]
28. Tsutsumi, S.; Fincato, R. Cyclic Plasticity Model for Fatigue with Softening Behaviour below Macroscopic Yielding. *Mater. Des.* **2019**, *165*, 107573. [CrossRef]
29. Tsutsumi, S.; Toyosada, M.; Hashiguchi, K. Extended Subloading Surface Model Incorporating Elastic Boundary Concept. *J. Appl. Mech.* **2006**, *9*, 455–462. [CrossRef]

30. *JIS G 3106:2008*; Rolled Steels for Welded Structure. Japanese Industrial Standards Committee: Tokyo, Japan, 2008.
31. A Theory of the Yielding and Plastic Flow of Anisotropic Metals. *Proc. R. Soc. Lond. Ser. A Math. Phys. Sci.* **1948**, *193*, 281–297. [CrossRef]
32. van den Boogaard, T.; Havinga, J.; Belin, A.; Barlat, F. Parameter Reduction for the Yld2004-18p Yield Criterion. *Int. J. Mater. Form.* **2016**, *9*, 175–178. [CrossRef]
33. Cazacu, O.; Barlat, F. Generalization of Drucker's Yield Criterion to Orthotropy. *Math. Mech. Solids* **2001**, *6*, 613–630. [CrossRef]
34. *JIS 2241:2011*; Metallic Materials-Tensile Testing-Method of Test at Room Temperature. Japanese Industrial Standards Committee: Tokyo, Japan, 2009.
35. Hashiguchi, K. *Elastoplasticity Theory*; Lecture Notes in Applied and Computational Mechanics; Springer: Berlin/Heidelberg, Germany, 2014; Volume 69, ISBN 978-3-642-35848-7.
36. Hashiguchi, K.; Tsutsumi, S. Elastoplastic Constitutive Equation with Tangential Stress Rate Effect. *Int. J. Plast.* **2001**, *17*, 117–145. [CrossRef]
37. Fincato, R.; Tsutsumi, S. Numerical Study of a Welded Plate Instability Using the Subloading Surface Model. *Mar. Struct.* **2017**, *55*, 104–120. [CrossRef]
38. Fincato, R.; Tsutsumi, S. An Overstress Elasto-Viscoplasticity Model for High/Low Cyclic Strain Rates Loading Conditions: Part I–Formulation and Computational Aspects. *Int. J. Solids Struct.* **2020**, *207*, 279–294. [CrossRef]
39. Fincato, R.; Tsutsumi, S. An Overstress Elasto-Viscoplasticity Model for High/Low Cyclic Strain Rates Loading Conditions: Part II–Numerical Analyses. *Int. J. Solids Struct.* **2021**, *208–209*, 247–261. [CrossRef]
40. Fincato, R.; Tsutsumi, S.; Sakai, T.; Terada, K. 3D Crystal Plasticity Analyses on the Role of Hard/Soft Inclusions in the Local Slip Formation. *Int. J. Fatigue* **2020**, *134*, 105518. [CrossRef]
41. Fincato, R.; Tsutsumi, S. Numerical Modeling of the Evolution of Ductile Damage under Proportional and Non-Proportional Loading. *Int. J. Solids Struct.* **2019**, *160*, 247–264. [CrossRef]
42. Fincato, R.; Tsutsumi, S. A Return Mapping Algorithm for Elastoplastic and Ductile Damage Constitutive Equations Using the Subloading Surface Method. *Int. J. Numer. Methods Eng.* **2018**, *113*, 1729–1754. [CrossRef]
43. Tsutsumi, S.; Kitamura, T.; Fincato, R. Ductile Behaviour of Carbon Steel for Welded Structures: Experiments and Numerical Simulations. *J. Constr. Steel Res.* **2020**, *172*, 106185. [CrossRef]
44. Tsutsumi, S.; Morita, K.; Fincato, R.; Momii, H. Fatigue Life Assessment of a Non-Load Carrying Fillet Joint Considering the Effects of a Cyclic Plasticity and Weld Bead Shape. *Frat. ed Integrità Strutt.* **2016**, *10*, 240–250. [CrossRef]
45. Tsutsumi, S.; Fincato, R.; Momii, H. Effect of Tangential Plasticity on Structural Response under Non-Proportional Cyclic Loading. *Acta Mech.* **2019**, *230*, 2425–2446. [CrossRef]
46. Fincato, R.; Tsutsumi, S. A Numerical Study of the Return Mapping Application for the Subloading Surface Model. *Eng. Comput.* **2018**, *35*, 1314–1343. [CrossRef]
47. The Society of Materials Science. *The Handbook of Fatigue Design*; The Society of Materials Science: Yokendo, Japan, 2015.
48. Iida, K. Micro-Crack Initiation Life and Micro-Fractographic Analysis in Strain Cycling Fatigue of a 60 kg/mm^2 High Strength Steel. *J. Soc. Nav. Archit. Japan* **1970**, *128*, a331–a342. [CrossRef]
49. Iida, K. Crack Initiation Life and Microfractographic Analysis in Strain Cycling Fatigue. *Trans. Jpn. Weld. Soc.* **1971**, *2*, 86–95.
50. Nieslony, A.; Dsoki, C.; Kaufmann, H.; Krug, P. New Method for Evaluation of the Manson–Coffin–Basquin and Ramberg–Osgood Equations with Respect to Compatibility. *Int. J. Fatigue* **2008**, *30*, 1967–1977. [CrossRef]
51. Muralidharan, U.; Manson, S.S. A Modified Universal Slopes Equation for Estimation of Fatigue Characteristics of Metals. *J. Eng. Mater. Technol.* **1988**, *110*, 55–58. [CrossRef]

Journal of
Marine Science and Engineering

MDPI

Article

Application of Equivalent Single Layer Approach for Ultimate Strength Analyses of Ship Hull Girder

Teguh Putranto [1,*], Mihkel Kõrgesaar [1] and Kristjan Tabri [1,2]

1 Department of Civil Engineering and Architecture, Kuressaare College, Tallinn University of Technology, Tallinna 19, 93819 Kuressaare, Estonia
2 Department of Civil Engineering and Architecture, School of Engineering, Tallinn University of Technology, Ehitajate tee 5, 19086 Tallinn, Estonia
* Correspondence: teguh.putranto@taltech.ee

Abstract: The objective of this paper is to present the application of equivalent single layer (ESL) approach for the ultimate strength assessment of ship hull girder in the context of numerical finite element (FE) simulations. In the ESL approach, the stiffened panel is replaced with a single plate, which has the equivalent stiffness of the original panel. Removal of tertiary stiffening elements from the numerical model facilitates time-savings in pre-processing and FE analysis stage. The applicability of ESL approach is demonstrated with two case studies, one compartment model and full-sized double hull tanker model in intact and damaged conditions. The damage extents are determined based on the international association of classification societies from common structural rules (IACS-CSR) for oil tanker. Ship hull girder is exposed to distributed pressure with the sinusoidal shape that bends the hull girder. This pressure load is applied separately to bottom and side structures to obtain the vertical and horizontal bending moments of the hull girder, respectively. Ultimate strength predictions obtained from ESL approach are compared to full three-dimensional finite element method (3D FEM) and IACS incremental-iterative method. The comparison between different methods is provided in terms of longitudinal bending moment and cross sectional stress distribution. Overall, ESL approach yields good agreement compared to the 3D FEM results in predicting the ultimate strength of ship hull girder while providing up to 3 times computational efficiency and ease of modeling.

Keywords: equivalent single layer; ultimate strength; hull girder; finite element method

Citation: Putranto, T.; Kõrgesaar, M.; Tabri, K. Application of Equivalent Single Layer Approach for Ultimate Strength Analyses of Ship Hull Girder. *J. Mar. Sci. Eng.* **2022**, *10*, 1530. https://doi.org/10.3390/jmse10101530

Academic Editor: Cristiano Fragassa

Received: 28 September 2022
Accepted: 15 October 2022
Published: 19 October 2022

1. Introduction

Due to extreme and accidental loads, a ship's hull girder can reach its ultimate load-carrying capacity. One of the fatal consequences of structural failure is that the ship may suffer progressive collapse due to internal and external loads during seafaring. To minimize the risk of such accidents, rules stipulate longitudinal strength assessment for all ships [1,2]. The objective of this assessment is to determine the ultimate strength capacity of ship hull girder when the ship is subjected to bending loads. In case of ship grounding or collision, damaged hull section reduces the bending strength further. Therefore, the ultimate strength assessment must be performed both in intact and damaged conditions to ensure that hull girder has sufficient strength reserve.

Several studies have been conducted to evaluate the ultimate strength of ship hull girder using simplified methods and are currently applied in the classification society and commercial software. Caldwell [3] proposed an equivalent thickness approach to replace stiffened panel and used strength reduction coefficients to consider buckling. Smith [4,5] further refined the strength reduction coefficient method and considered that the ultimate strength of a hull girder is dependent on the strength of individual elements reaching their limit at different times. The Smith's method has been adopted in the Common

Structural Rules for Bulk Carrier and Oil Tanker, but the loading is limited to the vertical bending moment. In parallel, Ueda and Rashed [6] proposed the idealized structural unit method (ISUM) which considers more loading scenarios. For example, ISUM can model the buckling response due to all possible hull girder sectional load components (i.e., vertical bending, horizontal bending, vertical shearing force, horizontal shearing force, and torsion). Several advanced applications of ISUM approach have been developed by Ueda et al. [7], Masaoka et al. [8], Fujikubo et al. [9], Paik et al. [10], as well as Lindemann and Kaeding [11] for ultimate strength analyses of stiffened panel structures in different loading conditions.

The ISUM is recognized as one of the most time efficient methods for progressive collapse analysis of ship hull girder [12]. The method has been implemented in the ALPS/HULL program within the MAESTRO FEM analysis code. The ISUM can deal with interaction between local and global failures [13] for a short section of the ship structure. However, larger models offer a number of advantages. To capture the compartment level buckling relevant for lightweight ship structures [14], a large-scale ship model should be considered. Larger models permit inclusion of actual pressure distributions and various load combinations. Additionally, the model length influences the post-buckling behavior, which ultimately determines the bending moment capacity [15]. Furthermore, full ship model are advocated in [16,17] to minimize the boundary effect which often lead to more conservative, heavier scantlings.

The full three-dimensional finite element method (3D FEM) is an effective tool used for performing progressive collapse analysis to obtain structural strength capacity of ship hull girder. The analysis can reflect the local failure of structural members, e.g., local plate buckling and stiffener tripping, if they are modeled in detail. However, the 3D FE simulation of entire ship structure requires enormous modeling and computational effort. To reduce these while maintaining the accuracy of 3D FEM, we propose the use of equivalent single layer (ESL) approach. In the context of ship structures, ESL has been used for analysis of buckling response of panels [18,19], vibration response of sandwich panels [20,21], and ultimate strength of stiffened panels [22,23]. However, the application examples for entire hull girder analysis are missing which this paper aims to fulfill. In the traditional FE modeling, a stiffened panel is modeled in detail composed of longitudinal stiffeners with its attached plating. Using ESL methodology, a stiffened panel is modeled as a plate without the stiffeners, but with the same stiffness as the original panel. Consequently, simplification of stiffened panels enables consideration of design alternatives without changing the FE mesh, and thus more efficient exploration of design space. Therefore, the main benefits of the ESL approach compared with 3D FE analysis are: (1) reduced modeling effort, (2) reduction in degree of freedom (DOF), and (3) reduced computational effort.

This paper presents the application of ESL approach for the ultimate strength assessment of ship hull girder and one compartment models. In the ESL model, the stiffeners are removed and shell properties are defined with equivalent stiffnesses composed of 6×6 membrane and bending stiffness matrices. The stiffness matrices are calculated by the first derivative of membrane force and bending moment obtained from a unit cell under six loading conditions. Two different ship-scale case studies are presented, first focusing on the compartment level analysis (Benson et al. [14]) and second, the full-scale analysis of ship hull girder including structural damage (Tabri et al. [24]). The ultimate strength of ship hull girder is analyzed in intact and damaged conditions due to grounding or collision. For the damage conditions, the damage extents are determined from the IACS-CSR and the structural members located in the damage extent area are removed. The distributed pressures are applied separately along the bottom and side structures of the ship to obtain the vertical and horizontal bending moment curves, respectively. For comparative purposes, analysis are also conducted with the incremental-iterative method from the IACS-CSR. The ultimate strength analysis are validated with the detailed 3D FE simulations.

2. Methods

In this paper, the ESL approach is used to predict the ultimate strength of hull girder. The theoretical framework and assumptions used in the ESL approach are explained in the following section.

2.1. Overview of the ESL Approach

The ESL approach employs the first-order shear deformation theory (FSDT) in which the Kirchhoff hypothesis is relaxed by removing the assumption of transverse normal. In the FSDT, the transverse normals do not remain perpendicular to the middle of the surface after plate is deformed, as can be seen in Figure 1. In this manner, transverse shear strains (γ_{xz}) are considered in the FSDT. Additionally, the rotation in the z direction is assumed to be zero. In the deformed plates, the transverse normals are displaced by u_0 and are rotated by ϕ_x from the undeformed position. Under the assumptions and restrictions used in the FSDT, the non-linear strains are expressed in Equation (1):

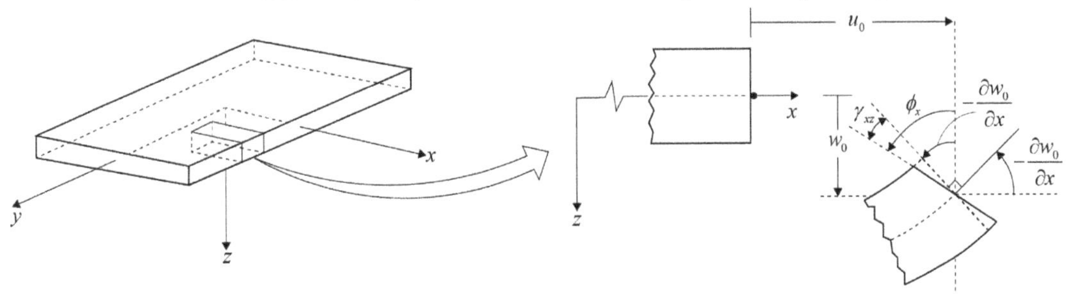

Figure 1. Undeformed and deformed geometries of an edge of a plate under the assumption of the first-order shear deformation theory (FSDT).

$$\begin{Bmatrix} \varepsilon_{xx} \\ \varepsilon_{yy} \\ \gamma_{yz} \\ \gamma_{xz} \\ \gamma_{xy} \end{Bmatrix} = \begin{Bmatrix} \varepsilon^0_{xx} \\ \varepsilon^0_{yy} \\ \gamma^0_{yz} \\ \gamma^0_{xz} \\ \gamma^0_{xy} \end{Bmatrix} + z \begin{Bmatrix} \varepsilon^1_{xx} \\ \varepsilon^1_{yy} \\ \gamma^1_{yz} \\ \gamma^1_{xz} \\ \gamma^1_{xy} \end{Bmatrix}, \tag{1}$$

where

$$\left\{\varepsilon^0\right\} = \begin{Bmatrix} \varepsilon^0_{xx} \\ \varepsilon^0_{yy} \\ \gamma^0_{yz} \\ \gamma^0_{xz} \\ \gamma^0_{xy} \end{Bmatrix} = \begin{Bmatrix} \frac{\partial u_0}{\partial x} + \frac{1}{2}\left(\frac{\partial w_0}{\partial x}\right)^2 \\ \frac{\partial v_0}{\partial y} + \frac{1}{2}\left(\frac{\partial w_0}{\partial y}\right)^2 \\ \frac{\partial w_0}{\partial y} + \phi_y \\ \frac{\partial w_0}{\partial x} + \phi_x \\ \frac{\partial u_0}{\partial y} + \frac{\partial v_0}{\partial x} + \frac{\partial w_0}{\partial x}\frac{\partial w_0}{\partial y} \end{Bmatrix}, \tag{2}$$

$$\left\{\varepsilon^1\right\} = \begin{Bmatrix} \varepsilon^1_{xx} \\ \varepsilon^1_{yy} \\ \gamma^1_{yz} \\ \gamma^1_{xz} \\ \gamma^1_{xy} \end{Bmatrix} = \begin{Bmatrix} \frac{\partial \phi_x}{\partial x} \\ \frac{\partial \phi_y}{\partial y} \\ 0 \\ 0 \\ \frac{\partial \phi_x}{\partial y} + \frac{\partial \phi_y}{\partial x} \end{Bmatrix}. \tag{3}$$

Note that the strains ($\varepsilon_{xx}, \varepsilon_{yy}, \gamma_{xy}$) are linear through the thickness, while the transverse shear strains (γ_{xz}, γ_{yz}) are constant through the thickness based on the FSDT. The strains (ε)

are composed of membrane (ε^0) and bending (ε^1) parts. Thus, the constitutive equations for the FSDT are obtained using the following relations:

$$\begin{Bmatrix} N_{xx} \\ N_{yy} \\ N_{xy} \\ M_{xx} \\ M_{yy} \\ M_{xy} \end{Bmatrix} = \begin{bmatrix} A_{11} & A_{12} & A_{13} & B_{11} & B_{12} & 0 \\ A_{21} & A_{22} & 0 & B_{21} & B_{22} & 0 \\ A_{31} & 0 & A_{33} & 0 & 0 & B_{33} \\ C_{11} & C_{12} & 0 & D_{11} & D_{12} & 0 \\ C_{21} & C_{22} & 0 & D_{21} & D_{22} & 0 \\ 0 & 0 & C_{33} & 0 & 0 & D_{33} \end{bmatrix} \begin{Bmatrix} \varepsilon^0_{xx} \\ \varepsilon^0_{yy} \\ \gamma^0_{xy} \\ \varepsilon^1_{xx} \\ \varepsilon^1_{yy} \\ \gamma^1_{xy} \end{Bmatrix}, \quad (4)$$

where N_{xx}, N_{yy} are the membrane forces, N_{xy} is the shear force, M_{xx}, M_{yy} are the bending moments, M_{xy} is the torsion, $\varepsilon^0_{xx}, \varepsilon^0_{yy}, \gamma^0_{xy}$ are the membrane strains, and $\varepsilon^1_{xx}, \varepsilon^1_{yy}, \gamma^1_{xy}$ are the curvatures. The **ABCD** stiffness matrices are obtained from the first derivative of membrane forces and bending moments of the unit cell (UC) simulations under six loading conditions, as can be seen in Figure 2. Compared to full 3D FEM, the generation of stiffness matrices requires an additional modeling and computational effort, which however can be made fairly automatic using programming.

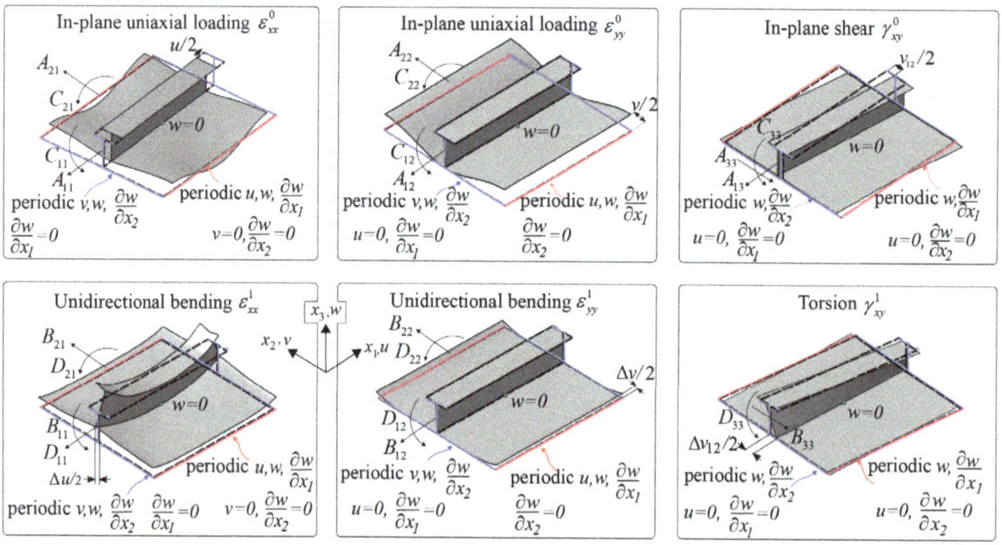

Figure 2. Six different unit cell (UC) configurations with boundary conditions needed for ABCD stiffness matrix definition. Forces and moments shown with arrows are associated with respective stiffness components. Boundary conditions for edges with the same color are identical.

The assumed UC is a periodic constituent of the full panel, thus periodic boundary conditions are imposed on UC, see Figure 2. The details of performing UC simulations are given in the previous papers by the authors [22,23]. Essentially, the same procedure was applied for each of the stiffened panel in case study analyses. Representative unit cell model was created for each panel with different topology (plate thickness and stiffener type) and 3D FE analyses were performed to obtain ABCD stiffness matrices. Compartment model (first case study) consisted only 1 stiffened panel configuration that was replaced by ESL. In a full ship hull (second case study), there were 45 different panel configurations, giving total of 270 (6 × 45) UC simulations. The entire procedure of UC analysis (pre- and post-processing) was made automatic using python scripts, which read the structural details from text file, prepare the models, and extract ABCD stiffness matrices.

Additionally, transverse shear strains are assumed to be constant based on the FSDT. The transverse shear force resultants (Q_x,Q_y) are calculated by multiplying the transverse shear stiffnesses (D_{Q_x},D_{Q_y}) and transverse shear strains (γ_{xz},γ_{yz}), as expressed in the following equation:

$$\begin{Bmatrix} Q_x \\ Q_y \end{Bmatrix} = \begin{bmatrix} D_{Q_x} & 0 \\ 0 & D_{Q_y} \end{bmatrix} \begin{Bmatrix} \gamma_{xz} \\ \gamma_{yz} \end{Bmatrix} \tag{5}$$

$$D_{Q_x} = k_{xz}\left(G_p t_p + G_w h_w + G_f h_f\right) \tag{6}$$

$$D_{Q_y} = k_{yz} G_p t_p \tag{7}$$

where k_{xz} is the longitudinal shear correction factor calculated by dividing the average shear stress ($\tau_{xz(avg)}$) to the maximum shear stress ($\tau_{xz(\max)}$), k_{yz} is the transverse shear correction factor as $5/6$. The shear moduli of the web (G_w) and flange (G_f) are the function of plate (G_p), web thickness (t_w), flange width (b_f), and stiffener spacing (s):

$$G_w = G_p(t_w/s\ ) \tag{8}$$

$$G_f = G_p\left(b_f/s\ \right) \tag{9}$$

2.2. Implementation of ESL in Abaqus

In ESL approach, stiffened panels are replaced by a single plate. An example is shown in Figure 3, where ship hull girder composed of stiffened panels is modeled using ESL approach. ESL properties are composed of 6 × 6 matrix considering membrane and bending stiffness components. In Abaqus software these ESL properties are given to elements by using shell general section option. This option allows reference to external Fortran-based VUGENS subroutine where the non-linear stiffness of ESL is calculated specific to loading stage. VUGENS subroutine is available starting from Abaqus 2022 and suitable for explicit analysis. To the best of authors knowledge, the analyses reported here are the first attempt to use VUGENS subroutine since in earlier versions of Abaqus only implicit version of shell general section definition could be used (UGENS). Note that in combination with explicit integration scheme the ESL approach could be potentially used in ship collision or grounding analysis. In non-ESL elements (frames, bracket, and other plates), standard isotropic properties are used by defining Young's modulus, Poisson's ratio, and material stress–strain curve.

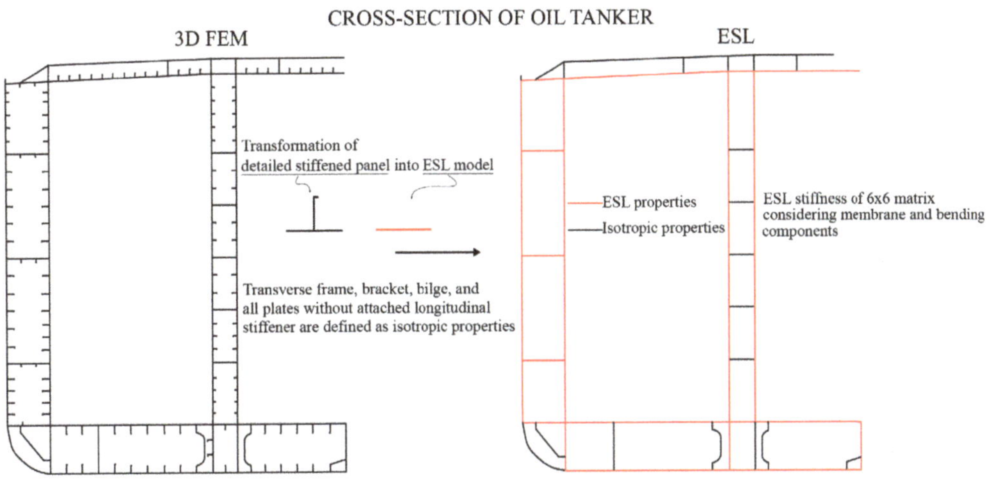

Figure 3. Application of ESL in a ship's hull girder.

3. Case Study of One Compartment Aluminium Box Girder

This paper examines the application of ESL to analysis of compartment level collapse, for details see Benson et al. [14]. The ultimate strength of one compartment box girder was analyzed and characterized by the bending moment versus curvature curve. The analyses were performed using the detailed 3D FEM and ESL approach. The detailed 3D FE analysis utilizes conventional modeling techniques with explicit modeling of stiffened plates further strengthened with larger transverse webframes with all parts given isotropic material properties. In the ESL model, longitudinally stiffened panels were represented with equivalent single layer plates having the same stiffness as the original plate with stiffeners while webframes were still explicitly modeled.

Modeled box girder with a 8.4 m length had a square cross section, stiffened on each side by 20 longitudinals spaced 400 mm apart. The panel configuration M1 was selected for analyses, see [14]. The longitudinal T-stiffeners had the web and flange dimensions of 120 × 55 mm and 55 × 7.7 mm, respectively. The transverse webframes were flat bars with the size of 180 × 10 mm spaced 1200 mm apart. There were six webframes between transverse bulkheads, which is sufficient to demonstrate buckling characteristics at the compartment level. To maintain the straightness of the compartment ends during bending, the bulkheads were modeled with a very large thickness. The mesh size used was 50 × 50 mm, which was determined by Benson et al. [14] through the mesh convergence analysis. The mesh size sensitivity study of ESL was performed using element sizes of 50 × 50 mm, 300 × 300 mm and 600 × 600 mm which are consisted of 24, 4 and 2 elements between webframes, respectively. The box girder was given aluminium 5083 properties with an elastic modulus of 70 GPa and a yield stress of 302 MPa. The non-linear stress–strain response of this material was characterized by the Ramberg–Osgood relationship, as expressed in the following equation:

$$\varepsilon = \frac{\sigma}{E} + 0.002\left(\frac{\sigma}{\sigma_{0.2}}\right)^{n} \tag{10}$$

where ε is the strain, σ is the applied stress, E is the elastic modulus, $\sigma_{0.2}$ is the 0.2% offset proof stress, n is the exponent.

Boundary conditions applied in the box girder are explained in Figure 4. At the reference point, a moment to the z-axis was applied resulting in compression on the top panel and tension on the bottom panel. This reference point was connected to the nodes at the boundary using kinematic coupling so that the section remained flat during rotation. At the opposite section, the clamped boundary condition was imposed. Only half of the girder was modeled by imposing symmetry on the center line.

The response of the box girder is compared in terms of bending moment vs. curvature curves and overall deformations obtained at the maximum bending moment, see Figure 5. Current 3D FE analysis gives slightly softer response compared with 3D FE results from Benson et al. [14]. Curvature in current study was determined by tracking the rotation at the reference point. The way curvature was obtained in analysis of Benson was not detailed, which possibly explains the slightly softer response obtained with current 3D FE analysis, while the overall behavior is well captured. Proceeding to analyze differences between current 3D FEM and ESL we also note close agreement in maximum bending moment. Even the coarsest (and stiffest) ESL model of 600 × 600 mm shows the difference of mere 4.57% compared with 3D FEM result. All studied element sizes can capture accurately compartment level (overall) buckling behavior and for overall efficiency we advocate to use the largest mesh of 600 × 600 mm or two elements between webframes. However, the ESL accuracy decreases if the stiffened panel is subjected to local buckling of the stiffener web or plate [22,23]. Therefore, further investigation should be performed what is the suitable element size for such simulations. In the post-ultimate stage, there is a sudden decrease in bending moment due to transition from interframe to overall buckling mode. ESL model cannot accurately trace this structural response at the post-ultimate stage.

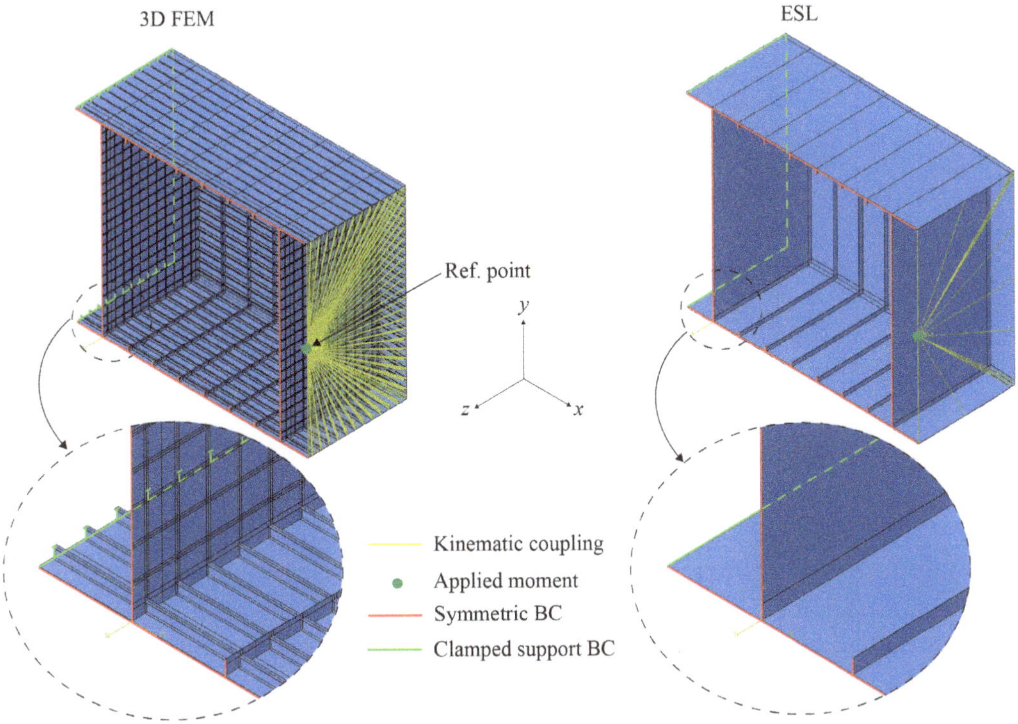

Figure 4. Boundary condition applied in 3D FEM and ESL of box girder.

In addition to providing a high level of accuracy with coarse elements of 600 × 600 mm, ESL approach also provides shorter computation times than 3D FEM. With ESL mesh of 600 × 600 mm the computation time is only 27.7% of full 3D model, see Figure 6. All simulations were ran with processor type of Intel(R) Xeon(R) CPU E5-2690 0 @ 2.90 GHz, 4 cores, 4 domains, and RAM of 24.0 GB.

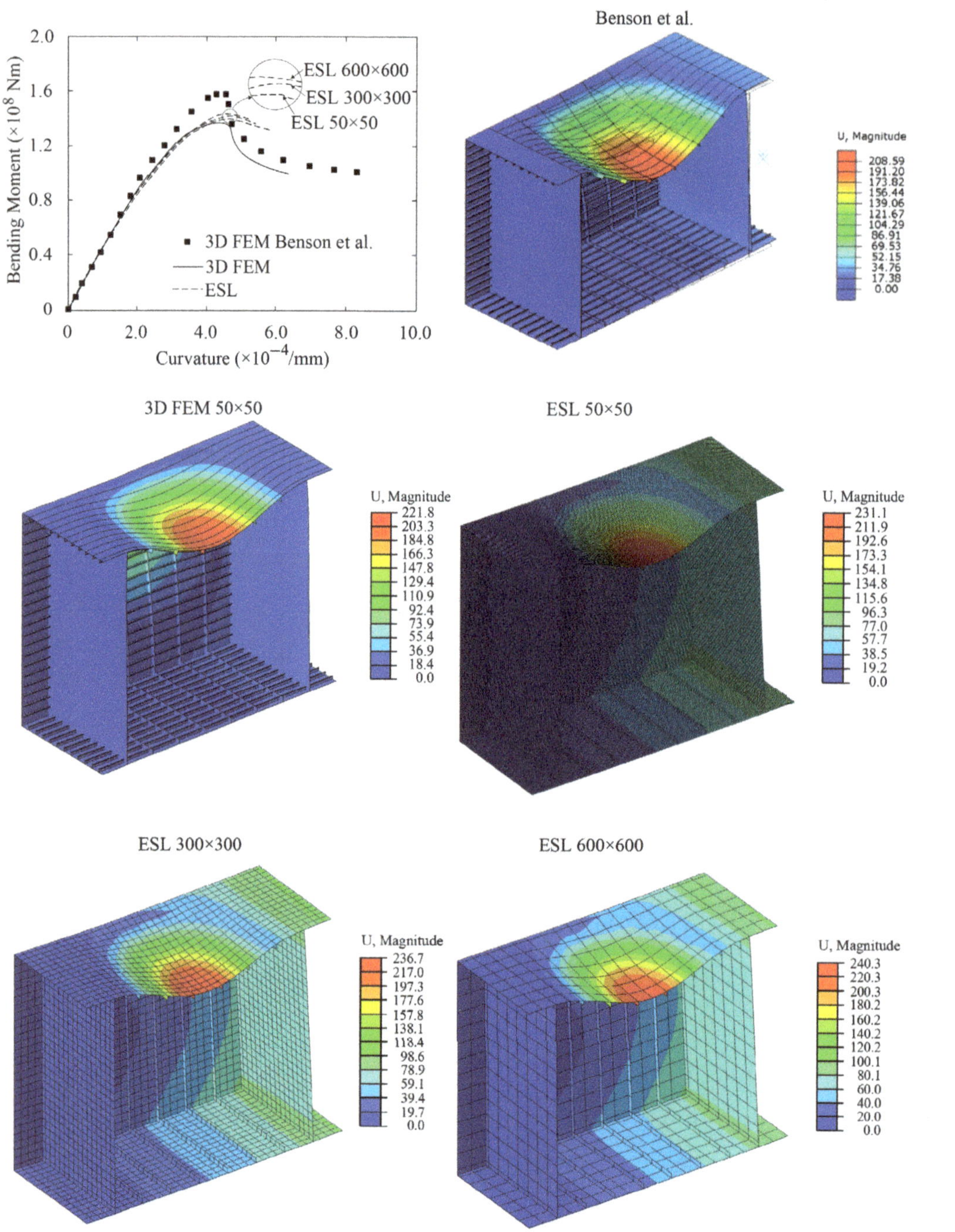

Figure 5. Bending moment–curvature curves and deformation shapes obtained from Benson et al. [14], 3D FEM and ESL models. U is in mm.

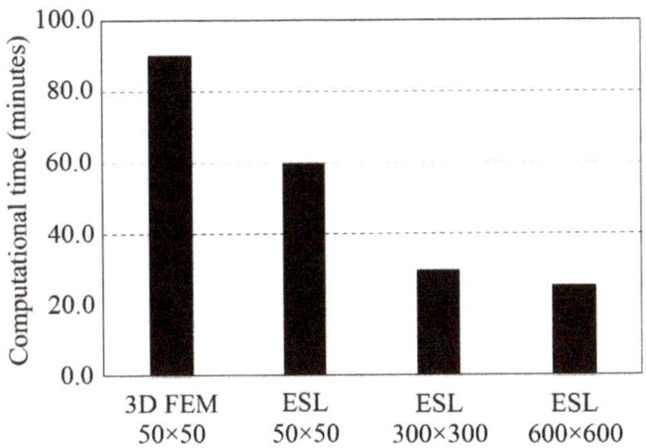

Figure 6. Computational time for 3D FEM and ESL with several mesh sizes.

4. Case Study of Full-Scale Steel Ship Structure

Ultimate strength analyses of ship hull girder in intact and damaged conditions were performed using the ESL approach, full 3D FEM, and incremental-iterative method from the IACS-CSR. A traditional ship hull can be considered as a light-weight thin-walled structure composed of an outer shell that is stiffened with framing members. In other words, ship hull girder is built from stiffened panels. From a design perspective, a detailed modeling of stiffened panel composed of plates and stiffeners is required for 3D FEM. However, ESL simplifies the modeling process as the plate elements are given stiffness properties representative of stiffened panels rendering explicit modeling of stiffeners unnecessary. In the IACS-CSR, ultimate strength is calculated based on a cross-section of hull girder between two adjacent transverse webframes. In Section 2, the theory and implementation of ESL approach was explained. In this section, the full 3D FEM and incremental-iterative method are described.

4.1. Ship Particulars

The case study structure is a chemical product tanker that was previously analyzed in Tabri et al. [24]. The midship section and bulkhead arrangement are given in Figure 7 and the main particulars in Table 1. The ship is designed from high strength steel (AH36) with Young's modulus of $E = 210$ GPa, Poisson ratio of $v = 0.3$, and yield stress of $\sigma_y = 355$ MPa.

Table 1. Main particulars of the chemical tanker.

Parameter	Symbol	Unit	Value
Overall length	L_{OA}	m	182.2
Length between perpendiculars	L_{PP}	m	175.3
Moulded breadth	B	m	32.2
Depth	H	m	15.0
Design draught	B	m	11.1
Displacement	∇	t	52,298
Double bottom height	H_{DB}	m	2.21

4.2. Loading and Boundary Conditions

Being a flexible thin-walled beam, the hull girder of the ship flexes globally when exposed to loads. The load components that act on the ship hull girder are the weight of the ship, its cargo, and the hydrostatic and hydrodynamic pressures (external load). The resultant of these load components can be treated with longitudinally distributed load applied on the hull girder, which can increase only by the external pressure due to waves. Therefore, global bending of the ship hull was achieved with longitudinally distributed pressure, which amplitude was gradually increased until ultimate strength was reached. The distributed pressure was applied either on ship bottom or side, depending on whether vertical or horizontal bending moment was determined, respectively. The sinusoidal shape of distributed pressure was kept unchanged during loading, as shown in Figure 8b. Although, in realistic situations, the distribution of load can play an important role, here only the simplified sinusoidal shape was considered. The direction of bending was controlled by the sign of the pressure amplitude (A), see the Equation in Figure 8b. This resulted in sagging/hogging in vertical bending and starboard/portside bending in horizontal bending. In damaged condition, the pressure was not applied in the damage opening. In Abaqus, the VDLOAD subroutine was invoked to apply the distributed pressure.

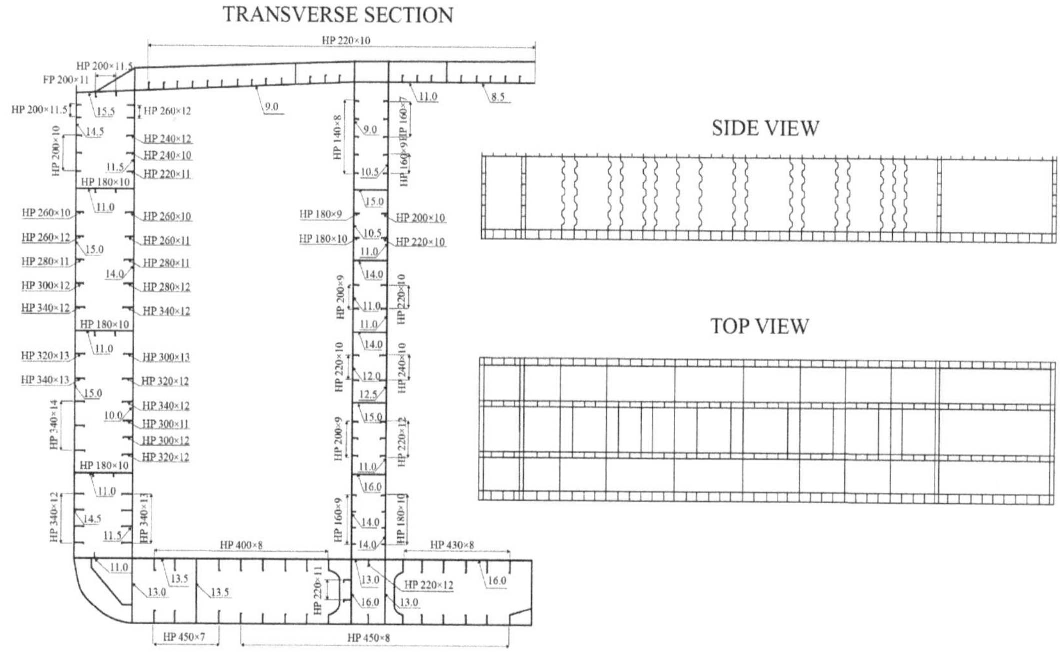

Figure 7. Design of hull structure for case study: midship section (**left**) and side view and top view (**right**).

A simply supported boundary condition was imposed on the ship at the mid nodes marked with •, see Figure 8a. The aft of the ship was pin constrained with all translations fixed while rotations were free. In the fore part, the constraints were the same except translation in longitudinal direction which was free to avoid an excessive stress concentration during deflection. Furthermore, rigid beams were modeled through the constraint nodes to keep the ends of the beam straight under increasing load and, thus, prevent local buckling. The surfaces where pressure was applied were specified with rectangles ABCD (for vertical bending moment) and EFGH (for horizontal bending moment), see Figure 8a. The reaction forces on the support is assumed to be very small since the distributed pressures produce the resultant forces of zero.

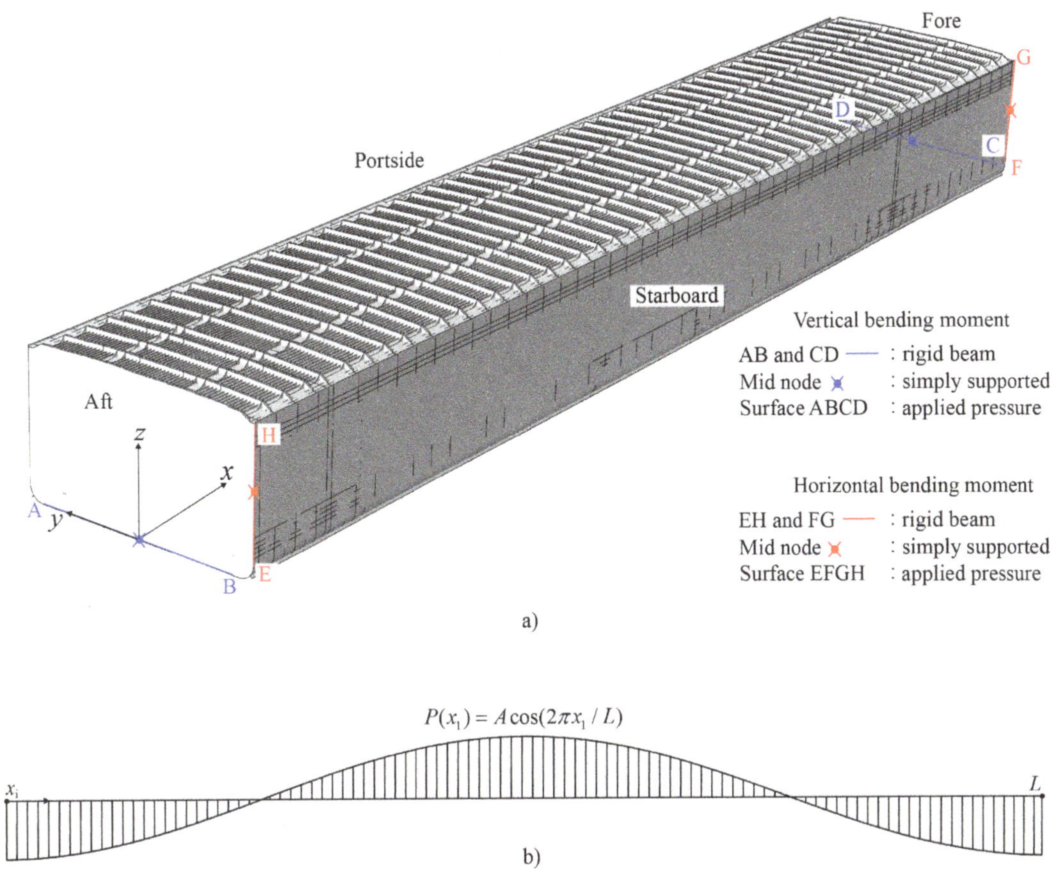

Figure 8. (**a**) Boundary condition applied in the tanker under vertical and horizontal bending moments. (**b**) Distributed pressure applied on the surface.

4.3. Damage Scenario

For post-accidental strength assessment of ship structure, both grounding and collision scenarios were considered. The extent of damage was determined based on the definition provided by the IACS-CSR. The damage was modeled by removing the plates, stiffeners, and frames that fall within the specified damage extent. The selected length of damage was 10 m with its center in the middle of the ship. This position was chosen since it significantly reduces the longitudinal bending strength of the ship. In case of collision damage, the transverse damage extent was equal to $B/16$ where B is the breadth of the ship. Therefore, respective structural elements in the parts of inner and outer skin and main deck were removed. The vertical damage extent was taken as $0.6D$ measured from the main deck, where D is the depth of the ship.

In case of grounding damage, the transverse damage extent was taken as $0.6B$, which is symmetric with respect to the centerline. The damage penetration height was equal to the depth of the double bottom (H_{DB}). Figure 9 shows the damage extents of the double hull oil tanker for grounding and collision scenario.

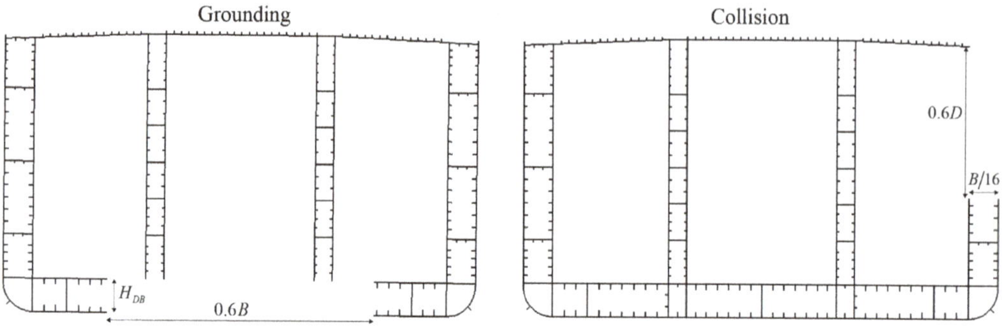

Figure 9. Assumed grounding and collision damage extents in tanker.

4.4. Full 3D Finite Element (3D FE) Model

Non-linear finite element method (NLFEM) is a sophisticated tool to solve the solid mechanics problem of complex engineering structures. Using NLFEM, the effect of material and geometrical non-linearities during progressive collapse of hull girder can be taken into account. A chemical tanker was modeled using Abaqus/Explicit software. All models were meshed using four-node shell elements (S4R), including plates, stiffener webs, frames, and girders. Stiffening was achieved with HP bulb profiles where the flanges were modeled using beam elements (B21). Stiffener webs were modeled using shell elements to capture the collapse mode, e.g., local buckling of stiffener web or stiffener tripping, expected during the progressive collapse of the hull girder. The mesh convergence study was conducted by Tabri et al. [24] to obtain the balance between numerical accuracy and simulation cost. Thus, the mesh density in the one compartment is illustrated in Figure 10 and can be summarized as follows: (1) plates between longitudinal stiffeners, 4 × 16 shells; (2) stiffener web plates, 1 × 16 (web height < 300 mm) shells and 2 × 16 (web height > 300 mm) shells; (3) side stringers and girders, 12 × 16 shells; and (4) corrugated bulkhead plates, 6 × 40 and 5 × 60 shells. The 3D FE model consisted of 3,160,000 nodes and 2,800,000 elements from which 2.5 M were shell and 0.3 M were beam elements. In the ESL model, stiffeners (shell elements related to web and beam element for flange) were removed so the total number of elements was reduced by 25% to 2,120,000 compared with the full 3D FE model.

4.5. Incremental-Iterative Method

The IACS-CSR provides the incremental-iterative method to calculate the ultimate strength of hull girder. Details of the method are given in CSR, which are briefly summarized here for entirety. The assumption is that hull girder collapse occurs in between two adjacent transverse webs. Accordingly, two-dimensional (2D) cross-section of the ship hull structure is divided into a series of structural elements, such as stiffeners, plates, and hard corners. The response of each of those elements is described with load-end-shortening curve compliant with prevalent collapse mode. Under compression, the stiffener element may experience the specific collapse mode, such as beam-column buckling, torsional buckling, or web buckling. For the plating element, the collapse mode of plate buckling may occur. Other element types under compression or tension may experience idealized elastic-plastic failure. Accordingly, the load-end-shortening curves for each structural element were obtained.

The calculation procedure of ultimate strength starts with estimation of neutral axis (NA) position. The iterative approach involves increasing the assumed curvature of the hull girder and calculating the new position of NA based on the moment equilibrium. Bending moment is obtained by integrating the load over the cross-section, while the load for each element is obtained from the load-end-shortening curves used as input. Once the updated location of the NA is obtained, the curvature is further increased and procedure is repeated until the maximum bending moment is achieved.

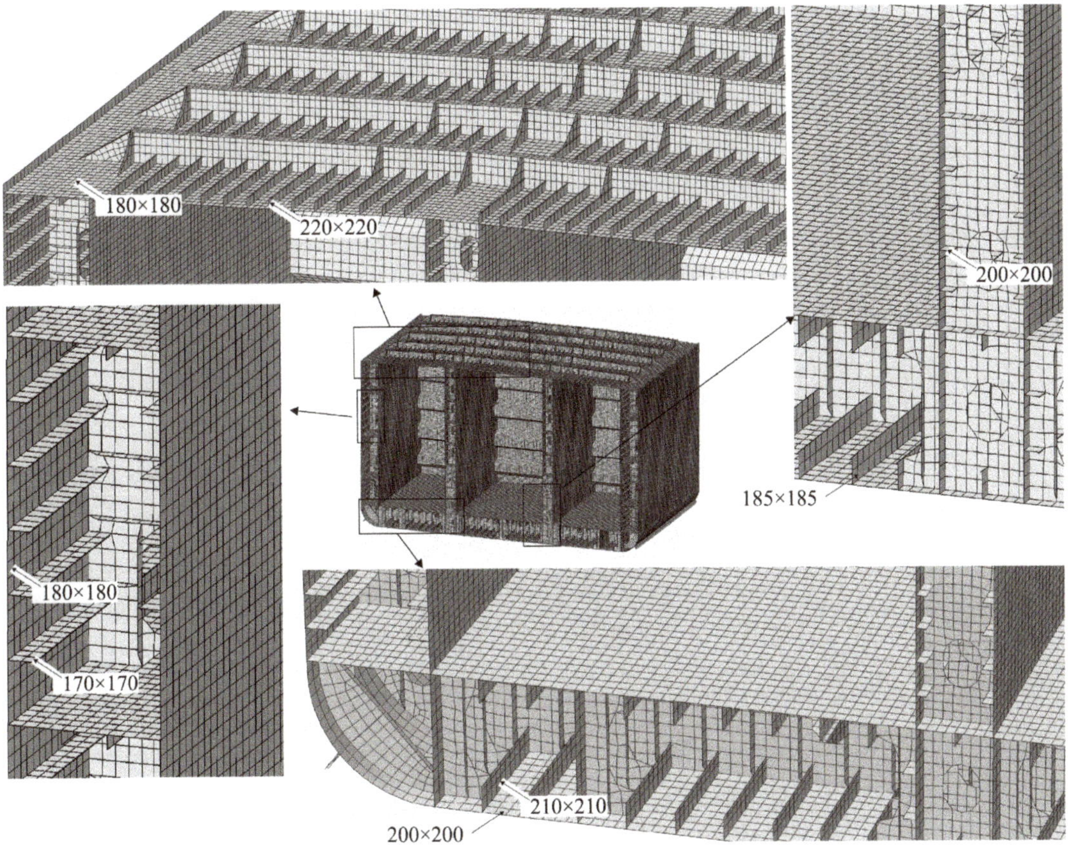

Figure 10. Mesh of the one compartment in the FE-model of chemical tanker.

5. Ultimate Strength Analysis

Ultimate strength of a chemical tanker was analyzed under vertical and horizontal bending moments. The results obtained from the ESL approach, 3D FEM, and incremental-iterative method are presented.

5.1. Vertical Bending Moment

Figure 11 shows the vertical deflection of intact chemical tanker along its length under hogging and sagging conditions. The horizontal bending moment is discussed in the next section. The deflections at the aft and fore are zero since the vertical translation of those parts is constrained. The maximum deflection occurs in the middle of the ship where the distributed pressure has maximum amplitude. In general, ESL deflection correlates well with the 3D FEM results.

The bending moment–deflection curves for 3D FEM, ESL, and incremental-iterative approach (hereinafter referred to as CSR method) for intact ship are presented in Figure 12a. The entire curve until maximum bending moment is accurately predicted by both, ESL and CSR method. Therefore, benefits of performing computationally more demanding ESL analyses are missing until more detailed information is desired. For instance, Figure 12b shows the longitudinal stress in vertical coordinate at the ship midsection under hogging condition. The constant tensile stresses in the upper decks (above side stringer S2) indicate to fully plastic condition. In the compressive side below the NA (slightly above side stringer S1) stress distribution is more complex due to non-linear buckling of structural elements. The stress provided by ESL approach has better agreement with the 3D FEM than CSR method. Furthermore, with 3D FE model available, similar stress distributions could be obtained in different cross-sections along the length, which is not possible with CSR.

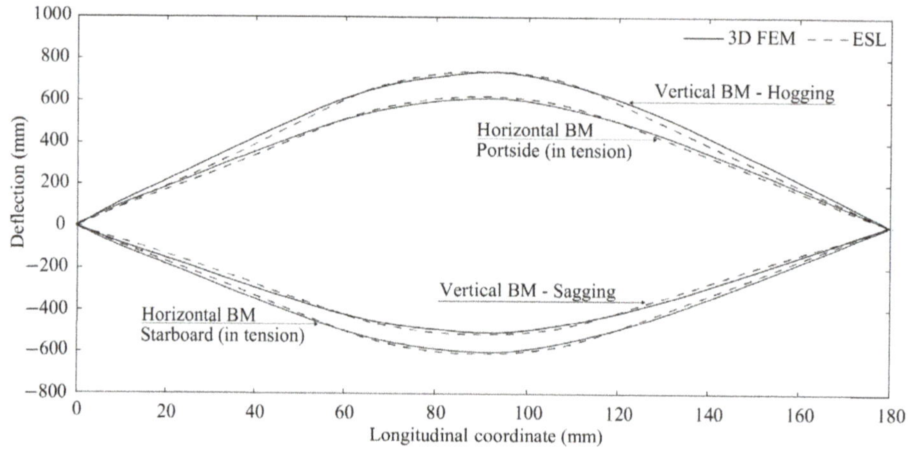

Figure 11. Deflection of ship's hull along its length at the ultimate stage under vertical and horizontal bending moments.

In the case of ship grounding, the bending moment–deflection curve is presented in Figure 13a. The initial stiffness is the same for all three curves. The bending moment increases linearly until gradual buckling of structural elements. The moment reduction under hogging condition is more significant since compressive loads are carried by damaged double bottom structure. The comparison of longitudinal stress distributions in Figure 13b shows that while ESL accuracy decreases compared with intact analyses, the stresses are still more accurately predicted compared with CSR.

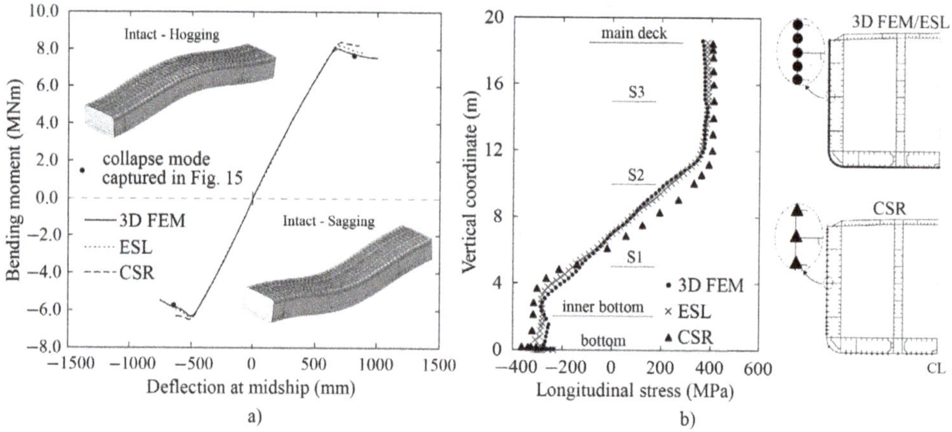

Figure 12. Response of intact ship under vertical bending moment. (**a**) Bending moment–deflection curves. (**b**) Longitudinal stress distribution to vertical coordinate at the midship under hogging condition.

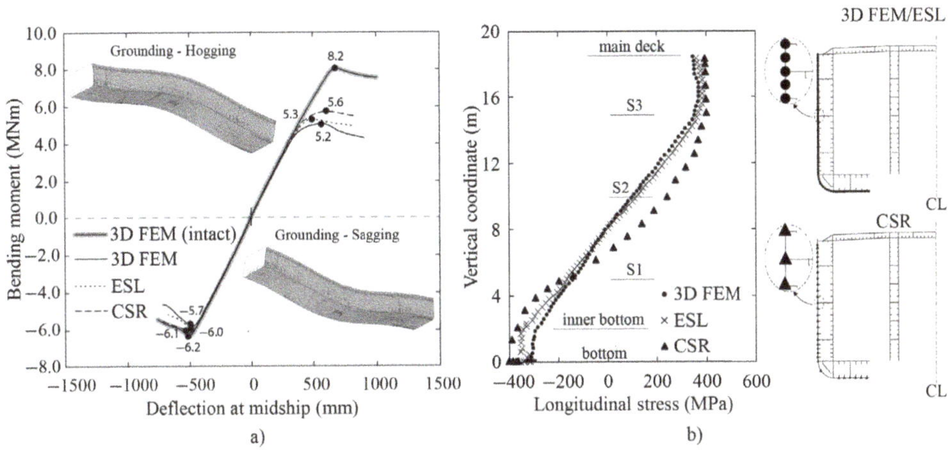

Figure 13. Response of ship grounding under vertical bending moment. (**a**) Bending moment–deflection curves. The maximum bending moments are used in calculating the histogram of moment reduction ratio. (**b**) Longitudinal stress distribution to vertical coordinate at the midship under hogging condition.

In the case of collision damage, the relationship between bending moment and deflection under sagging/hogging conditions is presented in Figure 14a. The longitudinal structure, especially the deck-side corner, greatly contributes to longitudinal strength. By removing the damaged side and deck structure, the ultimate bending moment decreases more than in case of grounding damage. However, this decrease in maximum bending moment is well captured by the ESL method which shows improved predictions compared with CSR. The reduced accuracy of CSR is explained by the effect of the removed deck-side corner. With respect to the longitudinal stresses in Figure 14b, the ESL method has similar accuracy as in grounding case, while maintaining the advantage over CSR.

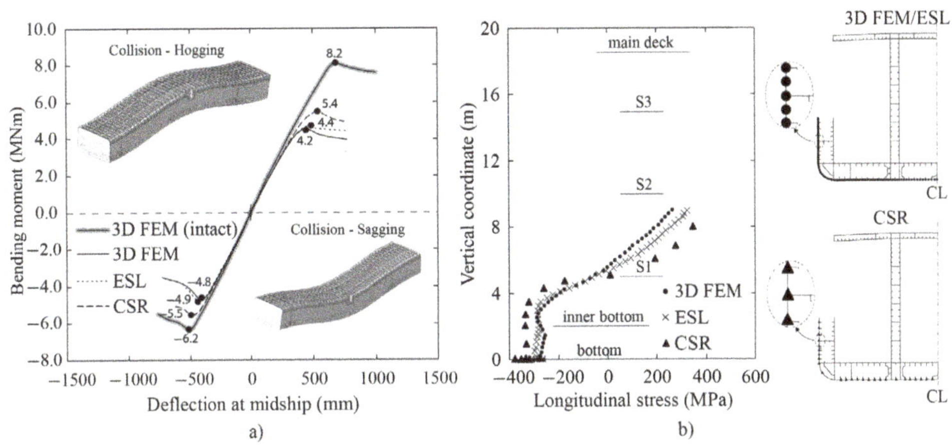

Figure 14. Response of ship collision under vertical bending moment. (**a**) The bending moment–deflection curves. The maximum bending moments are used in calculating the histogram of moment reduction ratio. (**b**) The longitudinal stress distribution to vertical coordinate at the midship under hogging condition.

The ESL approach captures the bending moment until ultimate stage with very good accuracy. In the post-ultimate stage bending moment is captured accurately under sagging, but not under hogging. Under hogging condition ESL fails to capture the local stiffener web buckling collapse in bottom structure as shown in Figure 15. It was also shown in analysis of [22,23] that beyond certain threshold slenderness ($\beta = 1.96$) the collapse mode becomes local which is not captured by ESL. In current analysis the plate slenderness in bottom structure is 2.12, while in deck structure slenderness is 1.55. In contrast, the slenderness of deck plates is much lower circumventing the local collapse under sagging condition, which ultimately leads to very good accuracy also in post-ultimate bending moment prediction, see Figure 14a.

The bending moment reduction with respect to intact condition obtained with each method (3D FEM, ESL, and CSR) are summarized in Figure 16. This reduction is calculated by dividing the maximum bending moment at damaged condition with the corresponding maximum from intact condition for each method concerned. Furthermore, the error in moment reduction obtained with ESL and CSR is calculated with respect to 3D FEM results and shown in the figure. The overall accuracy of ESL is excellent remaining in 5% in all cases. The compromise between accuracy and overall analyses cost favor CSR over ESL in grounding analyses, while accuracy of CSR reduces in collision analyses.

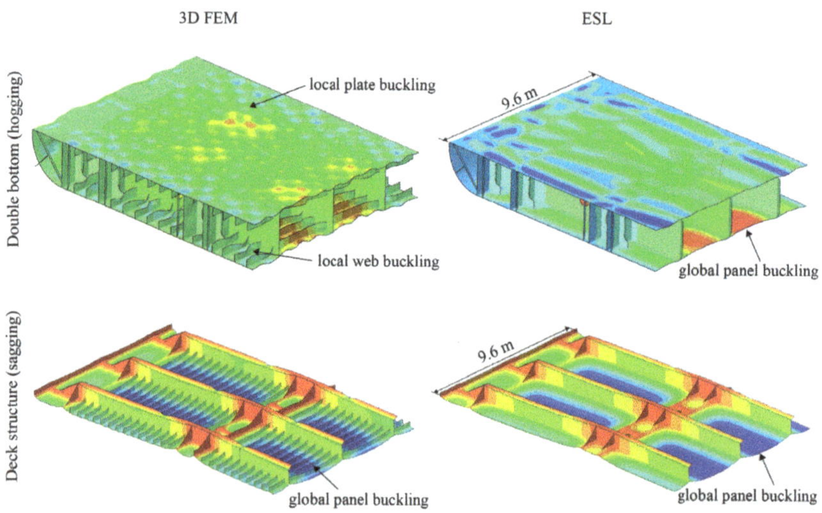

Figure 15. Collapse mode in double bottom and main deck structures under hogging and sagging conditions in post-ultimate stage, respectively. Deformation scaling factor is 5×.

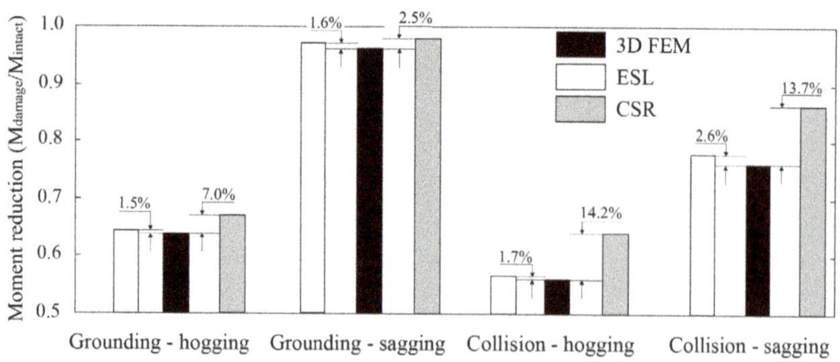

Figure 16. The moment reduction ratio under grounding and collision damage in the hogging and sagging conditions using the 3D FEM, ESL, and CSR methods.

5.2. Horizontal Bending Moment

The horizontal bending moment–deflection at midship curves are presented in Figure 17a–d for intact and damaged ship. Overall, a close agreement exists between the bending moment resulted by the ESL and 3D FEM until the ultimate stage. Since hull girder has greater depth than breadth, the ultimate horizontal bending moment is greater than vertical bending moment. For intact and grounding model, the cross-section of the ship is symmetric to the centerline so that the bending moment is independent whether the loading comes from port or starboard side. However, for collision model, the bending moment is dependent on the loading direction. For these cases, the ESL model can accurately capture the reduction in bending moment until the ultimate condition.

The moment reduction ratio under horizontal bending moment for the three methods are shown in Figure 18. The error percentage represents the difference between the analyzed method and 3D FEM. In all cases, the ESL results are consistently close to the 3D FEM.

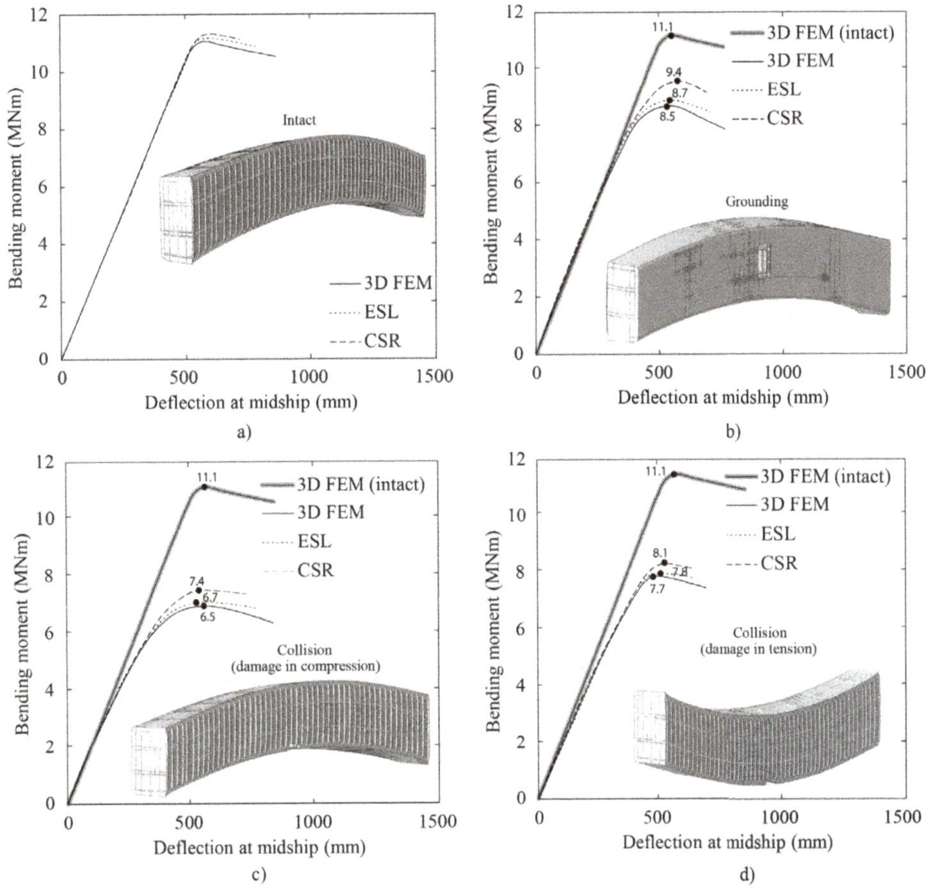

Figure 17. Horizontal bending moment-deflection relationships of tanker in the condition of (**a**) intact, (**b**) grounding, (**c**) collision with damage in compression, and (**d**) collision with damage in tension. The maximum bending moments are used in calculating the histogram in Figure 18.

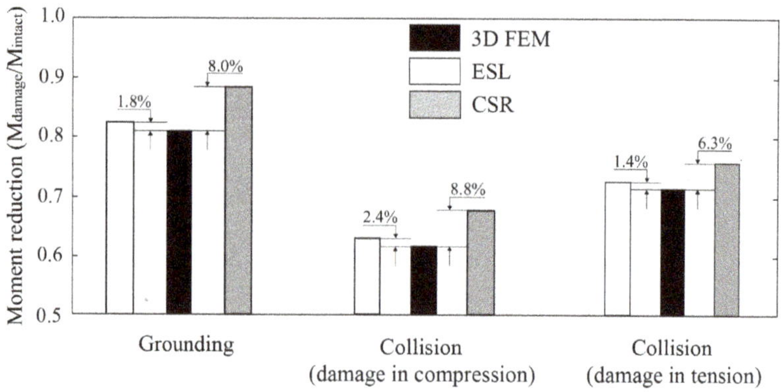

Figure 18. The moment reduction ratio of horizontal bending moment in grounding and collision cases for the 3D FEM, ESL, and CSR methods.

6. Conclusions

The ultimate strength analyses are carried out using the one compartment and full-scale ship models. The one compartment model is analyzed under vertical bending moment. The full-scale ship model in intact and damaged conditions is analyzed under vertical and horizontal bending moments. The analyses are performed with three methods, namely 3D FEM, ESL, and CSR. Overall, the ultimate strength predicted by the ESL approach give the results close to 3D FEM for all cases.

The ESL approach provides a more time-efficient way to analyze the ultimate strength compared to the detailed 3D FEM. In the ESL approach, plate with stiffeners is represented with an equivalent plate with equal stiffness so that stiffeners are not explicitly modeled. This reduces modeling effort as well as computational time and, thus, could be potentially used with great efficiency in structural optimization, which has seen increased popularity due to the advancements in computing power. Current analysis showed that in case of one compartment model, the analysis time was up to 3 times shorter when using the coarsest possible mesh. Moreover, the computational efficiency does not compromise the accuracy as the ultimate bending moment was captured with less than 5% error compared with 3D FEM in all analyzed cases. In contrast, the CSR results overestimate the collapse moment by up to 14.2% in collision damage scenario under hogging. This overestimation of bending moment obtained with CSR method is consistent with analysis in literature, see [25,26].

In addition to accurately visualizing the full-scale ship deflection, ESL model can capture the load-response of a ship structure until ultimate bending moment with very good accuracy. However, due to the interaction of structural members, the ESL approach cannot accurately account the local stiffener web buckling collapse in the post-ultimate stage. Furthermore, with ESL methodology one cannot visualize stresses in the post-processing stage. Both aspects need to be addressed in future investigations.

Author Contributions: Conceptualization, T.P., M.K. and K.T.; methodology, T.P. and M.K.; software, T.P.; validation, T.P.; formal analysis, T.P. and M.K.; investigation, T.P.; resources, T.P.; data curation, T.P.; writing—original draft preparation, T.P.; writing—review and editing, T.P., M.K. and K.T.; visualization, T.P.; supervision, M.K.; project administration, M.K. and K.T.; funding acquisition, M.K. and K.T. All authors have read and agreed to the published version of the manuscript.

Funding: The first author was financially supported by the European Regional Development Fund through the DORA scholarship for the doctoral students. The work has also been financially supported by the Estonian Research Council via grant PRG83 (Numerical simulation of the FSI for the dynamic loads and response of ships) and grant PSG754 (Coupled simulation model for ship crashworthiness assessment). These funding mechanisms are gratefully acknowledged.

Institutional Review Board Statement: Not applicable.

Informed Consent Statement: Not applicable.

Data Availability Statement: The data presented in this study are available on request from the corresponding author.

Conflicts of Interest: The authors declare no conflict of interest.

References

1. Det Norske Veritas, G.L. Rules for classification: Ships. *Ships Navig. Ice Det Nor.* **2016**, 3, 351–354.
2. IACS. Background Document—Section 9/1, Design Verification, Hull Girder Ultimate Strength. IACS Common Structural Rules for Double Hull Oil Tankers, January 2006. pp. 321–368. Available online: https://www.classnk.or.jp/hp/pdf/activities/csr/ECSR-T-TB002.pdf (accessed on 20 September 2022).
3. Caldwell, J. Ultimate Longitudinal Strength. *Trans. RINA* **1965**, 10, 411–430.
4. Smith, C.S. Influence of local compressive failure on ultimate longitudinal strength of a ship's hull. *Trans. PRADS* **1977**, 10, 73–79.
5. Tatsumi, A.; Iijima, K.; Fujikubo, M. A Study on Progressive Collapse Analysis of a Hull Girder Using Smith's Method—Uncertainty in the Ultimate Strength Prediction. Practical Design of Ships and Other Floating Structures, Singapore, 2021. Available online: https://www.sciencegate.app/document/10.1007/978-981-15-4672-3_7 (accessed on 20 September 2022).
6. Yukio, U.; Rashed, S.M.H. The idealized structural unit method and its application to deep girder structures. *Comput. Struct.* **1984**, 18, 277–293. [CrossRef]
7. Ueda, Y.; Rashed, S.M.H.; Abdel-Nasser, Y. An improved ISUM rectangular plate element taking account of post-ultimate strength behavior. *Mar. Struct.* **1994**, 7, 139–172. [CrossRef]
8. Masaoka, K.; Okada, H.; Ueda, Y. A rectangular plate element for ultimate strength analysis. In Proceedings of the 2nd International Conference on Thin-Walled Structures, Singapore, 2–4 December, 1998.
9. Fujikubo, M.; Kaeding, P.; Olaru, D.; Pei, Z. Development of ISUM Plate Element Considering Lateral Pressure Effects and Its Application to Stiffened Plate. Transactions of The West-Japan Society of Naval Architects The 109th West-Japan Society of Naval Architects Meeting (Joint Autumn Meeting of Three Societies of Naval Architects in Japan, 2004) (Transactions of The West-Japan Society of Naval Architects No. 109, Japan, 2005. Available online: https://www.jstage.jst.go.jp/article/jjasnaoe/3/0/3_0_285/_article (accessed on 20 September 2022).
10. Paik, J.K.; Seo, J.K.; Kim, D.M. Idealized structural unit method and its application to progressive hull girder collapse analysis of ships. *Ships Offshore Struct.* **2006**, 1, 235–247. [CrossRef]
11. Lindemann, T.; Kaeding, P. Application of the idealized structural unit method for ultimate strength analyses of stiffened plate structures. *Ship Technol. Res.* **2017**, 1, 15–29. [CrossRef]
12. Paik, J.K.; Thayamballi, A.K. *Ship-Shaped Offshore Installations: Design, Building, and Operation*; Cambridge University Press: Cambridge, UK, 2007.
13. Kim, D.K.; Kim, H.B.; Mohd, M.H.; Paik, J.K. Comparison of residual strength-grounding damage index diagrams for tankers produced by the ALPS/HULL ISFEM and design formula method. *Int. J. Nav. Archit. Ocean Eng.* **2013**, 5, 47–61. [CrossRef]
14. Benson, S.; Downes, J.; Dow, R.S. Compartment level progressive collapse analysis of lightweight ship structures. *Mar. Struct.* **2013**, 31, 44–62. [CrossRef]
15. Tekgoz, M.; Garbatov, Y.; Soares, C.G. Strength assessment of an intact and damaged container ship subjected to asymmetrical bending loadings. *Mar. Struct.* **2018**, 58, 172–198. [CrossRef]
16. Yoshikawa, T.; Bayatfar, A.; Kim, B.J.; Chen, C.P.; Wang, D.; Boulares, J.; Gordo, J.M.; Josefson, L.; Smith, M.; Kaeding, P. Report of ISSC 2015 Committee III. 1 Ultimate Strength. In Proceedings of the 19th International Ship and Offshore Structures Congress, Lisbon, Portugal, 7–10 September 2015.
17. Mohammed, E.A.; Benson, S.D.; Hirdaris, S.E.; Dow, R.S. Design safety margin of a 10,000 TEU container ship through ultimate hull girder load combination analysis. *Mar. Struct.* **2016**, 46, 78–101. [CrossRef]
18. Nordstrand, T. On buckling loads for edge-loaded orthotropic plates including transverse shear. *Compos. Struct.* **2004**, 46, 1–6. [CrossRef]
19. Jelovica, J.; Romanoff, J.; Ehlers, S.; Varsta, P. Influence of weld stiffness on buckling strength of laser-welded web-core sandwich plates. *J. Constr. Steel Res.* **2012**, 77, 12–18. [CrossRef]
20. Lok, T.S.; Cheng, Q.H. Free vibration of clamped orthotropic sandwich panel. *J. Sound Vib.* **2000**, 229, 311–327. [CrossRef]
21. Jelovica, J.; Romanoff, J.; Klein, R. Eigenfrequency analyses of laser-welded web-core sandwich panels. *Thin-Walled Struct.* **2016**, 101, 120–128. [CrossRef]
22. Putranto, T.; Kõrgesaar, M.; Jelovica, J.; Tabri, K.; Naar, H. Ultimate strength assessment of stiffened panel under uni-axial compression with non-linear equivalent single layer approach. *Mar. Struct.* **2021**, 78, 103004. [CrossRef]
23. Putranto, T.; Kõrgesaar, M.; Jelovica, J. Ultimate strength assessment of stiffened panels using Equivalent Single Layer approach under combined in-plane compression and shear. *Thin-Walled Struct.* **2022**, 180, 109943. [CrossRef]
24. Tabri, K.; Naar, H.; Kõrgesaar, M. Ultimate strength of ship hull girder with grounding damage. *Ships Offshore Struct.* **2020**, 15, 161–175. [CrossRef]

25. Shen, L.; Kyun, D.K. A comparison of numerical methods for damage index based residual ultimate limit state assessment of grounded ship hulls. *Thin-Walled Struct.* **2022**, *172*, 108854.
26. Paik, J.K.; Kim, D.K.; Park, D.H.; Kim, H.B.; Mansour, A.E.; Caldwell, J.B. Modified Paik-Mansour formula for ultimate strength calculations of ship hulls. *Ships Offshore Struct.* **2013**, *8*, 245–260. [CrossRef]

Journal of
Marine Science and Engineering

MDPI

Article

A New Study on the Fatigue Properties of SA Weld Joints by Considering the Effects of Welded Bead Shape

Dongpil Cho [1], Seongkil Nam [2], Chulho Cho [2], Dajung Lee [2], Eunyoung Jeong [2], Youngcheon Jeong [1], Jaeyik Lee [3], Marcus Cridland [4], Bongjin Kim [5], Kwangsoo Kim [6], Jihoon Kim [7] and Myunghyun Kim [7,*]

1 Material & Coating Research, Samsung Heavy Industry Co., Ltd., Geoje 53261, Korea
2 Welding Research, Samsung Heavy Industry Co., Ltd., Geoje 53261, Korea
3 Steel Business Division, POSCO, Incheon 21985, Korea
4 Corporate Chief Engineer's Office, ABS Class, Spring, TX 77389, USA
5 New Construction Team, BV Class, Busan 47351, Korea
6 Production Support Team, DNV Class, Busan 48120, Korea
7 Department of Naval Architecture and Ocean Engineering, Pusan National University, Busan 46241, Korea
* Correspondence: kimm@pusan.ac.kr

Abstract: Tandem SAW (Submerged Arc Welding) is one welding process that has been applied to maximize the welding productivity at the panel stage in ship building field. The weld bead profiles produced by Tandem SA welding exceed the acceptance criteria specified in some international regulations, such as AWS D1.1, ISO 5817 and NORSOK M-101. These regulations limit the applicable weld bead profiles, especially weld bead height, regardless of any consideration of design category. The fundamental reason for the limitation of weld bead profiles is related to the weldment fatigue properties. In this regard, we have investigated the effect of weld profiles on fatigue properties. The effect of weld bead profiles on fatigue properties has been experimentally verified and statistically analyzed, and new criteria for weld bead profiles which satisfy E curve as the design S-N curve are proposed for tandem SA welding.

Keywords: tandem SA welding; fatigue properties; weld bead profiles; angular distortion; E curve; circle

Citation: Cho, D.; Nam, S.; Cho, C.; Lee, D.; Jeong, E.; Jeong, Y.; Lee, J.; Cridland, M.; Kim, B.; Kim, K.; et al. A New Study on the Fatigue Properties of SA Weld Joints by Considering the Effects of Welded Bead Shape. *J. Mar. Sci. Eng.* **2022**, *10*, 1527. https://doi.org/10.3390/jmse10101527

Academic Editors: Sören Ehlers and Moritz Braun

Received: 29 September 2022
Accepted: 13 October 2022
Published: 18 October 2022

1. Introduction

Arc welding processes are predominantly used to join two steel members together in the fabrication of ships and offshore structures. Two adjacent steels, which will be joined by welding, are generally beveled by machining or cutting to the appropriate configuration in preparation for arc welding, as shown in Figure 1.

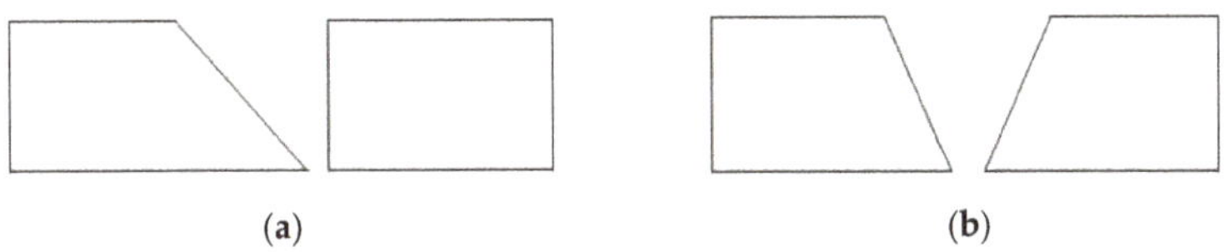

Figure 1. Typical weld joint configurations for arc welding: (**a**) single bevel; (**b**) single vee.

As the size of ships and offshore structures has increased, the thickness of the steel used in the construction has consequently increased. As the use of heavy thickness plates increases, welding productivity for the heavy thickness plate became one of the most important factors in the overall work processes. To respond to this, the tandem SA welding process has been improved and applied to the panel stage for welding of plate to plate at the beginning phase of all construction operations.

The maximized welding productivity can be achieved by tandem SA welding because superior deep penetration with a high welding speed can be obtained utilizing high welding current and controlled heat input. Therefore, weld joint configuration with increased root face dimension than that of other welding processes can be applied without back gouging. Figure 2 shows the typical weld joint configurations for tandem SA welding and Figure 3 shows the Tandem SAW equipment at the panel stage.

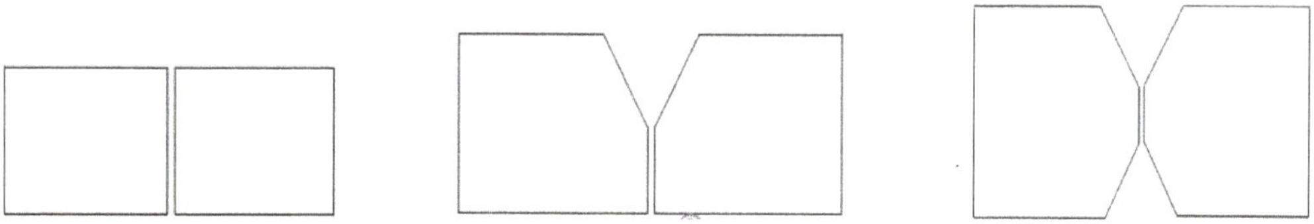

(a) (b) (c)

Figure 2. Typical weld joint configurations for Tandem SAW: (**a**) square groove; (**b**) Y (single V) groove; (**c**) double Y groove.

(a) (b)

Figure 3. Tandem SAW equipment at the panel stage: (**a**) tandem SAW equipment; (**b**) head of tandem SAW equipment.

Although tandem SA welding has many advantages, such as high productivity, high welding speed and high weld quality, its application is somewhat limited due to the regulation for weld bead profiles.

Some international regulations, such as AWS D1.1, ISO 5817, NORSOK M-101 and IACS Rec.47, limit the applicable maximum weld bead height and/or permissible weld profiles [1–4]. For example, AWS (American Welding Society) D1.1 specifies the maximum allowable bead reinforcement (height) of 3 mm for weld in base material thickness of 25 mm or less. However, the weld bead height of Tandem SAW generally exceeds 3 mm, especially at the second side welding.

It is well known that the fatigue behavior is influenced by the geometry of the welded joint, the mechanical properties of base material, the angular distortion by welding, the misalignment between members, and weld profiles, including weld toe angle and radius, weld bead height and so on. Numerous studies have been conducted to investigate the effect of each factor or multiple factors on fatigue behavior of the welded joints [5–11]. Based on the results of those studies, IIW (International Institute of Welding) recommends "Fatigue resistance S-N curve for steel, normal stress, standard application" as per Figure 4 [12].

The Class societies', the owners and designers adopt the same S-N curve as recommended by IIW.

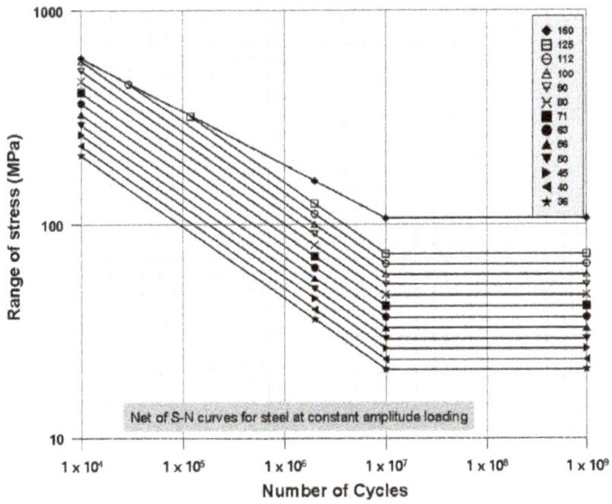

Figure 4. Fatigue resistance S-N curve proposed by IIW. The data was from Ref. [12].

Each S-N curve in Figure 4 is basically classified by structural detail and includes the influence of stress concentration by notch effect, metallurgical conditions, fusion welding processes and so on. However, the S-N curve does not represent a specific welding process and weld bead profiles, etc. that can influence the fatigue behavior.

The weld bead produced by the tandem SA welding process is much more uniform than other welding processes such as FCAW, GMAW and SMAW, with nearly complete uniformity. And there is almost no valley between weld ripples on the bead surface, which can increase the stress concentration by the notch effect [13]. This means that the fatigue behavior of weldments by tandem SA welding may be different from the S-N curve proposed by IIW.

As a result of checking the previous papers, the stress concentration factor (SCF) was analyzed through FE analysis after measuring the bead shape for fillet joint [14–17], and some papers also verified the relationship between fatigue properties and bead shapes [18,19]. In addition, other paper researched the relationship between fatigue properties and bead shapes for butt welding joints of laser welding on thin plates [20]. There was a paper that that compared the SCF, in accordance with bead shapes, with FE analysis results [21].

Therefore, in this study, it has been experimentally investigated to verify the effect of weld bead profiles produced by Tandem SA welding on fatigue behavior. The experimental research and statistical analysis for the test results were carried out to assess the correlation between weld profiles and fatigue properties.

This study was carried out through joint research project (JRP) together with class societies, such as ABS (America), BV (France), DNV (Norway) and Total (one of the major oil companies).

2. Experimental Details

2.1. Test Coupons and Welding Parameters

D, DH36 and EH36 grade steels specified in class society rule were used and the thicknesses selected were 20, 25, 30 and 40 mm. Welding test coupons were prepared based on base material thickness as indicated in Figure 5. That is, I groove for 20 and 25 mm thickness, Y groove for 30 mm thickness and double Y groove for 40 mm thickness.

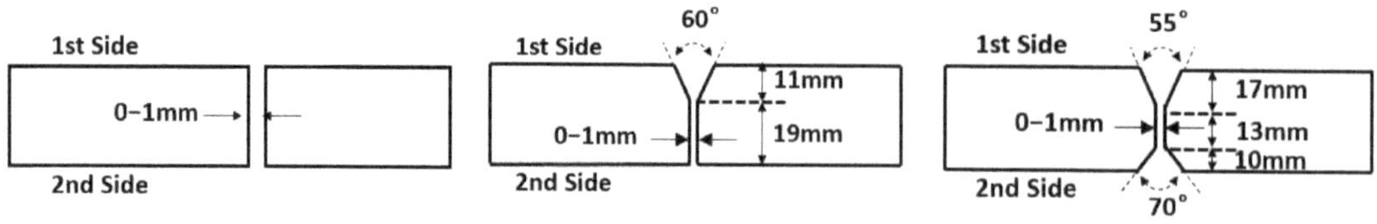

(a) (b) (c)

Figure 5. Joint configurations of the test coupons: (**a**) square groove for 20 and 25 mm; (**b**) Y (single V) groove for 30 mm; (**c**) double Y groove for 40 mm.

AWS classification of the welding consumables used in the study was F7A(P)8—EH14 of AWS A 5.17 and the welding wire diameter was 4.8 mm. Tandem SA welding with two welding wires was applied. A DC (+) power source for the leading wire and an AC power source for the trailing wire were used.

Twenty-three types of test coupons were prepared and welded with varying welding parameters to create the different weld bead profiles. The intention and the methods used for each test number are shown in Table 1.

Table 1. Summary of each test number.

Test No.	Base Material		Joint Config.	Intention	Methods for Intention
	Kind	Thick. (mm)			
1	EH36	20	I	Higher bead height	Lower voltage
2				Normal bead	Standard
3				Lower bead height	Higher voltage
4				Lower bead height	Lower ampere
5				Higher bead height	Higher ampere
6	DH36	20	I	Effect of base material	Change of material from EH36 to DH36 and D (same welding parameters as Test No. 2)
7	D	20	I		
8	DH36	30	Y	Higher bead height	Lower voltage
9				Lower bead height	Higher voltage
10		40	DY	Lower bead height	Higher voltage
11				Higher bead height	Lower voltage
P1		20	I	Higher angular distortion	Reverse pre-distortion during fit-up [(1)]
P2					
P3					
P4					
N1		25	I	Extremely higher bead	Higher flux burden
N2					Lower flux burden
N3				3 mm bead height	Machining [(2)]
N4				Extremely higher bead	Additional weld pass
N5					Preplaced filler wire
R1	EH36	20	I	Reproducibility of the tests of the P series	Reverse pre-distortion and same welding parameters as test No. 2, No.6 and No.7
R2	DH36	20	I		
R3	D	20	I		

Note: First trial: Test nos. 1 to 11, second trial: Test nos. P1 to P4 (for the angular distortion), third trial: Test nos. N1 to N5 (for an extremely higher bead by abnormal welding conditions), fourth trial: R1 to R3 (for the angular distortion). [(1)] P1 and P2 have the same welding parameters as Test no. 2. P3 and P4 have the same welding parameters as Test nos. 1 and 3, respectively. [(2)] N3: involves the removal of weld reinforcement by machining.

2.2. Measurement of Weld Bead Profiles

The main purpose of this study is to verify the effect of weld bead profiles on fatigue properties of welded joints. The weld profiles include the weld bead height and width, toe angle, angular distortion, trough depth at the weld toe and the radius at the weld toe.

Various tools and methods may be used to measure the weld bead profiles. The weld bead height and width can be easily measured but it is impossible or very difficult to measure them other than these two at the production stage. Therefore, various measuring methods were applied, as shown in Table 2, to verify the effective methods in production.

Table 2. Measuring methods.

Stage	Measuring	Measuring Method
For the weld on the test coupon	Weld bead height Weld bead width	Leg length gage Ruler
For fatigue specimen	Weld toe angle	SHI designed Templates Videometer
	Weld bead height Weld bead width Angular distortion Trough depth at the weld toe Min. radius at the weld toe	3D scan

The definition of each measuring item is shown in Figure 6.

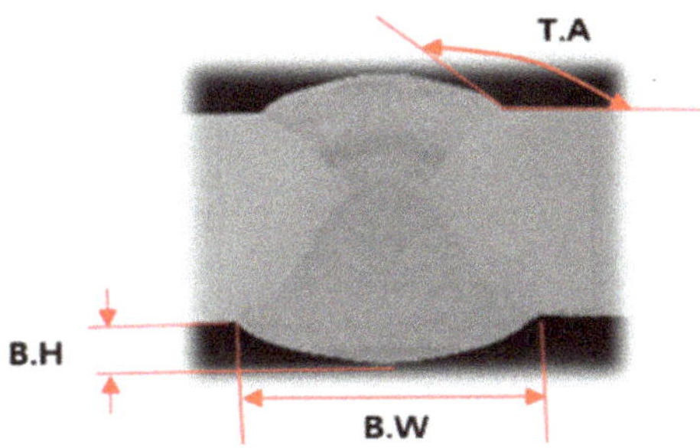

Figure 6. Definition of measuring items.

There is no internationally recognized standard that covers measuring methods of the weld toe angle, trough depth and radius at the weld toe and angular distortion. Therefore, new measuring methods were designed and applied for the accurate measurement and comparison between the newly designed measuring methods. Three kinds of measuring methods of a 3D Scan, templates and a videometer were applied to measure toe angle.

2.2.1. Measurement of Bead Height and Width on the Test Coupon

The weld bead height and width were measured on the test coupons, as shown in Figure 7.

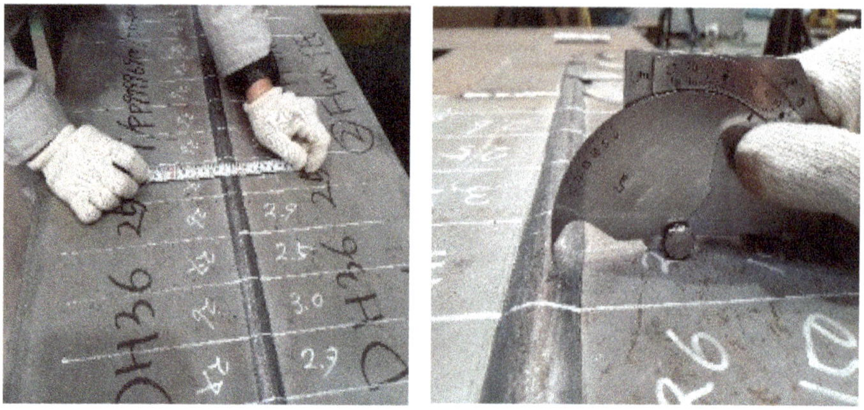

(**a**) (**b**)

Figure 7. Measurement of the weld bead height and width: (**a**) measurement of width; (**b**) measurement of height.

2.2.2. Measurement of Toe Angle by Using the Template

New templates were designed to measure the toe angle at eight points of each fatigue specimen prior to the fatigue test, as shown in Figure 8.

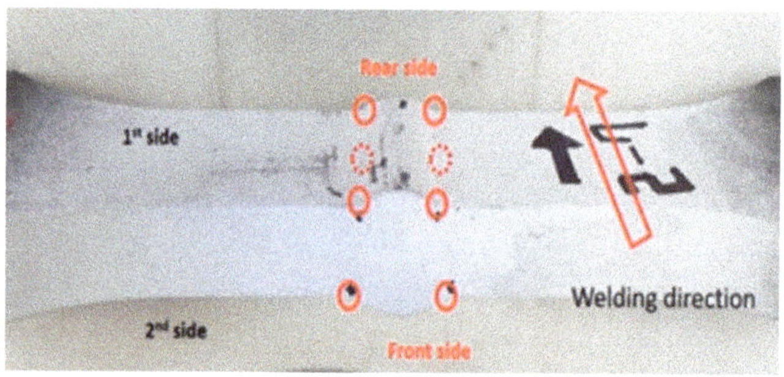

Figure 8. Locations for toe angle measurement using the template.

Nine kinds of templates were designed and developed to measure the toe angles between 115° to 155° with 5° intervals. Figure 9 shows the shape of the templates and the measuring methods.

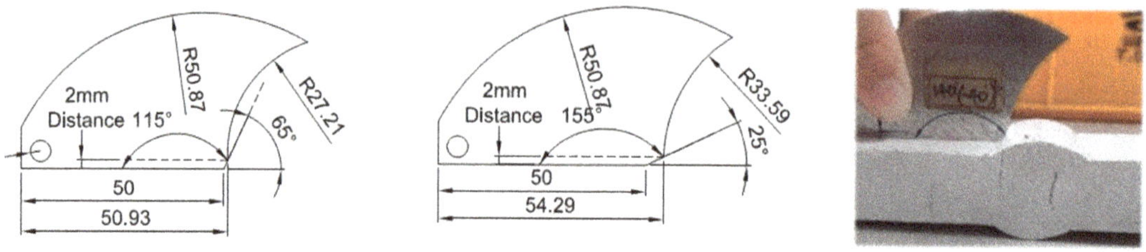

(**a**) (**b**) (**c**)

Figure 9. New templates for measuring the toe angle: (**a**) for 115°; (**b**) for 155°; (**c**) example of measurement.

The toe angles were measured with each newly designed template. The toe angle could not be measured accurately because the template was designed with 5° intervals. Therefore, the average value between two templates (a lower angle template and a higher angle than the actual toe angle) was calculated and used for analysis.

2.2.3. Measurement of Toe Angle by Using a Videometer

A videometer (Accura III Model, manufactured by the Seven Ocean Company) was used to measure the toe angle of the fatigue specimen at eight points prior to the fatigue test, as shown in Figure 8. The videometer and its specification are shown in Figure 10.

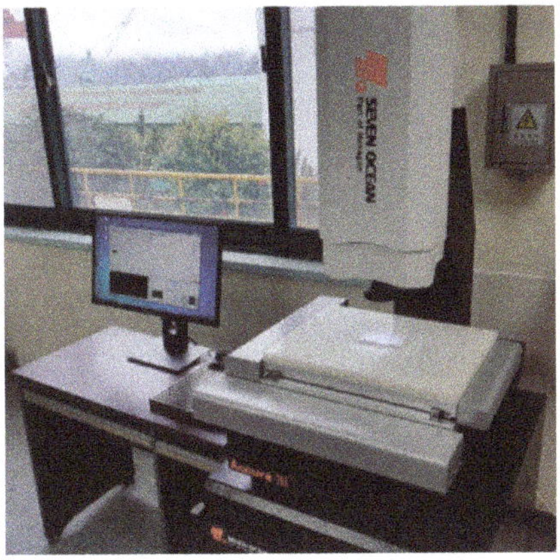

Figure 10. Videometer for measuring the toe angle.

Generally, it is impossible to predict the exact fatigue crack initiation point on the fatigue specimen prior to completing the actual fatigue test. In addition, the weld bead profiles at the fatigue crack initiation point cannot be confirmed after completing the fatigue test because the profiles are deformed during the fatigue test. Therefore, all fatigue test specimens were 3D scanned to record the overall weld bead profiles prior to the fatigue test.

The 3D scanner used for this study is the "EinScan-Pro+" model manufactured by SHINING 3D Company and the scanner is shown in Figure 11. The scanner has an accuracy of 0.05 mm.

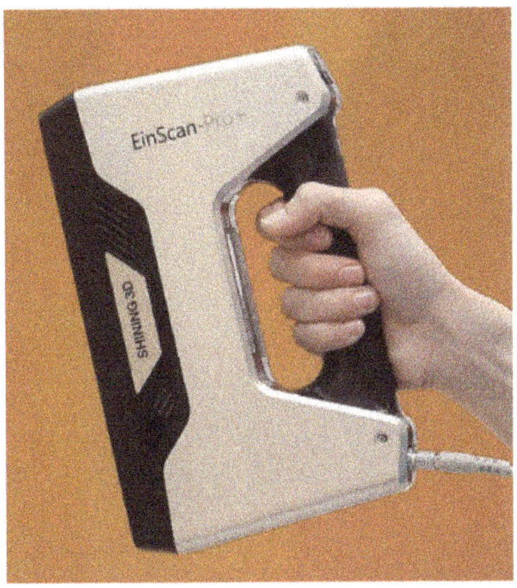

Figure 11. 3D Scanner used for this study.

2.2.4. Measurements by Using a 3D Scanner

The fatigue crack initiation point was determined by investigating the beach mark on the fractured surface, as shown in Figure 12.

Figure 12. Determination of the fatigue crack initiation point (the point marked by the red box denotes the initiation point of the fatigue crack).

The section profile at the fatigue crack initiation point was extracted from the data recorded using the 3D scanner and the weld profiles at the fatigue crack initiation point were then measured.

Establishment of a Reference Point for Measurements

The weld toe angle, trough depth and radius at the weld toe and angular distortion are known to influence the stress concentration [22]. On the other hand, the values of these parameters may differ according to the measuring methods applied which is further dependent on the reference point of each measuring item [1,23]. Therefore, determination of the reference point is very important.

Generally, it is difficult to determine an appropriate reference point on a weld cross-section because the transition area of the weld toe is quite complex and the inconsistent trough also exists at the toe area as shown in Figure 13. In addition, the surface of the base material is also uneven when enlarged.

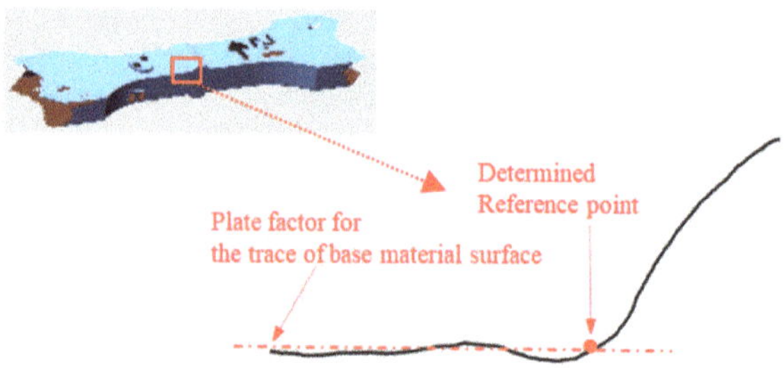

Figure 13. Changeable profile and the determination of the reference point.

Therefore, a "Plate factor" is defined as the arbitrarily straight line for base material surface calculated with the average tangent line and was drawn automatically by the software Geomagic Control-X for measuring and analyzing 3D scanned data. The intersection point between the "Plate factor" and the weld bead surface was determined as a reference point in this study as shown in Figure 13.

Weld Bead Height and Width

The weld bead heights and widths were measured from the reference point, as shown in Figure 14, because the extracted section profiles had some misalignment.

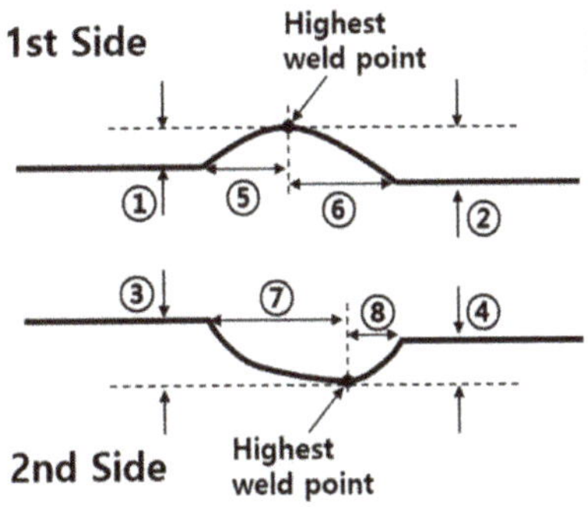

①	Left side weld bead height of 1st side
②	Right side weld bead height of 1st side
③	Left side weld bead height of 2nd side
④	Right side weld bead height of 2nd side
⑤	Left side weld bead width of 1st side
⑥	Right side weld bead width of 1st side
⑦	Left side weld bead width of 2nd side
⑧	Right side weld bead width of 2nd side
⑤+⑥	Weld bead width of 1st side
⑦+⑧	Weld bead width of 2nd side

Figure 14. Schematic diagram of the weld bead height and width measurement.

Toe Angle

A "Bead factor", which is defined as an arbitrarily straight line for the weld metal surface and is calculated with the average tangent line, was drawn automatically by the Geomagic Control-X program to measure the weld toe angle. Six kinds of bead factors for each weld toe were drawn for a weld bead height ranging from 0.5 mm to 3.0 mm with 0.5 mm interval.

The weld toe angle for each specific range was determined by measuring the angle between the "Plate factor" and the "Bead factor" using the Geomagic Control-X program, as shown in Figure 15.

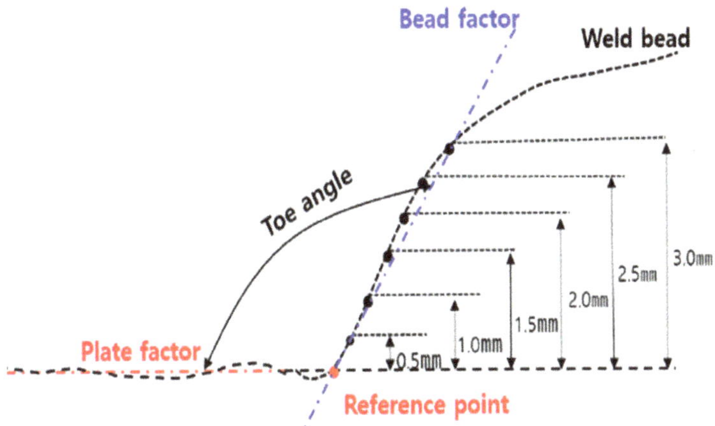

Figure 15. Schematic diagram of the toe angle measurement (toe ~ 3 mm).

Angular Distortion

The angular distortion generated by the welding and the fit-up of the test coupon was also measured from the extracted section profile at the fatigue-crack initiation point. A "Parallel line", which is the hypothetical line connecting two points on the base material, was drawn. These two points were 100 mm away from the reference point of each weld toe. The angular distortion for each point was determined by measuring the angle between the "Plate factor" and the "Parallel line". The sum of the two angles at the left and right points was determined to be the angular distortion of the side, as shown in Figure 16.

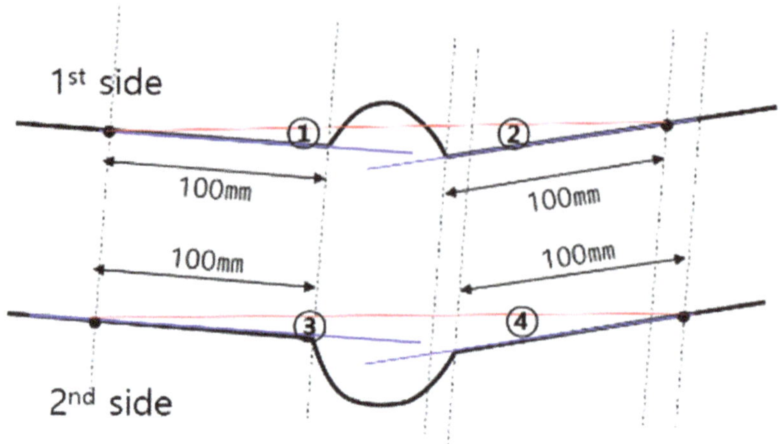

Figure 16. Schematic diagram of the angular distortion measurement.

Trough Depth at the Toe Area

Some troughs were found at the weld toe area when the extracted cross section considered as the fatigue crack initiation point was investigated with large magnification. However, these troughs had not been detected by a visual inspection and M.T. The trough depth at each weld toe was determined by measuring the vertical distance between the "Plate factor" and the deepest point of the trough, as shown in Figure 17.

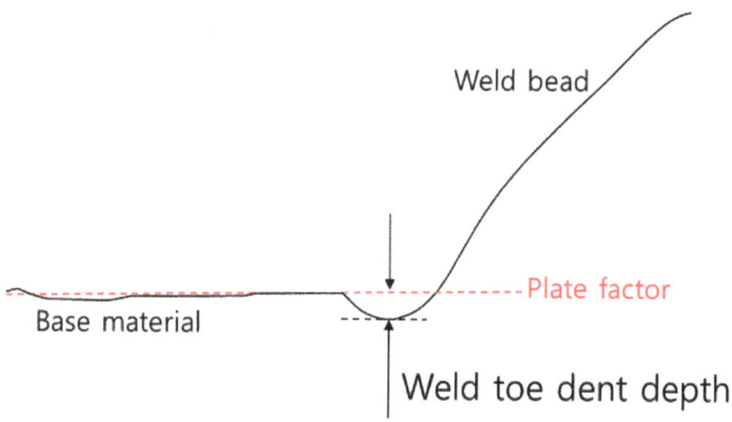

Figure 17. Schematic diagram of the trough depth measurement.

Minimum Radius at the Toe Area

The radius at the weld toe influences the fatigue behavior of the welded joint, but it is impossible or very difficult to measure the exact radius [13]. Therefore, this study drew the optimized circle automatically by the method of least squares. The radius of the circle was measured using the Geomagic Control-X program, as shown in Figure 18.

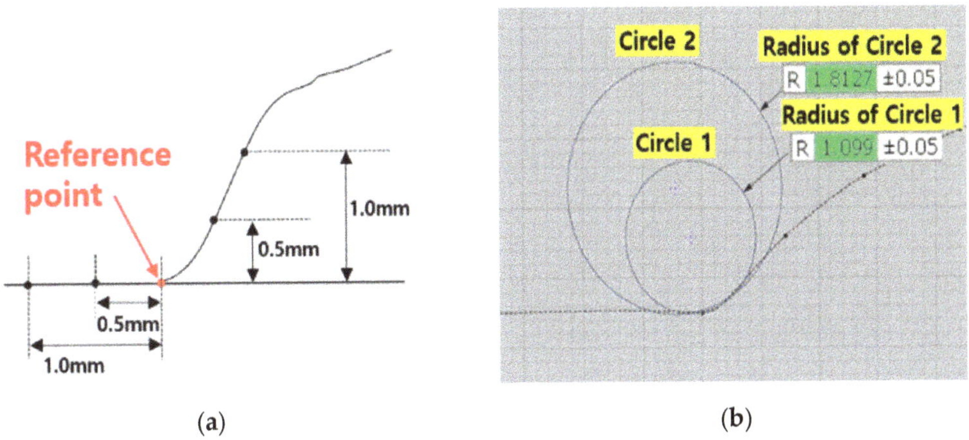

Figure 18. Schematic diagram and an example of radius measurement: (**a**) range for measuring radius at the reference point; (**b**) example of measured radius for two ranges.

The program automatically marks the point of a specifically designated distance to draw the optimized circle. For example, this study designated 0.5 mm and 1.0 mm distances along the base material surface and 0.5 mm and 1.0 mm heights along the weld metal surface from the reference point. The optimized circle was drawn automatically.

2.3. Fatigue Tests

2.3.1. Specimens for Fatigue Test

All test coupons were inspected by UT and MT and the fatigue test specimens were prepared by machining to the dimensions shown in Figure 19.

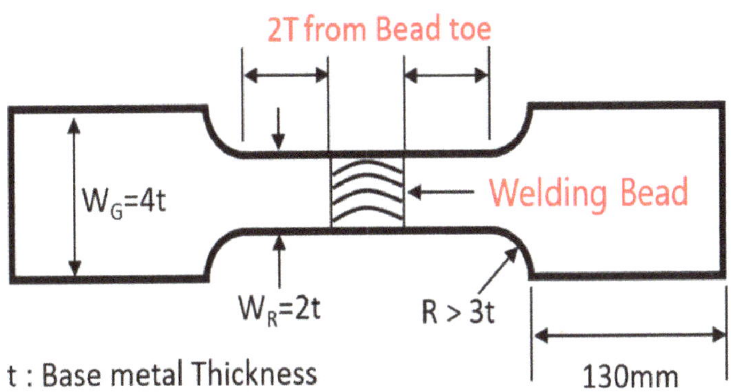

Figure 19. Shape and dimensions of the fatigue test specimens.

In total, 246 fatigue test specimens were prepared and the fatigue tests were carried out to estimate the fatigue strength for all test specimens.

2.3.2. Fatigue Test Machines

Instron 8803 (Capacity: 500 kN) manufactured by Instron was used to fatigue test for 20 and 25 mm test specimens, and MTS 311.41 (Capacity: 2500 kN) manufactured by MTS was applied to fatigue test for test specimens of 30 and 40 mm. Figure 20 shows the fatigue test machines.

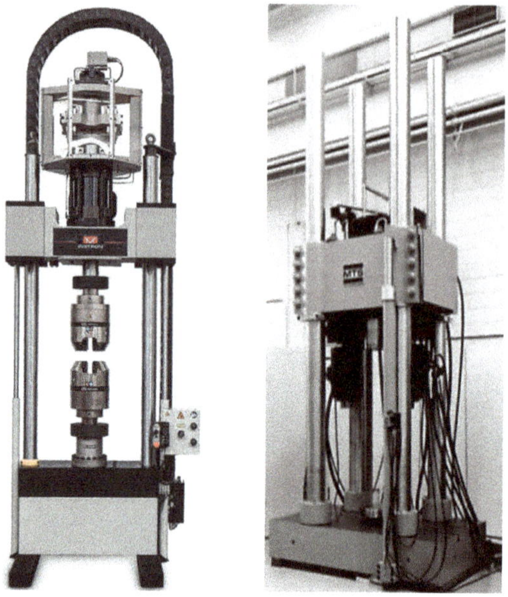

Figure 20. Fatigue test machine: (**a**) Instron 8803; (**b**) MTS 311.41.

2.3.3. Fatigue Tests

The fatigue tests were performed according to the conditions given in Table 3.

Table 3. Conditions for the fatigue tests.

Parameters	Condition			
Stress ratio (R)	0.1			
Applied stress	Below the yield strength			
Wave form	Sine wave			
Frequency (Hz)	10–40			
Thick. (mm)	20	25	30	40
Stress range (MPa)	112.5–292.5	67.5–247.5	65.0–315.0	112.5–292.5
Mean stress (MPa)	68.8–178.8	41.3–151.3	39.7–192.5	39.7–192.5
Amplitude (MPa)	56.3–146.3	33.8–123.8	32.5–175.5	32.5–157.5
Run-out (cycles)	2×10^6			

The S-N curve was plotted in accordance with Equation (1) and by using the test results for the range of $10^5 - 10^6$ cycles:

$$\log(\mathrm{N}) = \log(\mathrm{A}) - \mathrm{m}\log(\mathrm{S}) - \left(t_{dof}, \mathrm{a}\right)\mathrm{se} \quad (1)$$

where N, S, m, $\log(\mathrm{A})$, t_{dof} and se are the number of cycles, stress range, slope, intercept of $\log(\mathrm{N})$ axis, student's T-distribution value and standard deviation, respectively.

2.3.4. Criteria for Fatigue Test Results

Test results (fatigue limits: stress ranges at 10^7 cycles in air condition) were compared with the BV rule requirements, because the fatigue limit requirement of the BV rule is higher than that of the other five rules and codes as summarized in Table 4. The S-N curve required for a butt-welded joint of the plate-to-plate case is generally the E curve [12,24–28].

Table 4. Fatigue limits required in the rules and codes.

Rule or Code	S-N Curve	Fatigue Limit in Air (MPa)	Fatigue Limit at 10^7 Cycles Considering the Thickness Effect (MPa)			
			20 mm	25 mm	30 mm	40 mm
DNV	E curve	46.78	46.78	46.78	45.11	42.59
ABS		47	47.00	45.52	43.49	40.48
BV		46.96	46.96	46.96	45.28	42.75
ISO		47.13	44.58	42.16	40.28	37.48
BS		47	46.96	46.96	45.28	42.75
IIW	FAT 80	46.8	46.80	46.80	45.12	42.60

3. Test Results

3.1. Gage R&R (Repeatable and Reproducibility) for the Data Measured by Each Method

The correlation of data which were measured by the measuring methods of Table 2 for each weld profile was analyzed to verify the effectiveness of each measuring method statistically. Table 5 lists the results of this analysis.

Table 5. Summary of gage R&R for each weld profile.

Item	Comparing Object		Significance (*p*-Value)		R^2 Value
			F-Test	*t*-Test	
Bead height	Leg length gage	3D scan	0.779	0.599	0.947
Bead width	Ruler	3D scan	0.983	0.532	0.961
Toe angle	Template	Videometer	<0.05	<0.05	0.759
Toe angle	Template	3D scan	0.935	0.359	0.787

As the results of gage R&R for the bead height and width, the following was confirmed. The bead height measured by the leg length gage and the 3D scan and the bead width measured by the ruler and the 3D scan did not have a significant difference for the variance from the F-test and the mean value from the t-test.

Therefore, the leg length gage for the bead height measurement and the ruler for the bead width measurement can be used in production. Figure 21 shows the correlation for each measuring method of bead height and width.

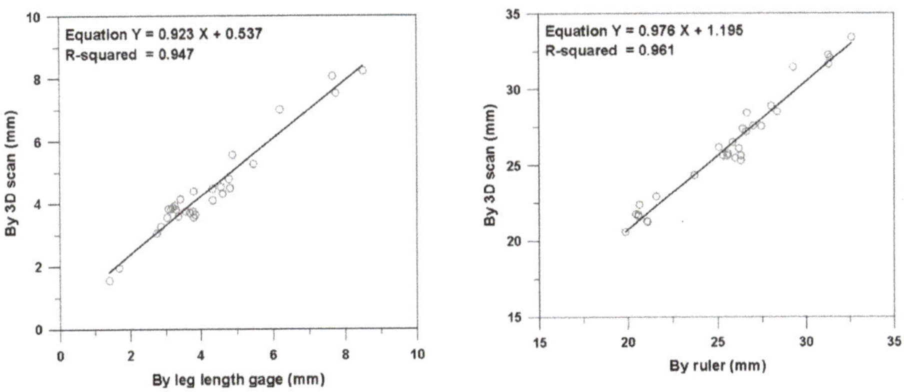

(a) (b)

Figure 21. Correlation for each measuring method of the bead height and width: (**a**) bead height; (**b**) bead width.

As the results of gage R&R for the toe angle, the following was confirmed. The toe angle measured by the template and the videometer showed a significant difference for the variance from the F-test and the mean value from the *t*-test. On the other hand, the toe angle, measured using the template and the 3D scan, did not show a significant difference for the variance from the F-test and the mean value from the *t*-test. Figure 22 shows the correlation for each measuring method of the toe angle.

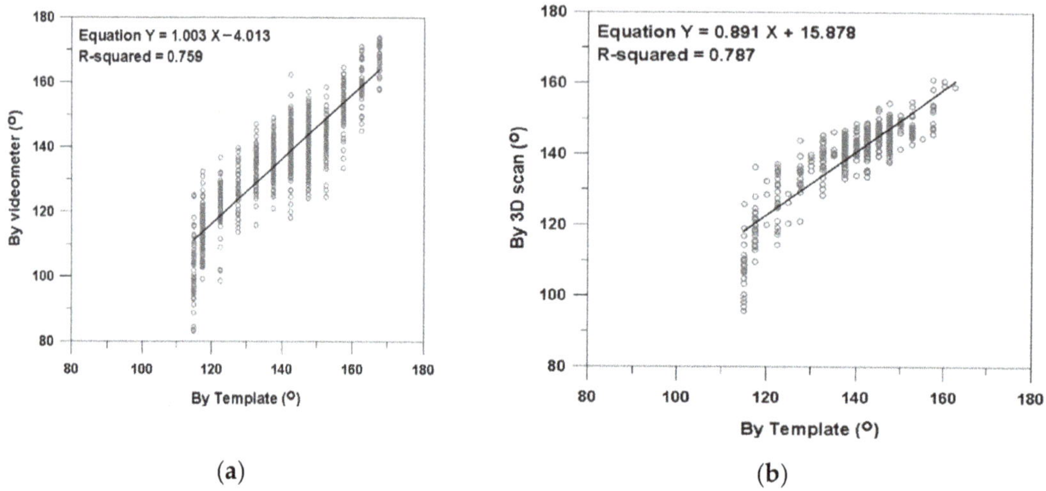

Figure 22. Correlation for each measuring method of the toe angle: (**a**) videometer; (**b**) 3D scan.

Some of the toe angles measured by the videometer and the 3D scan were less than 115°. Therefore, the toe angle less than or equal to 115° was assigned to the template angle of 115° for the analysis because the minimum angle of the template was 115°. The templates were manufactured with a 5° interval and the toe angles were determined from the average value of the two templates. In addition, it was difficult to determine the accurate reference tangent line on the weld bead to measure the toe angle. Therefore, there was a significant difference in the variance and the mean value between the toe angles measured using the template and the videometer.

Nevertheless, it was confirmed that the correlation of the toe angles measured using the template and the 3D scan was valid according to the result of the statistical analysis. It means that the measurements of the toe angle using the templates designed by SHI can be effectively used in production.

3.2. Fatigue Test Results

3.2.1. Weld Profiles and Fatigue Limits

The fatigue limit of each test number from the fatigue tests and the weld profiles measured at the fatigue initiation location of all fatigue test specimens using the 3D scan are summarized in Table 6. The parameters for weld profiles are expressed in terms of the average values for each test number.

Table 6. Summary of the measured weld profiles and fatigue limit for each test number.

Test No.	Min. Toe Angle (°)	Angular Distortion (°)	Bead Width (mm)	Bead Height (mm)	Toe Through (mm)	Min. Radius (mm)	Fatigue Limit (1) (mm)
1	138.53	0.14	22.49	4.60	−0.04	1.32	54.04
2	145.02	0.14	21.27	3.69	0.11	1.67	55.39
3	144.03	0.19	21.76	3.59	0.17	1.55	50.93
4	143.64	0.22	24.11	3.86	0.00	1.56	54.26
5	143.42	0.18	21.99	3.92	0.05	1.65	55.03
6 (2)	137.39	1.23	26.51	4.48	−0.06	1.31	46.34
7	140.46	0.22	25.63	4.53	−0.13	1.44	51.04
8	156.90	1.03	32.25	3.27	−0.08	5.18	53.79
9	155.09	0.99	32.04	3.06	−0.19	4.06	54.67
10	162.39	0.69	28.31	1.91	−0.19	5.93	59.70
11	159.76	0.89	28.85	1.95	−0.07	5.74	53.03
P1 (2)	137.80	1.42	27.10	4.75	−0.06	1.32	45.96
P2 (2)	138.14	1.51	26.10	4.63	−0.06	1.48	46.50
P3 (2)	133.65	1.17	23.10	5.27	−0.08	1.19	43.41
P4	142.61	1.65	26.47	4.44	−0.04	1.61	48.53
N1 (2)	110.43	1.42	25.72	7.51	−0.10	0.75	41.39
N2 (2)	113.75	1.49	27.38	6.99	−0.07	0.87	43.09
N3 (2)	116.24	1.27	26.06	3.25	−0.07	0.85	46.01
N4 (2)	117.43	1.50	27.53	8.22	−0.06	0.80	42.35
N5 (2)	105.49	1.23	28.39	8.05	−0.06	0.66	42.00
R1	142.53	1.27	26.61	4.63	−0.04	1.68	49.68
R2	143.06	1.09	26.33	4.38	−0.02	1.69	51.72
R3	143.07	0.93	26.45	4.34	−0.03	1.78	51.93

Note: 1 to 5: different welding parameters, 6 and 7: different base materials, 8 to 11: different welding parameters for the Y and double Y grooves, P1 to P4: Reverse pre-distortion, N1 to N5: Extremely higher bead, R1 to R3: Reproducibility of test nos. P1 to P4 (Refer to Table 1 for details). (1) Fatigue limit at 10^7 cycles. Table 7 lists the required fatigue limit of E curve for different base material thicknesses. (2) Failed to meet the E curve requirements of the BV rule [26].

Table 7. Fatigue test results for different base material thicknesses.

Base Material Thickness (mm)	Fatigue Limit at 10^7 Cycles (MPa)		The Ratio of the Test Result to the Requirement (%)
	BV Rule Requirement [26]	Test Results	
20	46.96	47.36	101
25	46.96	41.90	89
30	45.27	55.21	122
40	42.74	57.21	134

3.2.2. S-N Curve

S-N curves for each test number and base material thickness were drawn and the fatigue limits at 10^7 cycles were calculated according to the test results described in Section 2.3.2. Subsequently, the fatigue limits were compared with the E curve requirements of the BV rule [26].

Test numbers of 6, P1, P2, P3 and N1 to N5 did not meet the requirements, but the others met the requirements. The fatigue limits of test numbers 8, 9, 10 and 11 were much higher than the requirements when the notch effect due to the thickness factor is considered.

The fatigue limit of a 25 mm thick base material did not meet the requirements, but the fatigue limits of 20 mm, 30 mm and 40 mm thick base materials did meet the requirements. Table 7 presents the fatigue test results for each base material thickness.

Figure 23 presents the S-N curves obtained for each base material thickness. Extremely abnormal welding conditions were intended for the 25 mm thick base material to obtain the worse weld bead profiles, which may adversely affect the fatigue behavior of the welded joint.

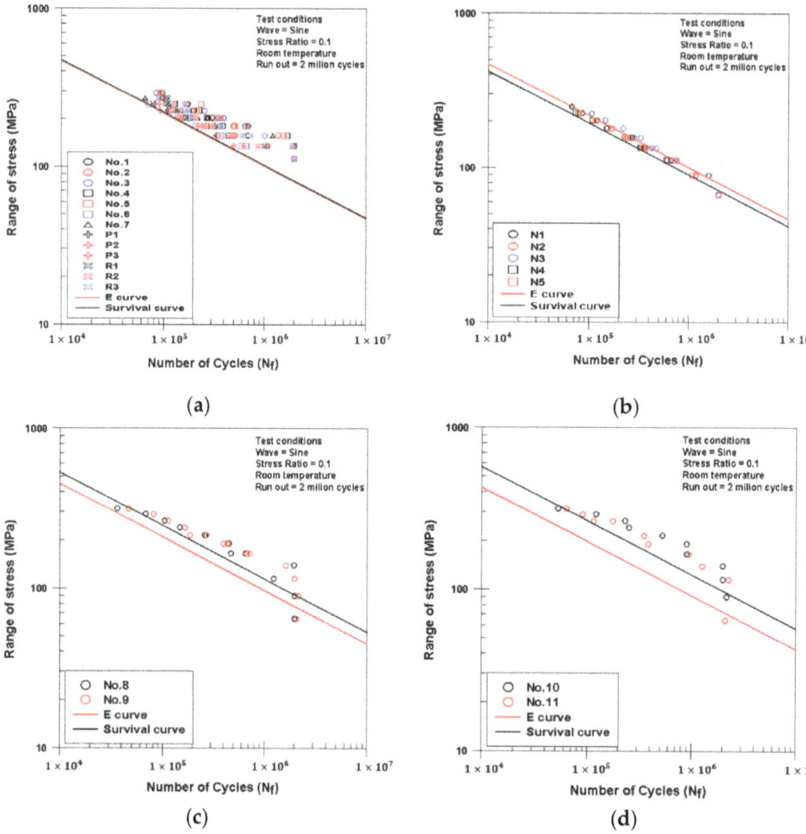

Figure 23. S-N curve for each base material thickness: (**a**) for 20 mm; (**b**) for 25 mm; (**c**) for 30 mm; (**d**) for 40 mm.

3.2.3. Brief Review of All Fatigue Test Results

Direction and Amount of Angular Distortion

Angular distortion occurred towards the second side for the square groove and the first side for the Y and double Y grooves. The angular distortion for each test number is shown in Table 6. The angular distortions of the Y and double Y grooves and the square groove with a reverse pre-distortion are higher than that of the normal square groove.

However, it is thought that the large angular distortion, such as written in Table 6, hardly occurs in production because of the size effect and self-weight of the plates welded. As a result of measuring the angular distortion of the actual block in production, it was confirmed that the angular distortion is much less than 0.7°.

Effect of Angular Distortion on the Fatigue Behavior of the Welded Joints

All specimens of test numbers 6, P1 to P3, N1 to N5 and 8 to 11, were fractured at the acute angle side of the two sides (first and second sides). On the other hand, fatigue fracture did not always occur at the acute angle side for the other test numbers. Therefore, the concordance rate of the fatigue fractured side and the acute angle sides were investigated based on the level of angular distortion and the results are shown in Table 8.

Table 8. Concordance rate of the fractured and acute angle sides.

Angular Distortion (°)	No. of Specimen (EA)	No. of Specimen with the Accordance Side	Concordance Rate (%)
exceed 1°	138	133	96
exceed 0.5°	169	160	95
exceed 0.2°	196	170	87

The larger the angular distortion, the more fractures occurred at the acute angle side. Some researchers considered that the fatigue fracture at the acute angle side is related to the secondary bending stress [23].

The fatigue limits of test numbers 6, P1 to P3 and N1 to N5 did not meet the E curve requirements of the BV rule [26]. It is thought that the stress concentration due to the large angular distortion may affect the lower fatigue limits of these test numbers.

On the other hand, the fatigue limits of the Y and double Y grooves met the E curve requirements of the BV rule, even though the level of angular distortions was relatively high. The angular distortions of the Y and double Y grooves occurred towards the first side and all specimens fractured at the first side. The weld bead profiles of the first side of the Y and double Y grooves were much better than those of the other test numbers and the second side of the Y and double Y grooves. It is thought that the lower stress concentration due to better weld bead profiles may affect the higher fatigue limits of the Y and double Y grooves.

4. Analysis of Test Results

Fatigue properties of the welded joints were analyzed with the data measured by the 3D scan. Fatigue cracks were not always generated on the worse weld bead profile side, which has a high stress concentration (SCF, stress concentration factor), or the acute angle side of angular distortion (Km, stress magnification factor). Therefore, an additional investigation was carried out using all weld profile parameters, such as weld bead height and width, toe angle, radius and trough depth at weld toe and angular distortion. All data used for the investigations were measured at the fatigue-fractured points and the investigations were carried out using regression analysis.

Misalignment and angular distortion in axially loaded joints lead to an increase of the stress in the welded joint because of secondary shell bending stresses. This stress increase is expressed using the stress magnification factor (Km) and is considered an absolute value. On the other hand, fatigue fracture did not always occur at the acute angle side.

The secondary bending stress that developed during the fatigue test has a negative effect (tensile stress) on the acute angle side and a positive effect (compressive stress) on the obtuse angle side of the specimen. Therefore, this study assigned the sign of positive (+) or negative (−) to the measured angular distortion considering the direction of distortion. Therefore, a negative sign (−) was assigned to the angular distortion at the acute angle side and a positive (+) sign was assigned to the angular distortion at the obtuse angle side.

4.1. *Effect of Weld Bead Profiles on the Fatigue Life of the Welded Joints*

It is well known that the fatigue properties of the welded joints are affected by the applied stress, weld bead profiles and joint configuration [13]. Therefore, the effect of the applied stress and weld bead profiles on the fatigue life of the welded joints was investigated by regression analysis. The applied stress was always included as a basic variable.

As the results of regression analysis, it was confirmed that the main variables affecting the fatigue life of the welded joints are the toe angle, bead height and angular distortion. Therefore, two different regression equations can be proposed to predict the fatigue life, as shown Equations (2) and (3):

$$(\text{F.L}) = -5101.220\text{X} - 68379.070\text{X}(\text{B.H}) + 51804.014\text{X}(\text{A.D}) + 1671724.645 \quad (2)$$

$$(\text{F.L}) = -5248.949\text{X} - 7046.702\text{X}(\text{T.A}) + 76949.256\text{X}(\text{A.D}) + 428025.683 \quad (3)$$

where F.L, X, (B.H), (A.D) and (T.A) mean fatigue life, stress, bead height, angular distortion and toe angle, respectively.

4.2. Effect of the Bead Profile on the Fatigue Limit at 10^7 cycles

The fatigue limit for each test number was calculated based on the fatigue test results corresponding to $10^5 - 10^6$ cycles of fatigue life. Subsequently, the effects of the weld bead profiles on the fatigue limit were investigated by regression analysis.

The regression analysis of Section 4.1 was carried out for the individual and the various combinations of variables of the weld bead profiles. These variables include the weld bead height and width, toe angle, angular distortion, trough depth at the weld toe, bead shape ratio (the ratio of the bead height to the bead width) and the radius at the weld toe, which are known to affect the fatigue properties of the welded joints. In total, 123 different regression analyses were carried out, and the regression results of 4 out of the 123 analyses showed a higher correlation. Table 9 lists the four different regression results from the analyses.

Table 9. Regression results for the weld bead profiles and the fatigue limit.

Factor X			Result of the Statistical Analysis			
X1	X2	X3	Adjust R^2	The p-Value for Each Factor		
				X1	X2	X3
Angular distortion	Bead shape ratio	-	0.87	<0.05	<0.05	-
Toe angle	Angular distortion	-	0.87	<0.05	<0.05	-
Bead height	Angular distortion	-	0.84	<0.05	<0.05	-
Bead height	Angular distortion	Bead width	0.87	<0.05	<0.05	<0.05

Four different regression equations can be proposed from the regression analysis results in Table 9, as follows:

$$(\text{F.L}) = 4.201\text{X}(\text{A.D}) - 51.330\frac{(\text{B.H})}{(\text{B.W})} + 62.430 \quad (4)$$

$$(\text{F.L}) = 0.238\text{X}(\text{T.A}) + 3.580\text{X}(\text{A.D}) + 19.930 \quad (5)$$

$$(\text{F.L}) = -2.096\text{X}(\text{B.H}) + 2.695\text{X}(\text{A.D}) + 61.974 \quad (6)$$

$$(\text{F.L}) = -1.763\text{X}(\text{B.H}) + 4.690\text{X}(\text{A.D}) + 0.460\text{X}(\text{B.W}) + 49.910 \quad (7)$$

where F.L, X, (A.D), (B.H), (B.W) and (T.A) mean fatigue limit, stress, angular distortion, bead height, bead width and toe angle, respectively.

It is concluded that the four different regression equations can be used to predict the fatigue limit based on the weld bead profiles.

If only one variable of the weld profiles exclusively affects the fatigue properties, the fatigue test specimens would be fractured at the side adversely affected by the variable dominantly. However, the fractured sides from the tests did not coincide with the side unfavorably affected by certain one variable. Therefore, the fatigue properties were investigated with combining two or more variables of weld profiles.

The p-value of one or more variables exceeds 0.05 when the toe angle is regression analyzed with the weld bead height and width or bead shape ratio. Based on the result, we can concludes that a specific reciprocal action between variables exists, which is explained in Section 5.

4.3. New Criteria of Weld Profiles for the E Curve

The four regression equations for the fatigue limit have angular distortions in common. Hence, the fatigue limit is dominantly influenced by the angular distortion. On the other hand, it is impossible or very difficult to measure all the angular distortions of the actual weld joints in production within a certain period of time for working speed. Therefore, some angular distortions for the weld joints of the actual production block were measured, and it is confirmed that the measured angular distortions were much lower than 0.7°.

When the angular distortion is 0.7° and required S-N curve is E curve of BV rule, the regression Equations (4)–(7) can be summarized as in Table 10.

Table 10. The regression equations for the conditions that the angular distortion is 0.7° and the required S-N curve is the E curve of the BV rule. The data was from Ref. [26].

Regression Equation	Weld Bead Profiles		
	Weld Bead Height (mm)	Weld Bead Width (mm)	Weld Toe Angle (°)
(4)	Refer to Table 11	Refer to Table 11	No restriction
(5)	No restriction	No restriction	Min. 124
(6)	Max. 6.2	No restriction	No restriction
(7)	Refer to Table 11	Refer to Table 11	No restriction

Table 11. Bead height and width meeting the E curve requirements of the BV rule when the angular distortion is 0.7°: (a) by Equation (4); (b) by Equation (7).

(a)		(b)	
Bead Width (mm)	Bead Height (mm)	Bead Width (mm)	Bead Height (mm)
20	4.9	20	5.0
21	5.1	21	5.3
22	5.4	22	5.6
23	5.6	23	5.8
24	5.9	24	6.1
25	6.1	25	6.3
26	6.3	26	6.6
27	6.6	27	6.9
28	6.8	28	7.1

The weld bead ratio (bead height/bead width) meeting the E curve requirements of the BV rule is 0.244 according to regression Equation (4) when the angular distortion is 0.7° and the weld bead height and width are shown in Table 11 (a) when the weld bead ratio is 0.244. In addition, the bead height and width which meet the required fatigue limit according to regression Equation (7) are shown in Table 11 (b) when the angular distortion is 0.7°.

From the results of this study, it is concluded that the fatigue limit meeting the E curve requirements of the BV rule can be controlled by ① only the weld bead height and width, ② only the weld toe angle or ③ only the weld bead height, as shown in Table 10. In addition, it is thought that a more robust weld quality could be achieved in production if the weld bead height and weld toe angle are restricted together.

A toe angle of 123.9° or a weld bead height of 6.26 mm was required to meet the E curve requirements (for a fatigue limit of 46.96 MPa) of the BV rule according to regression Equations (5) or (6), respectively, when the angular distortion of 0.7°. The toe angle and weld bead height to meet the E curve of the BV rule are shown in Figure 24.

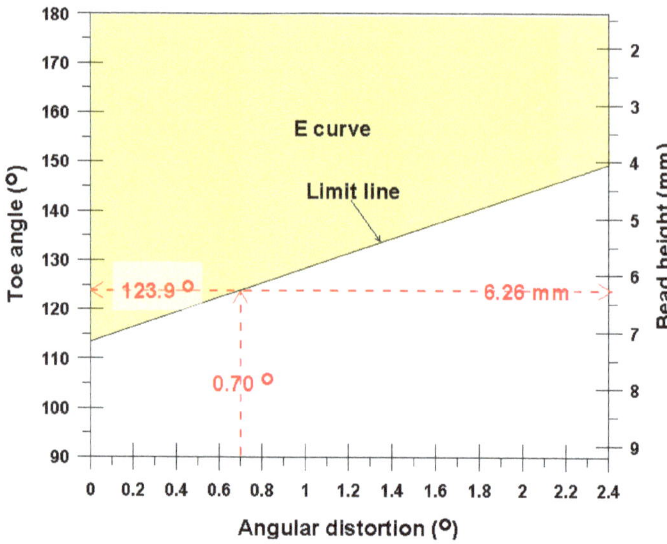

Figure 24. Toe angle and weld bead height satisfying the E curve requirements of the BV rule.

A graph like Figure 24 can be obtained according to the regression Equations (5) and (6) when the required fatigue limits are specified. The obtained toe angle and weld bead height can be used as new criteria for the weld profiles to meet the fatigue limits specified.

4.4. New Approach to Consider the Stress Concentration

It is known the fatigue failure initiation starts from the potential critical spots where geometric discontinuities and notches exist. The potential critical spots were explained by the K_t (SCF: Stress Concentration Factor) related to the weld bead profiles and the K_m (Stress Magnification Factor) related to the angular distortion or the misalignment of weld joints.

The K_t, which is the elastic stress concentration factor at the butt weld toe, can be calculated using the Lawrence equation in Equation (8) [29]:

$$K_t = 1 + 0.27 \tan \theta^{0.25} \left(\frac{t}{\rho}\right)^{0.5} \tag{8}$$

where θ, t and ρ are the toe angle in radius, thickness of the plate and toe radius, respectively.

The K_m, the structural stress concentration caused by misalignment, can be calculated with Equation (9) [30]:

$$K_m = 1 + \frac{\beta}{4}\frac{a_m S}{t} \tag{9}$$

where $\beta = 3$, a_m, S and t are the maximum angular misalignment between the flat plates, length of the plate and thickness of the plate, respectively.

The K_t and k_m calculated by Equations (8) and (9) using data at the fracture location of this study are shown in Figure 25.

It was not possible to determine the border line for K_t and K_m, nor whether values of SCF or K_m could meet the required fatigue limit or not. On the other hand, it can be concluded that the fatigue fracture will occur at the acute angle side when Km is equal to or exceeds 1.10 (angular distortion is equal to or exceed 0.4°).

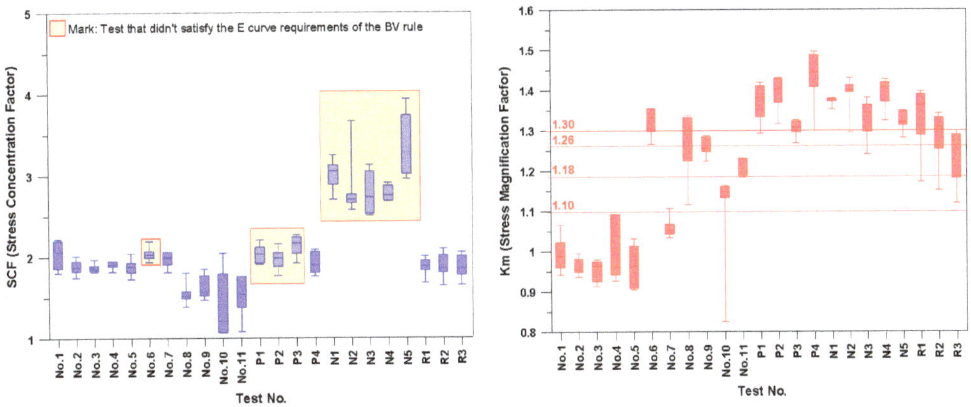

(**a**) (**b**)

Figure 25. K_t and K_m at the fracture location for each test number: (**a**) K_t; (**b**) K_m.

The combined stress concentration is expressed as a multiplication of K_t and K_m [30]. The $K_t \times K_m$ calculated by data at the fracture location of this study are shown in Figure 26.

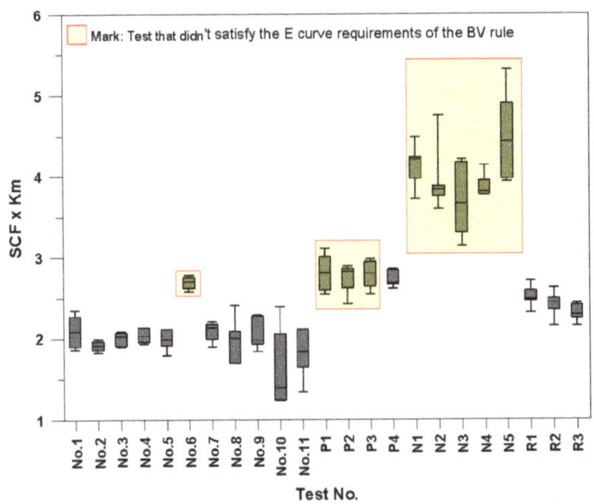

Figure 26. $K_t \times K_m$ at the fracture location for each test number.

A value of 2.6 for $K_t \times K_m$ can be used as a borderline SCF to confirm whether the fatigue limit (stress at 10^7 cycles) can satisfy the E curve requirements of the BV rule or not. It can also be concluded that the $K_m \times$ SCF may be used as a new index to decide whether the fatigue limit can meet a specific fatigue limit of the fatigue curve or not, especially the E curve requirements of the BV rule.

This research would like to name $K_t \times K_m$ as the S-SCF (Samsung-Stress Concentration Factor) when it is used as an index in the criteria for the fatigue limit.

5. Correlation between the Bead Height/Width and the Toe Angle

This study found that the weld bead shape of the SAW two-run method is similar to a circle, as shown in Figure 27.

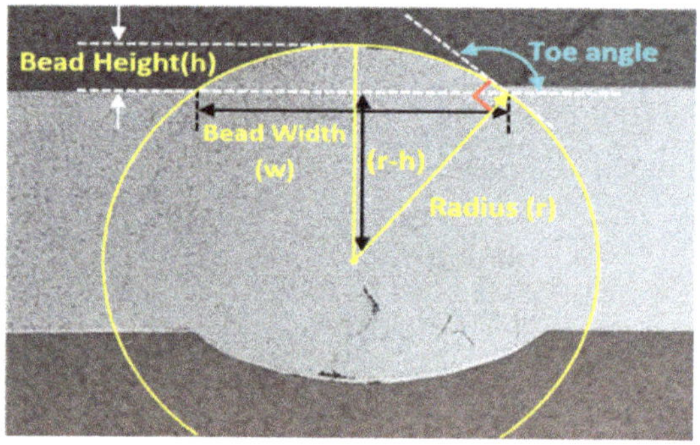

Figure 27. Weld bead shape of the SAW two-run method.

Suppose the actual weld bead can be converted to a segment of a circle made from the weld bead with a certain height and width, as shown in Figure 27. In that case, the radius (r) and toe angle (T.A) can be mathematically calculated from the measured weld bead height and width by the Equations (10) and (11), respectively:

$$r = \frac{h}{2} + \frac{w^2}{8h} \quad (10)$$

$$(\mathrm{T.A}) = 90^\circ + \left[\tan^{-1}\left(\frac{w}{4h} - \frac{h}{w}\right) \times \frac{180}{\pi}\right]^\circ \quad (11)$$

The radius was automatically measured using the Geomagic Control-X program. When the circle was drawn automatically, the radius was automatically calculated using the "Least squares method" by the Geomagic Control-X program depending on the actual weld bead profile. In addition, the toe angle was measured using the template according to the procedure in Section 2.2.2.

The theoretically calculated values were compared with the radius of the circle drawn automatically by the Geomagic Control-X program and the toe angle, measured using the template to verify the effectiveness of theoretical calculations were verified by reflecting the confidence level of 95%. In particular, the average value of the toe angles measured using the template was used for comparison. Figure 28 shows the regression analysis results.

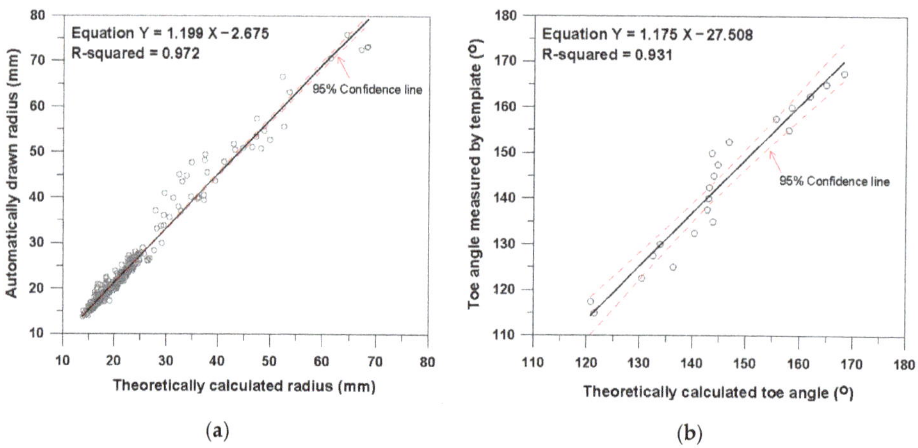

Figure 28. Regression analysis results for the radius and toe angle of the weld bead: (**a**) radius; (**b**) toe angle.

The radius and toe angle of the weld bead can be calculated using regression Equations (12) and (13), respectively, based on the bead height and width measured in production:

$$\text{Radius of the weld bead} = 1.199 \times \text{Equation (10)} - 2.675 \tag{12}$$

$$\text{T.A of the weld bead} = 1.175 \times \text{Equation (11)} - 27.508 \tag{13}$$

6. Conclusions

This study experimentally verified the effects of weld bead profiles and angular distortion on the fatigue properties of the welded joint. The following are important conclusions from the study.

- The fatigue limit of the SA welded joint can be predicted using the proposed regression equation based on the angular distortion, the weld bead height and the toe angle.
- The new criteria for the weld bead profiles, which can achieve more robust fatigue properties for the E-curve of the SA welded joints, can be proposed as a minimum toe angle of 130° and a maximum bead height of 6 mm.
- The weld bead of SA welding can be converted to a segment of a circle. Hence, the expected toe angle can also be obtained by measuring the bead height and width in production.
- The new templates designed by SHI are effective tools to measure the toe angle of a welded joint.

Author Contributions: Conceptualization, D.C. and S.N.; methodology, S.N., C.C., D.L. and E.J.; validation, D.C., Y.J. and J.L.; formal analysis, D.C., Y.J. and J.L.; investigation, M.C., B.K. and K.K.; data curation, D.C.; writing—original draft preparation, D.C.; writing—review and editing, S.N., J.K. and M.K.; visualization, D.C.; supervision, S.N. and M.K.; project administration, D.C. and S.N. All authors have read and agreed to the published version of the manuscript.

Funding: This research received no external funding.

Institutional Review Board Statement: Not applicable.

Informed Consent Statement: Not applicable.

Data Availability Statement: The data presented in this study are available on request from the corresponding author.

Acknowledgments: This research was supported by Samsung Heavy Industry Co., Ltd.

Conflicts of Interest: The authors declare no conflict of interest.

References

1. *AWS D1.1/D1.1M*; Structural Welding Code—Steel. American Welding Society: Miami, FL, USA, 2020.
2. *ISO 5817*; Welding—Fusion-Welded Joints in Steel, Nickel, Titanium and Their Alloys (Beam Welding Excluded)—Quality Levels for Imperfections. International Organization for Standardization: Geneva, Switzerland, 2014.
3. *NORSOK M-101*; Structural Steel Fabrication. Norwegian Petroleum Industry: Stavanger, Norway, 2011.
4. *IACS No.47*; Shipbuilding and Repair Quality Standard. International Association of Classification Societies: London, UK, 1996.
5. Pang, J.C.; Li, S.X.; Wang, Z.G.; Zhang, Z.F. Relations between fatigue strength and other mechanical properties of metallic materials. *Fatigue Fract. Eng. Mater. Struct.* **2014**, *37*, 958–976. [CrossRef]
6. Han, J.W.; Han, D.K.; Han, S.H. Stress intensity factors for three-dimensional weld toe cracks using weld toe magnification factors. *Fatigue Fract. Eng. Mater. Struct.* **2014**, *37*, 146–156. [CrossRef]
7. Livieri, P.; Tovo, R. Overview of the geometrical influence on the fatigue strength of steel butt welds by a nonlocal approach. *Fatigue Fract. Eng. Mater. Struct.* **2020**, *43*, 502–514. [CrossRef]
8. Ohta, A.; Maeda, Y.; Suzuki, N. Effect of yield strength on the basic fatigue strength of welded joints. *Fatigue Fract. Eng. Mater. Struct.* **1993**, *16*, 473–479. [CrossRef]
9. Ninh, N.T.; Wahab, M.A. The effect of residual stresses and weld geometry on the improvement of fatigue life. *J. Mater. Process. Technol.* **1995**, *48*, 581–588. [CrossRef]
10. Cho, J.H.; Baek, S.J.; Shin, S.C.; Lee, G.H.; Kang, M.G.; Jo, Y.R. Effect of Aluminum Arc Butt Joint Weld Geometry on Fatigue Strength. *J. Weld. Join.* **2021**, *39*, 103–108. [CrossRef]

11. Wang, Z.M.; Chang, K.H.; Muzaffer, S. Fatigue Analysis of the Effects of Incomplete Penetration Defects on Fatigue Crack Initiation Points in Butt-Welded Members. *J. Weld. Join.* **2020**, *38*, 543–550. [CrossRef]
12. *IIW-1823-07*; Recommendations for Fatigue Design of Welded Joints and Components. International Institute of Welding: Paris, France, 2008.
13. Schork, B.; Kucharczyk, P.; Madia, M.; Zerbst, M.; Hensel, J.; Bernhar, J.; Tchuindjang, D.; Kaffenberger, M.; Oechsner, M. The effect of the local and global weld geometry as well as material defects on crack initiation and fatigue strength. *Eng. Fract. Mech.* **2018**, *198*, 103–122. [CrossRef]
14. Renken, F.; Polach, R.U.F.B.; Schibnell, J.; Jung, M.; Oswald, M.; Rother, K.; Ehlers, S.; Braun, M. An algorithm for statistical evaluation of weld toe geometries using laser triangulation. *Int. J. Fatigue* **2021**, *149*, 106293. [CrossRef]
15. Stenberg, T.; Barsoum, Z.; Astrand, E.; Ericson Oberg, A.; Schneider, C.; Hedegard, J. Quality control and assurance in fabrication of welded structures subjected to fatigue loading. *Weld. World* **2017**, *61*, 1003–1015. [CrossRef]
16. Hultgren, G.; Barsoum, Z. Fatigue assessment in welded joints based on geometrical variations measured by laser scanning. *Weld. World* **2020**, *64*, 1825–1831. [CrossRef]
17. Hultgren, G.; Myren, L.; Barsoum, Z.; Mansour, R. Digital Scanning of Welds and Influence of Sampling Resolution on the Predicted Fatigue Performance: Modelling, Experiment and Simulation. *Metals* **2021**, *11*, 822. [CrossRef]
18. Wang, Y.; Luo, Y.; Tsutsumi, S. Parametric Formula for Stress Concentration Factor of Fillet Weld Joints with Spline Bead Profile. *Materials* **2020**, *13*, 4639. [CrossRef] [PubMed]
19. Niederwanger, A.; Warner, D.H.; Lener, G. The utility of laser scanning welds for improving fatigue assessment. *Int. J. Fatigue* **2020**, *140*, 105810. [CrossRef]
20. Riofrio, P.G.; Ferreira, J.A.M.; Capela, C.A. Imperfections and Modelling of the Weld Bead Profile of Laser Butt Joints in HSLA Steel Thin Plate. *Metals* **2021**, *11*, 151. [CrossRef]
21. Braun, M.; Neuhausler, J.; Denk, M.; Renken, F.; Kellner, L.; Schubnell, J.; Jung, M.; Rother, K.; Ehlers, S. Statistical Characterization of Stress Concentrations along Butt Joint Weld Seams Using Deep Neural Networks. *Appl. Sci.* **2022**, *12*, 6089. [CrossRef]
22. Radaj, D.; Sonsino, C.M.; Frick, W. *Fatigue Assessment of Welded Joints by Local Approaches*, 2nd ed.; Woodhead Publishing: Cambridge, UK, 2006; p. 183.
23. Fricke, W. Recent developments and future challenges in fatigue strength assessment of welded joints. *J. Mech. Eng. Sci.* **2015**, *229*, 1224–1239. [CrossRef]
24. *DNV-RP-C203*; Fatigue Design of Offshore Steel Structures. Det Norske Veritas: Oslo, Norway, 2016.
25. American Bureau of Shipping. *Guide for Fatigue Assessment of Offshore Structures*; ABS: New York, NY, USA, 2003.
26. *BV NI 611*; Guidelines for Fatigue Assessment of Ships and Offshore Units. BV: Neuilly-sur Seine, France, 2020.
27. *ISO 19902*; Petroleum and Natural Gas Industries—Fixed Steel Offshore Structures. ISO: Geneva, Switzerland, 2007.
28. *BS 7608*; Guide to Fatigue Design and Assessment of Steel Products. British Standard Institution: London, UK, 2015.
29. Maddox, S.J. *Fatigue Strength of Welded Structures*, 2nd ed.; Woodhead Publishing: Cambridge, UK, 1991.
30. Yung, J.Y.; Lawrence, F.V. Analytical and graphical aids for the fatigue design of weldments. *Fatigue Fract. Eng. Mater. Struct.* **1985**, *8*, 223–241. [CrossRef]

Journal of
Marine Science and Engineering

MDPI

Article

Estimation of Fracture Toughness of API 2W Gr.50 Steel in Ductile to Brittle Transition Behavior Using Master Curve Approach

Jihoon Kim [1], Jeongyeol Park [2] and Myunghyun Kim [1,*]

1 Department of Naval Architecture and Ocean Engineering, Pusan National University, Busan 46241, Korea; wlgns3417@pusan.ac.kr
2 Energy Component & Material R&DB Group, Korea Institute of Industrial Technology, Busan 46938, Korea; jypaark@kitech.re.kr
* Correspondence: kimm@pusan.ac.kr

Abstract: Welding is used as the main joining method in various industries, including the shipbuilding industry. In the case of welded structures, structural integrity assessment is essential to ensure the safety of the structure because many defects inevitably exist during the manufacturing process. The value of reliable fracture toughness is required for structural integrity assessment. It is obtained by the fracture toughness test, but the fracture toughness test requires a lot of time and effort. Therefore, many studies have been conducted on efficient methods to evaluate fracture toughness. Among the various studies that estimate fracture toughness, some have been conducted using the Charpy impact test, which is relatively simple compared to the fracture toughness test. This study conducted a series of experimental investigations on API 2W Gr.50 steel applied with different welding conditions. Based on the Charpy impact test results, the fracture behavior was well estimated in the ductile to brittle transition region according to the temperature. However, there was a difference in the accuracy of predicting fracture behavior depending on the welding process. Therefore, additional consideration reflecting the various welding conditions is required to ensure the safety of welded structures.

Keywords: Charpy impact energy; fracture toughness; crack tip opening displacement; ductile to brittle transition

Citation: Kim, J.; Park, J.; Kim, M. Estimation of Fracture Toughness of API 2W Gr.50 Steel in Ductile to Brittle Transition Behavior Using Master Curve Approach. *J. Mar. Sci. Eng.* **2022**, *10*, 1066. https://doi.org/10.3390/jmse10081066

Academic Editors: Sören Ehlers, Moritz Braun and José A.F.O. Correia

Received: 4 July 2022
Accepted: 27 July 2022
Published: 3 August 2022

1. Introduction

As the size of ships and offshore structures gradually increases, careful verification of safety and integrity is required. In particular, as the structure becomes larger, the possibility of defects such as cracks increases during manufacturing. In the case of welded structures, this tendency increases and inevitably includes welded defects. Under this background, many studies have been conducted on the evaluation of the fracture toughness and structural integrity of members with cracks [1]. Fracture toughness is essential to assess the integrity of structures with cracks. It can be obtained by performing the fracture toughness tests using the standard test specimen [2–5]. However, the fracture toughness test requires a lot of time and cost, such as inserting fatigue pre-crack and manufacturing test specimens. Therefore, an efficient method to evaluate the fracture toughness is required, and there is a method using Charpy impact energy [6]. The Charpy impact test requires only a small impact specimen and an impact test machine. Using Charpy impact energy, BS 7910 provides a procedure for estimating fracture toughness values [7]. However, the procedure is very conservative because it is not based on the fracture mechanics theory but only on the correlation between the impact energy and the fracture toughness obtained through the experiment. For this reason, BS 7910, modified in 2019, seeks to achieve minimal conservatism through the yield strength of the material and Charpy upper shelf energy [8].

Pisarski and Bezensek reviewed the Charpy-fracture toughness correlations in Annex J of BS 7910: 2013 together with the 2019 revision of the standard [9]. As the results indicated, the use of an improved equation to determine T_0, which applies the yield strength and Charpy upper shelf energy, provides a better prediction of the fracture toughness transition curve than the existing equation in BS 7910. In addition, it was found that T_{27J} can be estimated from the Charpy impact data obtained at a single temperature in the absence of a Charpy transition curve. Lee et al. investigated the validity of the master curve approach considering the groove shapes and sample location of the impact test specimen. It was confirmed that the tendency of fracture toughness varies depending on the specimen sample location. In addition, these studies required additional analysis according to the welding conditions, such as heat input and welding method [6,10].

As an extension of previous studies, the target of this study is to evaluate the fracture toughness in the ductile–brittle transition region of API 2W Gr.50 steel using the master curve approach and to validate this approach. Therefore, a series of experimental investigations were conducted on API 2W Gr.50 steel by considering heat input and welding methods. To this end, this study compared the derived master curve and the crack tip opening displacement (CTOD) test results. From the results, it can provide a lower limit of fracture values when it is not available.

2. Fracture Toughness Estimation

2.1. Correlation between Charpy Impact Energy and Fracture Toughness

Over the years, many studies have been conducted to develop the empirical correlations between Charpy impact energy and fracture toughness [11–13]. Barsom and Rolfe developed a relationship between fracture toughness and Charpy impact energy [14]. This relationship describes the correlation between K_{Ic} and Charpy impact energy in the transition range for several steels of different compositions and strengths. Wallin reanalyzed the existing relationship between Charpy impact energy and fracture toughness [15]. It was found that the relationship is influenced by the material yield strength and Charpy upper shelf energy. In addition, he suggested a revised equation for reference temperature, T_0, and it is included in the proposal for revision BS 7910. Lucon et al. established a method to estimate T_{27J} from incomplete transition curve or only single temperature data. When the Charpy transition curve is not available, and the results are given only at a single temperature, this method can be useful [16].

Although many empirical correlations are established and have simplicity and efficiency, there are significant differences between the Charpy impact test and fracture toughness test [17]. The Charpy impact test measures the energy of both fracture initiation and propagation for high-strain rate conditions in a short blunt notch as shown in Figure 1 [18]. On the other hand, fracture toughness test uses the standard fracture specimen having a deep sharp crack as shown in Figure 2 and measures the fracture resistance at cleavage instability under quasi-static conditions [4]. Moreover, the correlations are not based on the fracture mechanics but only established by the experimental results for some specific materials. As a result, the correlations do not explain the difference in the material contents and production process. Thus, the correlations are established very conservatively.

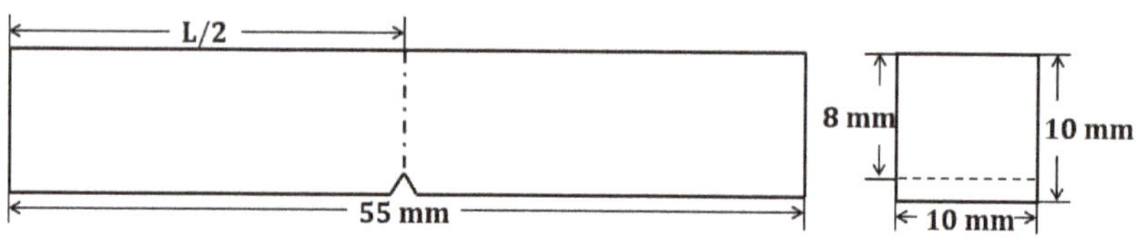

Figure 1. Charpy impact test specimen. Reprinted with permission from Ref. [18]. 2018, ASTM International.

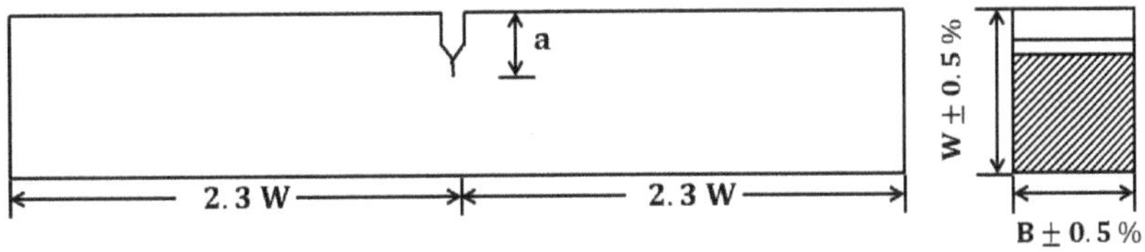

Figure 2. CTOD test specimen. Reprinted with permission from Ref. [2]. 1991, BSI.

2.2. Master Curve Approach

BS 7910, which uses Charpy transition curve, proposes a method for estimating the fracture toughness in ductile to brittle transition behavior. If the specifications of the Charpy impact test specimen suggested in ASTM E23 are satisfied, the fracture toughness dependent on temperature is expressed as Equation (1) [8].

$$K_{mat} = 20 + \{11 + 77\exp[0.019(T - T_0)]\}\left(\frac{25}{B}\right)^{0.25}\left[\ln\left(\frac{1}{1 - P_f}\right)\right]^{0.25} \quad (1)$$

In Equation (1), the B is the thickness of the specimen and the P_f value of 0.05 (5%), which means the probability of K_{mat} being less than the estimated value is recommended. In this equation, T_0 means the reference temperature with the fracture toughness of 100 MPa$\sqrt{m}$ for a 25 mm thickness specimen. It can be calculated by Equation (2) using the material yield strength, σ_Y, and Charpy upper shelf energy, CV_{US} [9]. Charpy upper shelf energy is the maximum amount of impact energy in the ductile fracture mode.

$$T_0 = T_{27J} - 87 + \frac{\sigma_Y}{12} + \frac{1000}{CV_{US}} \quad (2)$$

For the purpose of validating the master curve approach, fracture toughness in terms of J-integral or CTOD can be converted to the equivalent stress intensity factor, K_J. At this time, the critical value of J is limited by the $K_{J(lim)}$. This $K_{J(lim)}$ means the maximum fracture toughness at which fracture takes place under small-scale yielding conditions and is defined as a function of specimen geometry and mechanical property. Equation (3) represents the maximum fracture toughness where b_0 is the specimen ligament ($W - a_0$) [8].

$$K_{J(lim)} = \left[\frac{E b_0 \sigma_Y}{(1 - v^2)}\right]^{0.5} \quad (3)$$

3. Experimental Details

In this study, the base metal used to manufacture the test specimen is API 2W Gr.50 steel for marine structures. The base metal has 345 MPa of yield strength and 448 MPa of tensile strength at room temperature. FCAW (Flux Cored Arc Welding) and SAW (Submerged Arc Welding) were applied with a X-groove configuration. FCAW was applied with two heat input conditions, overheat input and normal heat input, and SAW was applied with normal heat input conditions. Heat treatment was carried out for 48 h at 150 °C to minimize the effects of hydrogen on the time difference between welding and testing. The weldment was cut to have a cross-sectional area of 45 × 90 mm^2 and polished using Labo System of Struers corporation. Then, etching was performed using a 3 % Nital solution. The main reason for observing macrostructures in this study is to examine the intervention of impurities in the section and the degree of uniformity of the weldment and weld defects. As a result of macrostructure observation, no impurity was found in the weldment. Table 1. lists the chemical composition of the test materials, and Figure 3 shows the macrostructure of the weldment.

Table 1. Chemical composition of the base metal and weld metal.

(wt.%)	C	Si	Mn	p	S	Cu	Ni	Mo	Ti	Al	Cr	Ce	V
B.M.	0.16	0.33	1.38	0.03	0.01	0.35	0.75	0.08	0.02	0.012	-	0.41	-
FCAW [1*]	0.043	0.371	1.330	0.012	0.012	0.038	1.331	0.008	0.043	-	0.038	-	0.017
FCAW [2*]	0.047	0.301	1.218	0.014	0.017	0.051	1.204	0.012	0.040	-	0.024	-	0.016
SAW [2*]	0.077	0.307	1.362	0.015	0.011	0.122	0.151	0.005	0.014	0.019	0.029	-	0.005

[1*] overheat input, [2*] normal heat input.

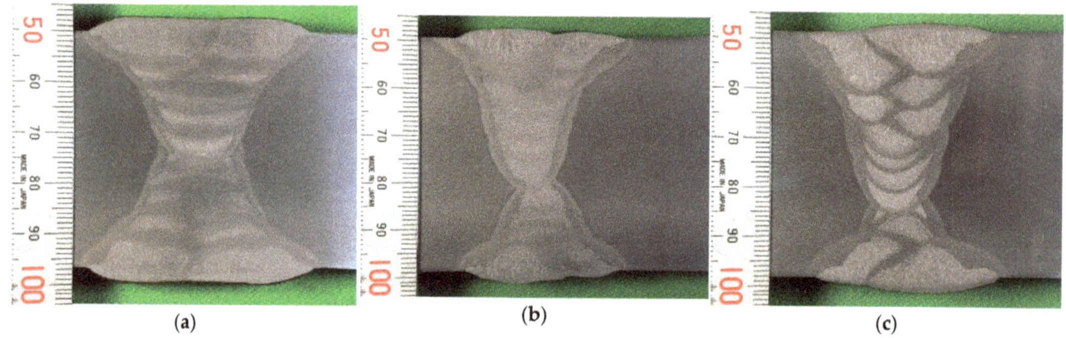

Figure 3. Macrostructure of weldment: (**a**) FCAW with overheat input condition; (**b**) FCAW with normal heat input condition; (**c**) SAW with normal heat input condition.

The main weld parameters used for FCAW, overheat input conditions are: (1) average heat input > 45 kJ/cm, (2) welding current 380 A, and (3) welding voltage 40 V. The parameters used for FCAW, normal heat input conditions are: (1) average heat input 25~30 kJ/cm, (2) welding current 320 A, and (3) welding voltage 36 V. For the SAW process, the main parameters are (1) average heat input 30~45 kJ/cm, (2) welding current 660 A, and (3) welding voltage 32 V. Table 2 lists the main parameters used for the welding process.

Table 2. Welding parameters used for welding process.

Welding Method	Groove Angle (°)	Root Face (mm)	Root Gap (mm)	Heat Input (kJ/cm)	Welding Current (A)	Welding Voltage (V)
FCAW [1*]	60	0~3	6	>45	380	40
FCAW [2*]	60	0~3	0~3	25~30	320	36
SAW [2*]	60	Min. 6	0~2	30~45	660	32

[1*] overheat input, [2*] normal heat input.

The welded plate has sizes of 1000 mm, 500 mm, and 45 mm, respectively, in length, width, and thickness. Figure 4 shows the schematic diagram of the welded plate and sample location for test specimen.

3.1. Tensile Test

Mechanical tensile tests were performed using a standard test specimen with a diameter of 8 mm and a length of 40 mm in the center parallel section at room temperature [19]. The test equipment used in this study is a servo-hydraulic testing machine (IMT 8803 of Instron Corporation). The tensile test specimens for weldment were machined with their longitudinal axes parallel to the welding direction. The specimens were sampled from the top and bottom of the weldment, and the average values of yield strength and tensile strength were used in this study. When fracture toughness has been obtained in CTOD, the values are required to be converted to the critical stress intensity factor, K_{mat}. In this procedure, yield strength and tensile strength determined at the fracture toughness test temperature are required. The following Equations (4) and (5) can be applied to obtain yield strength and tensile strength corresponding to the temperature [8,9].

$$\sigma_Y = \sigma_{Y,R} + \frac{10^5}{(491 + 1.8T)} - 189 \quad (4)$$

$$\sigma_U = \sigma_{U,R} \times \left[0.7857 + 0.2423exp\left(-\frac{T}{170.646}\right)\right] \quad (5)$$

where, σ_Y and σ_U mean the yield strength and tensile strength at test temperature in MPa and $\sigma_{Y,R}$ and $\sigma_{U,R}$ mean the yield strength and tensile strength at room temperature in MPa. T means the test temperature in °C.

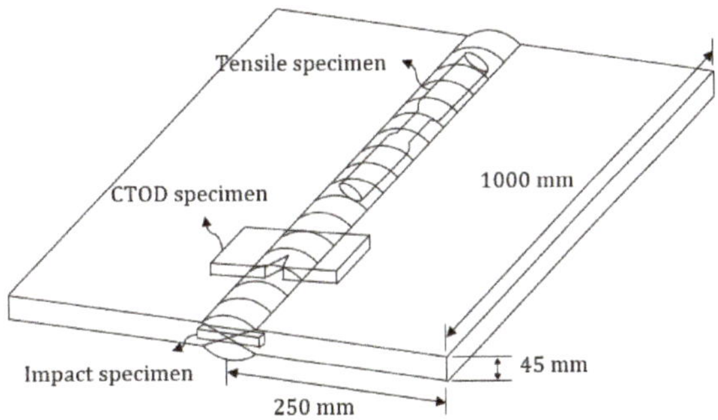

Figure 4. Schematic diagram of the welded panel and test specimen.

Table 3 provides the tensile characteristics of API 2W Gr.50 according to the welding conditions at room temperature.

Table 3. Mechanical properties of API 2W Gr.50 steel at room temperature.

Material	Yield Strength (MPa)	Tensile Strength (MPa)	Elastic Modulus (GPa)
B.M	345	448	205
FCAW [1*]	576.4	621.7	195
FCAW [2*]	539.4	601.3	209
SAW [2*]	541.8	583.8	210

[1*] overheat input, [2*] normal heat input.

3.2. Charpy Impact Test

Charpy impact specimens were sampled from the weldment with their longitudinal axes perpendicular to the welding direction [17]. The thickness of welded plates is thicker than that of impact specimen. Thus, the sample locations are divided into three regions: First, Root and Second, according to the welding sequence, in areas 2 mm from the top and bottom. Figure 5 shows the schematic diagram of the sample location for the Charpy impact specimen.

A total of 111 impact specimens were tested using TM-CIMC of TEST MATE Corporation. The impact tests were performed between the temperatures −90 °C and 60 °C, and Figure 6 represents the test results of absorbed energy according to the temperature. Figure 6a represents the comparison of the test results between overheat input and normal heat input conditions with FCAW. As a result, it was confirmed that the transition temperature was low in the overheat input condition except for the root location. Additionally, Figure 6b represents the comparison of the test results between FCAW and SAW with normal heat input conditions. As shown in Figure 6b, for the same welding heat input, SAW showed lower transition temperature at all sample locations compared to

FCAW. In Figure 6, the lines represent the hyperbolic tangent curve fitting expressed as Equation (6) [20].

$$CVE = A + Btanh\left(\frac{T - T_t}{C}\right) \quad (6)$$

where, T, T_t, and CVE mean assessment temperature, transition temperature, impact energy for assessment temperature, respectively. Using the function, the impact energy values according to temperature can be calculated, and the temperature corresponding to impact energy 27J used for reference temperature calculation can be obtained. Table 4 lists the coefficients of the hyperbolic tangent function and the temperature corresponding to impact energy 27J.

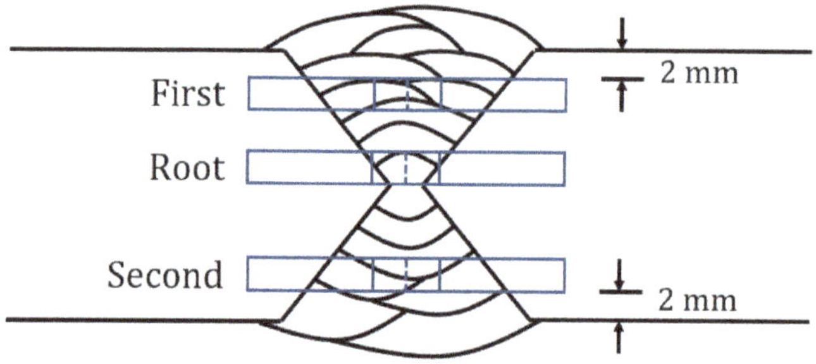

Figure 5. Schematic diagram of sample location for Charpy impact specimen.

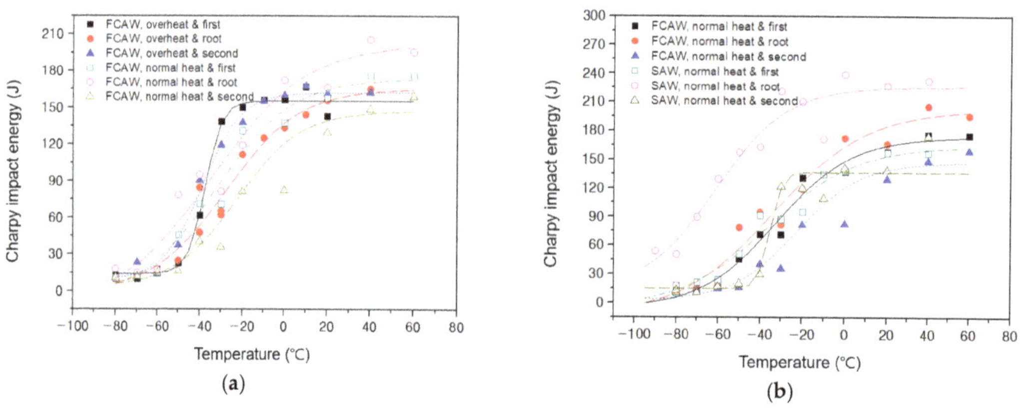

Figure 6. Charpy impact test results: (**a**) Charpy transition curve according to heat input; (**b**) Charpy transition curve according to welding method.

Table 4. Hyperbolic tangent curve fit coefficient of impact test results according to sample location.

Materials	Sample Location	A	B	C	T_t	T_{27J}
	First	84.7	70.5	7.9	−37.5	−46.6
FCAW [1*]	Root	82.2	83.2	34.4	−27.5	−55.0
	Second	85.8	76.2	18.0	−38.8	−57.3
	First	84.0	89.7	36.7	−32.2	−59.0
FCAW [2*]	Root	94.8	107.5	44.1	−33.5	−66.3
	Second	75.0	72.1	29.9	−22.5	−46.5
	First	79.9	84.0	41.0	−36.0	−66.4
SAW [2*]	Root	119.0	107.4	33.9	−61.1	−104.6
	Second	75.9	61.1	5.7	−35.2	−41.4

[1*] overheat input, [2*] normal heat input.

3.3. Fracture Toughness Test (CTOD)

The CTOD test specimens are manufactured in the shape of SENB with a rectangular cross section of B×2B in accordance with BS 7448 [2,3]. The approximate dimensions of the specimens are 42 mm thickness, 84 mm width, and 336 mm span length. For the purpose of evaluating the fracture toughness of weldment, the specimens are machined with their notch parallel to the welding direction. Furthermore, in order to reduce the influence of the notch shape, the radius of the notch was manufactured to be less than 0.1 mm, and fatigue pre-crack was generated so that the a/W became 0.5.

All CTOD test results and load types are included in Appendix A. Figure 7 represents the CTOD test results according to the temperature. Figure 7a represents the comparison of the test results between overheat input and normal heat input conditions with FCAW. Additionally, Figure 7b represents the comparison of the test results between FCAW and SAW with normal heat input conditions. In the comparison according to the heat input condition, the transition temperature of normal heat input is about −10 °C, which is lower than that of overheat input of 0 °C. Before the transition temperature, the fracture toughness of overheat input condition are higher than that of normal heat input condition. However, there was no difference in fracture toughness at temperatures lower than the transition region. In cases of the welding process, the fracture toughness of SAW is higher than that of FCAW. The difference in transition temperature between the two conditions was about 40 °C. Table 5 lists the coefficient and transition temperature of the CTOD test results.

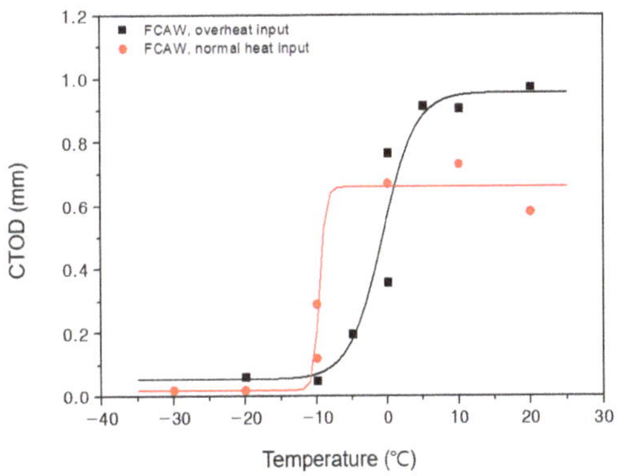

(a)

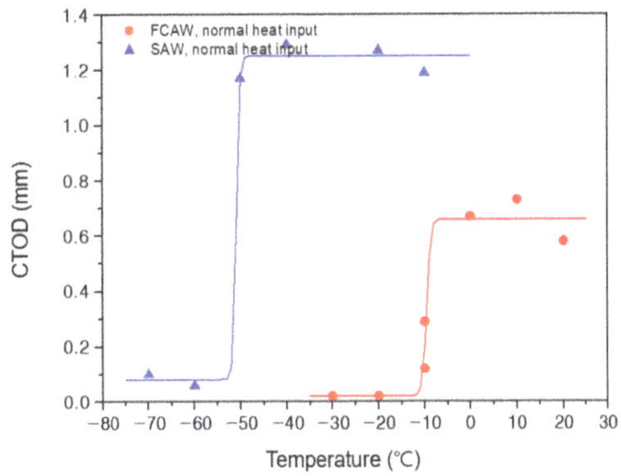

(b)

Figure 7. CTOD test results: (**a**) CTOD transition curve according to heat input; (**b**) CTOD transition curve according to welding method.

Table 5. Hyperbolic tangent curve fit coefficient of CTOD test results.

Materials	A	B	C	T_t
FCAW [1*]	0.51	0.45	4.68	−0.7
FCAW [2*]	0.34	0.32	0.90	−9.6
SAW [2*]	0.67	0.59	0.69	−50.9

[1*] overheat input, [2*] normal heat input.

3.4. Comparison between Charpy and CTOD Transition Curve

Although the test specimens are manufactured using the same material and welding conditions, the difference in transition temperature between the two tests is observed. This difference in transition temperature occurs due to the difference in conditions such as the specimen geometry, strain rate, and notch shape. Therefore, in this study, the transition temperature derived through the CTOD results and the Charpy impact test results of the root were compared.

In Figure 8, the solid symbols and lines represent the Charpy impact test results of weld root. Additionally, hollow symbols and dash lines represent the CTOD test results. Regardless of the welding conditions, the ductile to brittle transition temperature derived from the CTOD test is higher than the value derived from the Charpy impact test. A maximum difference in FCAW with overheat input condition is 27 °C. Based on the test results, Figure 9 shows the relationship between the ductile to brittle transition temperature derived from the two tests. The transition temperature derived from the two tests shows a linear relationship, and it is possible to predict the CTOD transition temperature based on the Charpy impact test results. The linear relationship of two transition temperatures can be expressed as Equation (7).

$$y = 1.495x + 40.437 \tag{7}$$

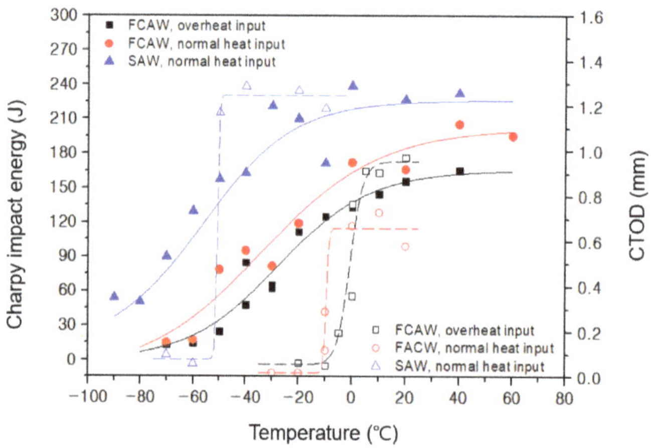

Figure 8. Comparison of ductile to brittle transition behavior according to test method.

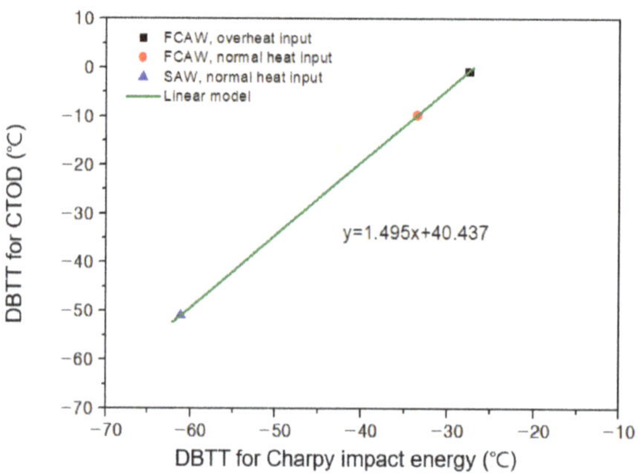

Figure 9. Relationship of DBTT between CTOD and Charpy impact tests.

4. Master Curve Analysis

In this study, the thickness of the welded plate is 45 mm which is thicker than that of the Charpy impact specimen. Therefore, the tendency of master curves varies according to the sample location. Figures 10–12 show the estimated median values of fracture toughness according to the sample location. In case of FCAW with overheat input condition, the estimated fracture toughness of the first location is the lowest. However, in case of other conditions, the second location represents the lowest fracture toughness. Based on the estimated fracture toughness at CTOD transition temperature, the difference according to the sample location is about 36 MPa$\sqrt{m}$ under FCAW with overheat input condition, and a maximum difference is approximately 140 MPa$\sqrt{m}$ under SAW with normal heat input condition. Therefore, consideration of the sample location is required if the welded plate is thicker than the Charpy impact specimen.

The CTOD test results are compared with the master curve to verify the validity of the master curve approach. When fracture toughness has been obtained in terms of CTOD, BS7910 suggests Equation (8) for converting CTOD values to the equivalent stress intensity factor. In the equation, σ_Y, δ, E, and v are yield strength, CTOD value, elastic modulus and Poisson's ratio, respectively. Additionally, m is the constant expressed as Equation (9) for steel [8].

$$K_{mat} = \sqrt{\frac{m\sigma_Y \delta E}{1-v^2}} \tag{8}$$

$$m = 1.517\left(\frac{\sigma_Y}{\sigma_U}\right)^{-0.3188} \tag{9}$$

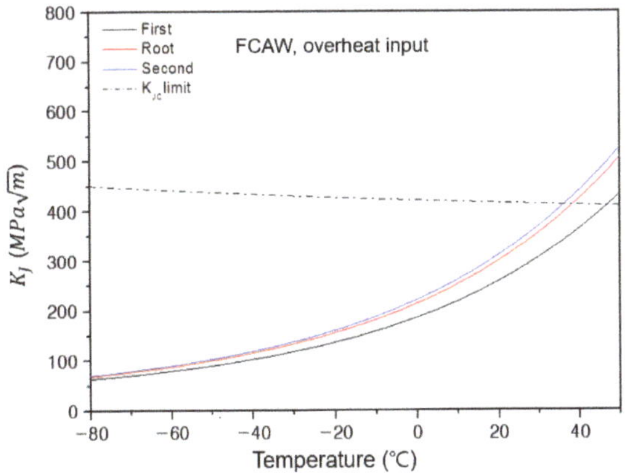

Figure 10. Master curve for FCAW, overheat input condition according to sample location.

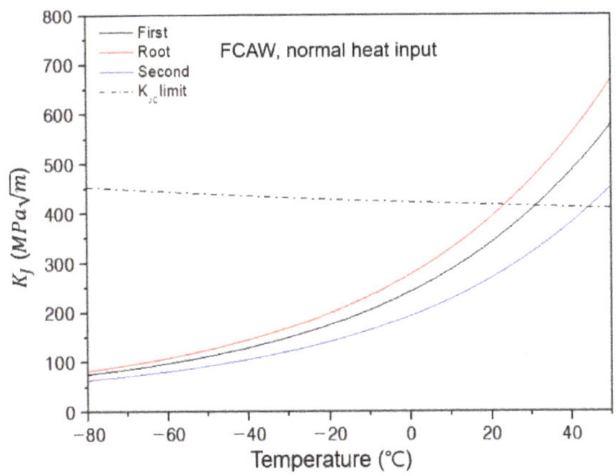

Figure 11. Master curve for FCAW, normal heat input condition according to sample location.

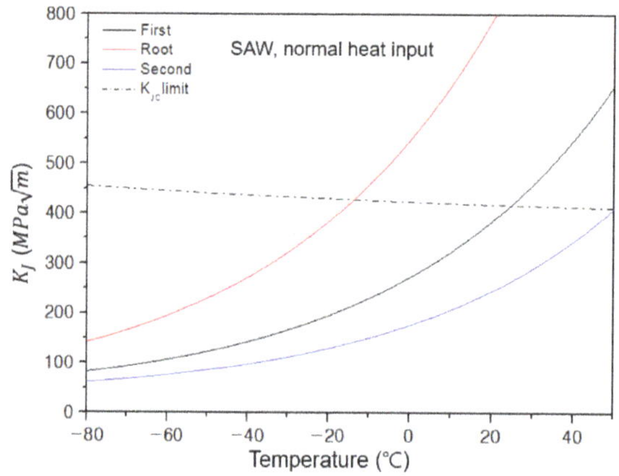

Figure 12. Master curve for SAW, normal heat input condition according to sample location.

Figures 13–15 represent the master curve derived from the Charpy impact test results of weld root. In the figures, the solid lines are the median fracture toughness of the master curve, and dash lines are 5 % and 95 % tolerance bounds. The symbols are the equivalent stress intensity factor converted from the CTOD value, and the dash-dot lines represent the maximum fracture toughness value depending on temperature. In Figures 13 and 14, it is confirmed that the fracture toughness values converted from CTOD were distributed above the median curve in the region above the fracture toughness transition temperature. In contrast, the fracture toughness values converted from CTOD are distributed below the median curve after transition temperature. Although there are some points in which the fracture toughness values deviated from the tolerance bound, the master curve by the Charpy impact test predicted fracture toughness relatively well in the region close to the transition temperature.

In Figure 15, the master curve derived from the Charpy impact test results conservatively evaluates the fracture toughness compared to the CTOD test results. In particular, the master curve has difficulties predicting the tendency of fracture toughness in the region above −60 °C. One of the reasons for the difficulty in predicting fracture toughness may be that the master curve is an empirical correlation based on experimental data. Therefore, additional considerations reflecting various welding conditions are required for accurate fracture toughness estimation.

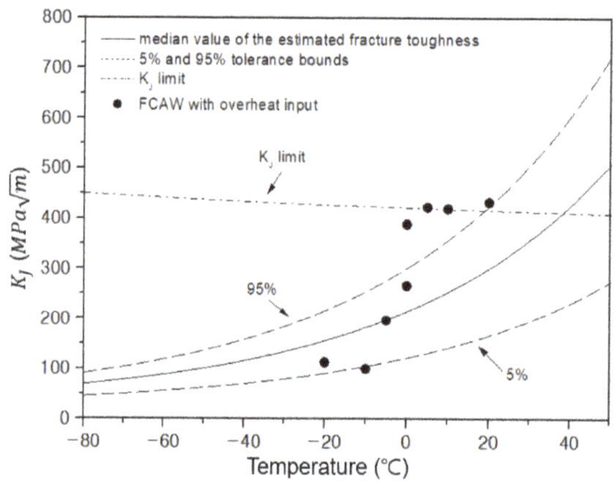

Figure 13. Master curve for FCAW, overheat input condition with CTOD test results.

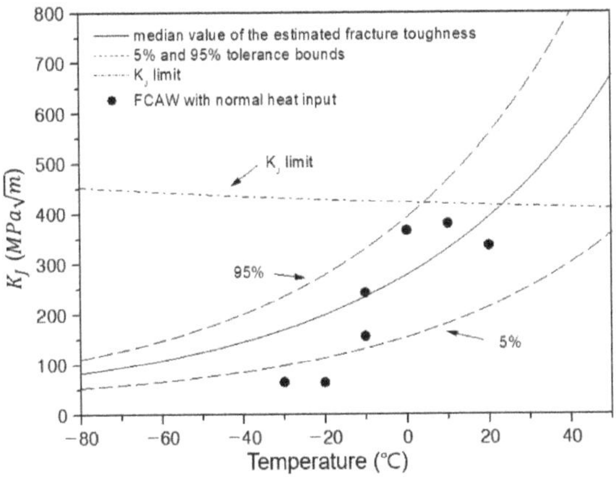

Figure 14. Master curve for FCAW, normal heat input condition with CTOD test results.

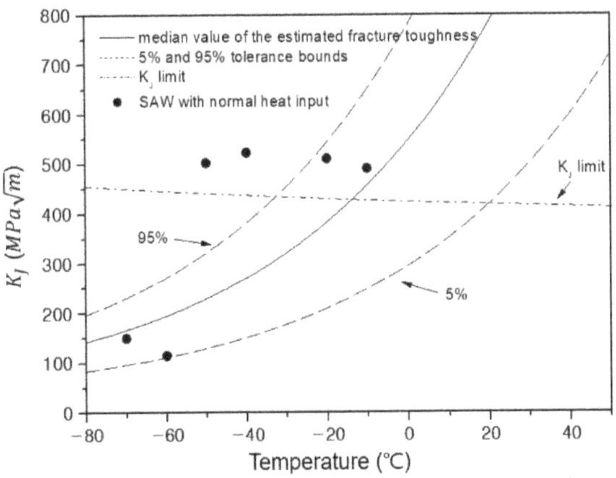

Figure 15. Master curve for SAW, normal heat input condition with CTOD test results.

5. Conclusions

This study used the master curve approach to estimate the fracture toughness of API 2W Gr.50 steel weldment in the ductile to brittle transition region. The test specimens were manufactured using various welding methods (FCAW, SAW) and heat input conditions (overheat input, normal heat input). The reference temperature was calculated using the Charpy impact test results, and the master curve was determined based on the impact energy. Finally, the determined master curve was validated by comparing the CTOD results obtained through the fracture toughness test. Based on this study, the following conclusions are drawn:

- In this study, the thickness of the welded plate is thicker than that of the Charpy impact specimen. Therefore, the impact test specimens are manufactured in three locations, and the DBTT and master curve according to the sample location are different. Based on the master curve by the CTOD transition temperature, the difference in estimated fracture toughness is the largest when the SAW process is applied. As confirmed in this study, if the thickness of the plate is much thicker than that of the standard specimen of the Charpy impact test, it is essential to examine the effect of the sample location.
- The DBTT derived from the CTOD test is higher than that derived from the Charpy impact test. In addition, the CTOD transition temperature was calculated using

the Charpy impact transition temperature. In case of transition temperature, the relationship between CTOD and Charpy impact tests represents the linear line.
- When the FCAW process is applied, master curves derived from the impact test results provide appropriate fracture toughness values in the transition region. Although there are some points outside the tolerance bounds, fracture toughness can be predicted simply from the impact test results, and the transition behavior can be efficiently evaluated.
- When the SAW process is applied, the master curve conservatively evaluates fracture toughness and shows a significant difference from the CTOD test result. Therefore, additional considerations reflecting various welding conditions are required for accurate fracture toughness estimation.

Author Contributions: Conceptualization, J.K. and M.K.; methodology, J.K. and J.P.; validation, J.P.; formal analysis, J.K.; investigation, J.P.; data curation, J.K.; writing—original draft preparation, J.K.; writing—review and editing, J.K. and M.K.; visualization, J.K.; supervision, M.K.; project administration, M.K. All authors have read and agreed to the published version of the manuscript.

Funding: This research received no external funding.

Institutional Review Board Statement: Not applicable.

Informed Consent Statement: Not applicable.

Data Availability Statement: The data presented in this study are available on request from the corresponding author.

Acknowledgments: This research was supported by Korea Institute for Advancement of Technology (KIAT) grant funded by the Korea Government(MOTIE) (P0017006, The Competency Development Program for Industry Specialist).

Conflicts of Interest: The authors declare no conflict of interest.

Nomenclature

B	thickness (mm)
b_0	the size of uncracked ligament (mm)
CV_{US}	the Charpy upper shelf energy (J)
δ	crack tip opening displacement (mm)
E	elastic modulus (MPa)
$K_{J(lim)}$	the maximum fracture toughness at which fracture takes place under small-scale yielding conditions (MPa$\sqrt{m}$)
K_{mat}	the estimate of the fracture toughness (MPa$\sqrt{m}$)
m	constant expressed as the ratio of yield strength to tensile strength for steel
P_f	the probability of K_{mat} being less than estimated value
σ_U	tensile strength
$\sigma_{U,R}$	tensile strength at room temperature
σ_Y	yield strength (MPa)
$\sigma_{Y,R}$	yield stress at room temperature
T	temperature (°C)
T_0	temperature for a median toughness of 100 MPa$\sqrt{m}$ in 25 mm thick specimen (°C)
T_{27J}	the temperature for 27 J measured in a standard Charpy impact specimen (°C)
T_t	transition temperature (°C)
ν	Poisson's ratio
CTOD	Crack Tip Opening Displacement
CVE	Charpy V-notch Energy
DBTT	Ductile to Brittle Transition Temperature
FCAW	Flux Cored Arc Welding
SAW	Submerged Arc Welding
SENB	Single Edge Notched Bend

Appendix A

Table A1. CTOD test results.

Materials	Force Type	Temperature (°C)	CTOD (mm)
FCAW [1*]	F_c	−20	0.06
	F_c	−10	0.05
	F_c	−5	0.20
	F_c	0	0.36
	F_m	0	0.77
	F_m	5	0.91
	F_m	10	0.91
	F_m	20	0.97
FCAW [2*]	F_c	−30	0.02
	F_c	−20	0.02
	F_c	−10	0.12
	F_u	−10	0.29
	F_m	0	0.67
	F_m	10	0.73
	F_m	20	0.58
SAW [2*]	F_c	−70	0.10
	F_c	−60	0.06
	F_m	−50	1.17
	F_m	−40	1.29
	F_m	−20	1.27
	F_m	−10	1.19

[1*] overheat input, [2*] normal heat input.

References

1. Lee, J.S.; Kim, M.H. Strain-Based Failure Assessment Based on a Reference Strain Method for Welded Pipelines. *J. OMAE* **2020**, *142*, 041701. [CrossRef]
2. *BS 7448-1:1991*; Fracture Mechanics Toughness Tests-Part1: Method for Determination of K_{IC}, Critical CTOD and Critical J Values of Metallic Maeterials. British Standard Institution: London, UK, 1991.
3. *BS 7448-2:1997*; Fracture Mechanics Toughness Tests-Part 2: Method for Determination of K_{IC}, Critical CTOD and Critical J Values of Welds in Metallic Materials. British Standard Institution: London, UK, 1997.
4. *ASTM E1921-19A*; Standard Test Method for Determination of Referece Temperature, T_0, for Ferritic Steels in the Transition Range. ASTM International: West Conshohocken, PA, USA, 2019; T0.
5. ISO 15653:2018; Metallic Materials—Method of Test for the Determination of Quasistatic Fracture Toughness of Welds. ISO: Geneva, Switzerland, 2018.
6. Lee, J.H.; Shin, Y.T.; Kim, M.H. Validation of the Master Curve Approach Based on Charpy Impact Test with Groove Shapes. *J. Weld. Join.* **2019**, *37*, 539–546. [CrossRef]
7. BS 7910: 2013; Guide to Methods for Assessing the Acceptability of Flaws in Metallic Structures. British Standard Institution: London, UK, 2013.
8. BS 7910: 2019; Guide to Methods for Assessing the Acceptability of Flaws in Metallic Structures. British Standard Institution: London, UK, 2019.
9. Pisarski, H.; Bezensek, B. Estimating Fracture Toughness from Charpy Data. In Proceedings of the ASME 2019 38th International Conference on Ocean, Offshore and Arctic Engineering, Glasgow, UK, 9–14 June 2019.
10. Lee, J.H.; Kim, J.H.; Kim, M.H. Validation of the Master Curve Approach with Various Welding Conditions—Groove Shapes, Heat Inputs and Welding Processes. In Proceedings of the ASME 2020 39th International Conference on Ocean, Offshore and Arctic Engineering, Fort Lauderdale, FL, USA, 28 June–3 July 2020.
11. Chao, Y.J.; Ward, J.J.D.; Sands, R.G. Charpy impact energy, fracture toughness and ductile-brittle transition temperature of dual-phase 590 Steel. *Mater. Des.* **2007**, *28*, 551–557. [CrossRef]
12. Barbosa, V.S.; Ruggieri, C. Fracture toughness testing using non-standard bend specimens—Part II: Experiments and evaluation of T_0 reference temperature for a low alloy structural steel. *Eng. Fract. Mech.* **2018**, *195*, 297–312. [CrossRef]
13. Hadley, I.; Pisarski, H. Materials propertied for Engineering Critical Assessment: Background to the advice given BS 7910:2013. *Int. J. PVP* **2018**, *168*, 191–199.
14. Barsom, J.M.; Rolfe, S.T. Correlation between K_{IC} and Charpy V-Notch Test Results in the Transition-Temperature Range. In *ASTM STP 466, 28, Impact Testing of Metals*; American Society for Testing and Materials: Philadelphia, PA, USA, 1970; pp. 281–302.

15. Wallin, K. A simple fracture mechanics based Charpy-V impact energy criterion for plastic collapse. *Eng. Fract. Mech.* **2020**, *237*, 107247. [CrossRef]
16. Lucon, E.; Wallin, K.; Langenberg, P.; Pisarski, H. The Use of Charpy/Fracture Toughness Correlations in the Fitnet Procedure. In Proceedings of the OMAE2005 24th International Conference on Offshore Mechanics and Arctic Engineering, Halkidiki, Greece, 12–17 June 2005.
17. Barbosa, V.S.; Godois, L.A.C.; Bianchi, K.E.; Ruggieri, C. Charpy impact energy correlation with fracture toughness for low alloy structural steel welds. *Theor. Appl. Fract. Mech.* **2021**, *113*, 102934. [CrossRef]
18. ASTM E23-18; Standard Test Method for Notched Bar Impact Testing of Metallic Materials. ASTM International: West Conshohocken, PA, USA, 2018.
19. ASTM E8/E8M-16a; Standard Test Methods for Tension Testing of Metallic Materials. ASTM International: West Conshohocken, PA, USA, 2016.
20. Oldfield, W. Fitting curves to toughness data. *J. Test. Eval.* **1979**, *7*, 326–333.

Journal of
Marine Science and Engineering

MDPI

Article

Optimizing the Voce–Chaboche Model Parameters for Fatigue Life Estimation of Welded Joints in High-Strength Marine Structures

Alice Petry [1,*], Pasquale Gallo [1], Heikki Remes [1] and Ari Niemelä [2]

[1] Department of Mechanical Engineering, Aalto University, 00076 Aalto, Finland; pasquale.gallo@aalto.fi (P.G.); heikki.remes@aalto.fi (H.R.)
[2] Meyer Turku Oy, Hull Basic Design, Telakkakatu 1, 20240 Turku, Finland; ari.niemela@meyerturku.fi
* Correspondence: alice.petry@aalto.fi

Abstract: This work studies the Voce–Chaboche (V–C) material model parameter optimization for high-strength steel welded joints subjected to cyclic loading. The model parameters of each material zone in a S690 steel butt-welded joint were determined using an optimization algorithm based on the Newton trust region (NTR) method and an accumulated true strain parameter. The model parameters were fitted to stress–strain histories from uniaxial strain-controlled cyclic tests. To validate the model, fully-reversed variable amplitude fatigue experiments were performed under load control. The experimental results were then compared to numerical results from a finite element analysis. When the elastic modulus is optimized as a V–C parameter, the results indicate that the V–C model slightly underestimates the strain range, leading to conservative fatigue life estimates. However, the results can be improved by using an elastic modulus obtained experimentally. In this case, the resulting material model slightly overestimates the strain range, leading to a non-conservative, but more accurate, fatigue life estimation. It can be concluded that the NTR-based accumulated true strain approach successfully determined the V–C model parameters for different material zones in the welded joint, and closely estimated the strain range and the fatigue life for a variable amplitude load history.

Keywords: Voce–Chaboche; welded joints; high-strength steel; fatigue life; marine structures; cyclic plasticity

Citation: Petry, A.; Gallo, P.; Remes, H.; Niemelä, A. Optimizing the Voce–Chaboche Model Parameters for Fatigue Life Estimation of Welded Joints in High-Strength Marine Structures. *J. Mar. Sci. Eng.* **2022**, *10*, 818. https://doi.org/10.3390/jmse10060818

Academic Editors: José A.F.O. Correia, Sören Ehlers and Moritz Braun

Received: 23 May 2022
Accepted: 10 June 2022
Published: 14 June 2022

1. Introduction

In many civil applications, welded joints introduce geometric discontinuities and material nonlinearities [1]. This is particularly relevant in ship structures, as a cruise ship has around 1000 km of welded joints [2]. Welds are often located at structural discontinuities, such as connections of structural elements, and they can include notches and undercuts acting as stress raisers [3–5]. Additionally, the local microstructure surrounding a weld is altered during the welding process, leading to material nonlinearities. For high-performing welds, this material effect is as damaging as any geometrical effect [6]. Improvements in manufacturing technologies have led to a decrease in the size of defects present in welded joints. Due to these developments, the most relevant fatigue failure process changed from macro-crack propagation (defect size of about 0.1 mm) towards fatigue crack initiation [6]. This paradigm shift allows the utilization of high-strength steels and higher loads, leading to larger material plasticity around stress concentrations. However, current design guidelines for welded joints assume only linear elastic material behavior [7]. Consequently, the recommended fatigue life assessment methods are unable to accurately estimate the fatigue life of high-performing welded joints in lightweight ship structures [8]. One of the existing challenges is modeling the material plasticity and stress–strain behavior of the weld, which can change under cyclic loads [9], e.g., loads caused by wave–ship interaction. Improved fatigue life estimation methods for high-performing welded joints need to consider both the elastic and plastic material response, including transient material effects, such as isotropic and kinematic hardening [8].

Transient material effects can be considered by the Voce–Chaboche (V–C) cyclic plasticity model. Among its applications, it has been used to study the cyclic behavior of welded steel joints subjected to low cycle fatigue [10–12], and the redistribution of residual stresses during elastic shakedown [13] or after high-frequency mechanical impact (HFMI) treatment [14]. Finally, the V–C model has also been used to study the effect of seismic loading on steel structures such as moment-resisting frames [15,16], buckling-restrain braces [17], or pressure-retaining metal components in nuclear power plants [18]. One shortcoming of the V–C model is the lack of a universally accepted model parameters optimization algorithm [19]. In the original paper, Chaboche et al. [20] determined the model parameters for 316 stainless steel through manual curve-fitting of different expressions to five stabilized cyclic hysteresis loops. However he acknowledged that "automatic identification procedures could be used". Nonetheless, many researchers still choose to determine the V–C model parameters using manual curve-fitting methods applied to stabilized hysteresis loops [21–23] or to cyclic stress–strain curves [24]. This process is often reduced to an inefficient trial-and-error approach [25,26]. Other popular methods of determining the V–C model parameters use genetic algorithms [19,27,28] or particle swarm optimization (PSO) [25].

Recently, de Castro e Sousa et al. [15] developed a new model parameter estimation approach. This process uses a Newton trust region (NTR) method to minimize an error function that is defined by the true stress and the accumulated true strain. This formulation of the inverse problem presents a more robust alternative to genetic algorithms and particle swarm optimization methods [15]. The optimization procedure is implemented in a Python library named RESSPyLab (Resilient Steel Structures Laboratory Python Library) [29]. The proposed methodology consists of three phases: application-relevant load protocol sampling, material and testing procedures, and parameters determination. This methodology has successfully determined the V–C model parameters of several different types of structural steel [15], and of the base material and the HAZ of welded S690 QL high-strength steel [11] under low cycle fatigue.

This paper focuses on the experimental validation of the V–C material model applied to high-strength steel welded joints subjected to cyclic loading. The aim is to apply the V–C material model to estimate strains of different material zones of welded joints subjected to fatigue-relevant loads. This study uses the NTR-based accumulated true strain approach [15] to optimize the V–C model parameters for the base metal (BM), the heat-affected zone (HAZ), and the weld metal (WM). The considered cyclic loads are representative of the fatigue-relevant loads acting on a cruciform welded joint located in an upper deck longitudinal deck girder of a large cruise ship. To the best of the authors' knowledge, no other studies have used the V–C model to represent the cyclic material behavior limited to the transition region between elastic and plastic material behavior. Neither has the NTR-based accumulated true strain been applied to determine V–C model parameters to describe the material response around the yield stress of a material. Therefore, the parameter optimization methodology outlined in [15] is modified to provide the best results for the material behavior expected in welded joints in ship structures. The fitness of the resulting V–C model is determined by comparing numerical results from a finite element analysis to experimental results. The evaluation criteria focus on the strain range per load cycle, as this is a relevant quantity for fatigue life estimation using a strain–life approach.

The structure of the paper is as follows: Section 2 introduces the V–C cyclic plasticity model and the chosen model parameter estimation method. Section 3 presents the methodology for the V–C model parameter estimation and the finite element analysis (FEA). Section 4 presents the results of the material model parameter estimation experiments and the validation experiments. Section 5 presents the discussion. The main findings of the paper are summarized in Section 6, the conclusion.

2. Theoretical Background

This section presents the theoretical framework for the uniaxial V–C model in the J2 plasticity framework. First, the general equations of the V–C model are introduced. Second, the material model parameter optimization method developed by de Castro e Sousa et al. [15] is summarized briefly.

2.1. Uniaxial Voce–Chaboche Model

The V–C model is a nonlinear isotropic and kinematic hardening model that combines the Voce isotropic hardening law [30] and the Chaboche kinematic hardening law [20]. In the Haigh–Westergaard stress space, the isotropic hardening law describes how the yield surface (defined by the von Mises criterion) expands or contracts as a function of the equivalent plastic strain ε^p_{eq},

$$\sigma_y = \sigma_{y0} + Q_\infty\left(1 - e^{-b\varepsilon^p_{eq}}\right), \tag{1}$$

where σ_{y0} is the initial yield stress of the material, Q_∞ is the maximum saturated value by which the yield stress can change, and b is the exponential saturation parameter that describes the rate of change of the yield stress.

The kinematic hardening law describes how the origin of the yield surface shifts in the stress space as a function of the equivalent plastic strain rate $\dot{\varepsilon}^p_{eq}$. This shift is called the backstress α. In the Chaboche hardening law, the total backstress α is calculated as the sum of the individual backstresses α_i. Each backstress α_i is described by an Armstrong–Franklin rule [31],

$$\dot{\alpha}_i = \mathrm{sign}(\sigma - \alpha)C_i\dot{\varepsilon}^p_{eq} - \gamma_i\alpha_i\dot{\varepsilon}^p_{eq}, \tag{2}$$

where C_i is the *ith* kinematic stress parameter and γ is the *ith* kinematic saturation parameter. The total backstress is calculated as

$$\alpha = \sum_{i=1}^{N}\alpha_i, \quad \text{i=1, 2, ..., N}, \tag{3}$$

where N is the total number of backstresses. In the uniaxial load case, the equivalent plastic strain rate $\dot{\varepsilon}^p_{eq}$ can be calculated as

$$\dot{\varepsilon}^p_{eq} = |\dot{\varepsilon}^p|. \tag{4}$$

2.2. Accumulated True Strain Approach and RESSPyLab

The methodology proposed by de Castro e Sousa et al. [15] uses service-relevant stress–strain histories to determine the V–C model parameters for a specific material. In their original work, the proposed strain-based load protocols were representative of the deformation in steel moment-resisting frames subjected to seismic loads. Tensile test coupons were subjected to these load protocols in order to collect the applied load signal F and the corresponding controlled strain signal ε_{eng}. These signals were then converted to true stress and true strain using the well-known relations

$$\varepsilon_{\text{true}} = \ln(1 + \varepsilon_{\text{eng}}), \text{ and} \tag{5}$$

$$\sigma_{\text{true}} = \frac{F}{A_0}(1 + \varepsilon_{\text{eng}}), \tag{6}$$

where A_0 is the initial cross-sectional area of the test coupon. In the case of cyclic loading, a stress value σ_{true} can be attributed to multiple strain values $\varepsilon_{\text{true}}$ due to cycling-induced changes in the mechanical properties of the material. To link each stress value to an unique strain value, the accumulated true strain parameter ε^* was introduced.

$$\varepsilon^* = \int_0^t |\dot{\varepsilon}_{\text{true}}| d\tau. \tag{7}$$

Consequently, the true stress σ_{true} can be expressed as a function of the accumulated true strain ε^*. The goal of the material parameter optimization routine is to find the best-fitting stress signal σ_{model} for an input accumulated true strain signal ε^*. The error function that is to be minimized during the parameter estimation process is

$$\varphi(\varepsilon^*;\theta) = \sum_i \frac{\int_0^{\varepsilon^*} \left[\sigma_{\text{model}}\left(\varepsilon_i^*;\theta\right) - \sigma_{\text{Test}}\right]^2 d\varepsilon^*}{\int_0^{\varepsilon^*} d\varepsilon^*}, \quad i \in \{1, \ldots, n_{\text{Test}}\}, \tag{8}$$

where n_{Test} is the number of considered input signals. The vector θ contains all of the parameters that need to be determined during the parameter estimation process, namely,

$$\theta = \{E, \sigma_{y0}, Q_\infty, b, C_1, \gamma_1, \ldots, C_i, \gamma_i\} \tag{9}$$

Each component of the vector θ is restricted within $\theta_i \leq 1 \times 10^{-4}$. The error function $\varphi(\varepsilon^*;\theta)$ is minimized using the NTR approach. For more information on the optimization procedure, see [15].

3. Materials and Methods

This section presents the methodology for each experiment performed as a part of this study: (a) V–C model parameter estimation, (b) experimental V–C model validation, and (c) numerical V–C model validation using a finite element analysis. Since the material parameter estimation experiments and the model validation experiments follow similar methodologies, they are presented together in Sections 3.1 and 3.2. Section 3.3 is dedicated to the finite element analysis.

3.1. Material and Weld Samples

The test specimens were manufactured from butt-welded S690 steel plates provided by the shipyard Meyer Turku Oy. Figure 1 shows a macro sample of the weld and the corresponding hardness profile. The base material consists of two 22 mm thick plates made of S690 SHIP QT from SSAB. The base material has a reported yield stress of 854 MPa, an ultimate tensile stress of 874 MPa, and an area reduction of 16%. The butt-welds were generated from an X-groove through metal active gas (MAG) welding. The filler material was OUTERSHIELD 690-H 1.2 mm. Its reported as-welded mechanical properties are a yield stress of 780 MPa, an ultimate tensile stress of 810 MPa, and an area reduction of 18%. The chemical composition of both materials is presented in Tables 1 and 2. The quality of the weld was assessed through radiography and ultrasonic inspection.

Table 1. Chemical composition of S690 Ship QT in mass percent % as specified by the manufacturer.

C	Si	Mn	P	S	Al	Nb	V	Ti	Cu	Cr	Ni	Mo	N	B	CET
0.159	0.19	1.42	0.007	0.002	0.036	0.016	0.021	0.009	0.012	0.25	0.04	0.498	0.003	0.0013	0.36

Table 2. Chemical composition of Outershield 690-H in maximum mass percent % (except for C) as specified by the manufacturer.

C	Si	Mn	P	S	Cr	Ni	Mo	Cu	V	Ti	Al	As	Sn	Pb	N
0.805	0.01	0.01	0.001	0.001	0.01	0.01	0.01	0.01	0.01	0.001	0.001	0.001	0.001	0.001	0.001

Hardness tests were performed using a Struers Duramin-40 tabletop hardness testing machine. Hardness measurements were taken with a load of 1 kgf and a Vicker's hardness indenter. The optical resolution was set according to ISO standards [32]. The Vicker's hardness (HV) results were converted to the Brinell hardness (HB) scale as proposed by [33]. The average Brinell hardness was used to estimate the coefficients of the Coffin–Manson–Basquin strain–life relationship according to [34].

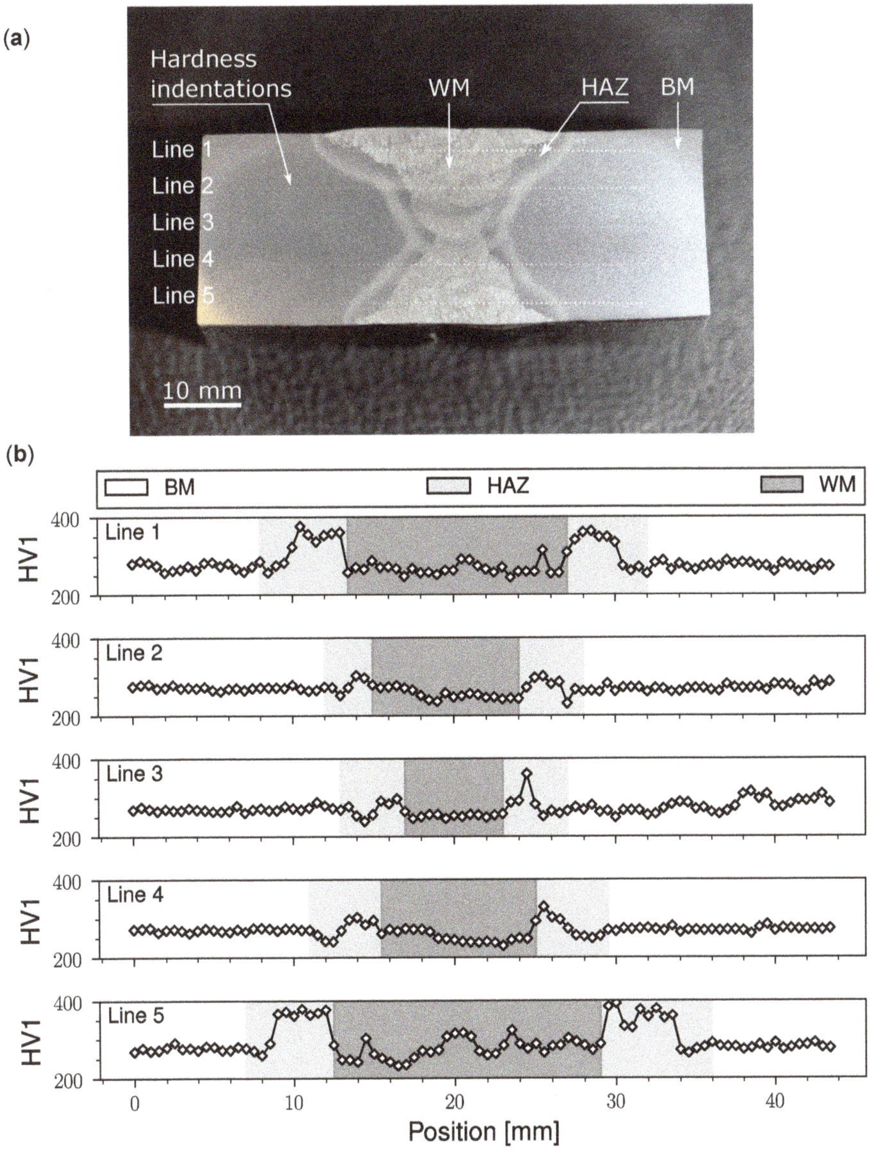

Figure 1. Macro sample of the welded S690 steel with micro hardness indentations: (**a**) annotated macro sample, (**b**) corresponding hardness profile.

The standard deviation σ_{SD} of the hardness measurements is provided. The average hardness values at the center of the macro sample were 271 HV1 (257 HB) in the base material ($\sigma_{SD} = 6$ HV1) and in the HAZ ($\sigma_{SD} = 12$ HV1), and 254 HV1 (241 HB) in the weld material ($\sigma_{SD} = 14$ HV1).

Figures 2 and 3 present the geometries of the test specimens manufactured from the welded plates and used in the material parameter estimation experiments and the validation experiments. The material parameter estimation experiments were performed using round dog-bone specimens designed according to ASTM standards [35]. The validation experiments were performed on plates with a circular hole at the center of the test section.

The circular hole is placed in the weld to introduce a stress concentration effect comparable to that of a micro notch at the weld toe of a cruciform-welded joint [36]. The specimen design was validated against buckling through finite element analysis. To manufacture the dog-bone specimens, the original welded plate was cut into individual strips. Each strip was machined into a test specimen using CNC machining. To manufacture the flat specimens, wider sections of the original welded plates were cut. These plates were then milled to the outer dimensions of the specimen. The test section and the circular notch were realized using electrical discharge machining (EDM). Both types of test specimens had a surface roughness of 0.2 μm in the test section.

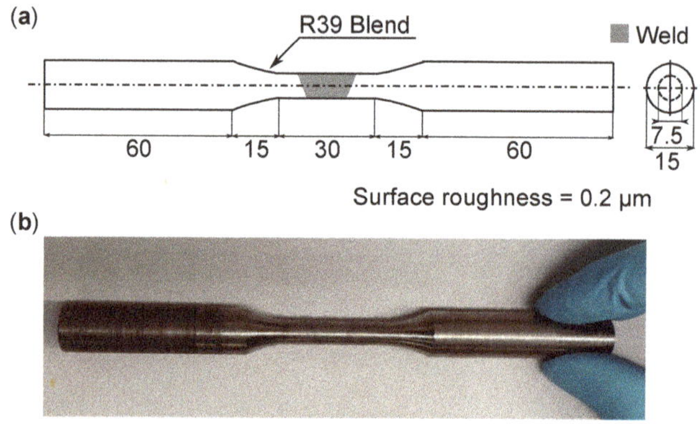

Figure 2. Test specimen used in the material model parameter estimation experiments: (**a**) dimensions of the dog-bone specimen, and (**b**) manufactured dog-bone specimen. The dimensions are provided in mm unless stated differently. The schematic is not drawn to scale.

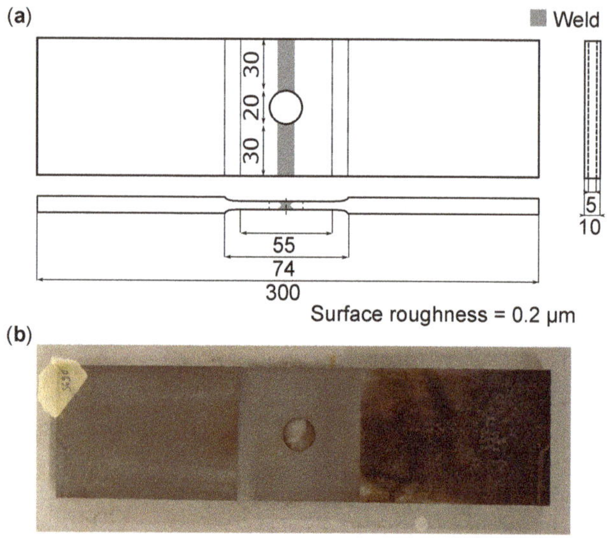

Figure 3. Test specimen used in the validation experiments: (**a**) dimensions of the flat specimen, and (**b**) manufactured flat specimen. The dimensions are provided in mm unless stated differently. The schematic is not drawn to scale.

3.2. Experimental Test Setup

Two different experimental test setups were used for the material parameter estimation experiments and the validation experiments. The strain-controlled material parameter estimation experiments were performed using the MTS-810 servo-hydraulic test system in combination with an MTS Model 632.11F-20 extensometer with a gauge length of 25 mm. The dog-bone test specimens were held in place using the 646 Hydraulic Collet Grip. The load-controlled validation experiments were performed using the MTS 370.50 servo-hydraulic test system with a load capacity of 250 kN. The flat specimens were held in place using wedge grips of type 647 Hydraulic Wedge Grips with smooth clamping wedges. An L-shaped guide was used to align the flat specimens. Each experimental test setup is presented in Figure 4.

(a) (b) (c) (d)

Figure 4. Experimental test setup used in the material model parameter estimation experiments and the validation experiments: (**a**) MTS-810 test system with a dog-bone specimen specimen, (**b**) instrumented dog-bone specimen with extensometer, (**c**) MTS-370.50 test system with a flat specimen, (**d**) instrumented flat specimen.

The load protocols used in the material parameter estimation experiments and the validation experiments considered the recommendations by [15] and fatigue-relevant load cycles as proposed in the DNV standard [37]. In the present study, three different load protocols were used: (a) one fully-reversed load cycle and tension until failure, (b) fully-reversed constant amplitude cyclic loading, and (c) an incremental step test (see Figure 5). The latter two protocols are standard material characterization tests [38]. The controlled strain range was chosen to be representative of the expected service loads and to guarantee stable structural behavior. The validation load protocol consisted of a load block with 21 fully-reversed, variable amplitude load cycles (see Figure 5d). The load block is representative of fatigue-relevant stresses experienced in high-performing welds in large cruise ship structures. The reader is referred to [36] for additional information on the choice of load amplitudes. Each load protocol was repeated until one of the strain gauges failed. Here, failure is defined as measuring a strain that exceeded the measurement range of 5%. Table 3 presents the experimental test matrix.

The test section of each specimen was etched using Nital 2% to highlight the location of the different material zones in the weld. Three strain gauges were then applied to each specimen in the BM, the HAZ and the WM, respectively. The strain gauges were KYOWA KFGS-2-120-C1-11 L1M2R with a base area of 2.8×6.3 mm^2 and a grid area of

1.2×2.0 mm^2. The error in strain measurement due to axial misalignment was limited to less than 1%. The location of the strain gauges was consistent across different specimens. In the dog-bone test specimens, the WM strain gauge was located in the thinnest section of the WM because preliminary cyclic tests showed that cracks initiated in this area. The HAZ strain gauge was placed in the area where the HAZ was the widest to ensure that only the deformation of the HAZ was measured. The BM strain gauge was placed such that its placement did not interfere with the remaining strain gauges. In the flat specimen with a hole, the WM strain gauge was placed as close to the location of the highest stress concentration as possible to measure the stress concentration effect. The HAZ strain gauge was also placed close to the location of the highest stress concentration; however, the width of the HAZ was also taken into account. The BM strain gauge was again placed such that its placement did not interfere with the remaining strain gauges, while still being located in an area with an elevated stress field. The strain gauges on the dog-bone test specimens were covered with polyimide adhesive tape to protect them from the sharp support edges of the extensometer. In both experimental setups, the input voltage of the strain gauges was set to 1.25 V as this allowed to measure strains up to 5%. The strain-controlled experiments were performed with a constant strain rate of 1.25×10^{-4} s^{-1} according to ISO standards [39]. The load-controlled experiments were performed with a constant cycling frequency of 5 Hz according to ISO standards [40]. The controlled signals were defined by a triangular waveform to avoid smearing around the peak. In the case of the material parameter estimation experiments, the load and the controlled strain were recorded by the testing machine, while the strain from the strain gauges was recorded using the data acquisition system QuantumX MX410B. Each signal was recorded with a sampling frequency of 50 Hz. In the case of the validation experiments, the applied load and the strain measured by each strain gauge were recorded at a sampling rate of 100 Hz using the data acquisition system QuantumX MX410B. The data of both experimental setups were output using the data acquisition software Catman Easy.

Table 3. Overview of the material parameter estimation experiments and the validation experiments. Each load protocol is fully reversed. The strain amplitude ε_a is given in percent %.

Load Protocol	Specimen [#]	ε_a [%]	Analyzed Signal
One cycle and tension	2	1.0	Whole signal
One cycle and tension	1	2.0	Whole signal
Constant amplitude	5	0.5	First five cycles
Constant amplitude	6	0.5	First five cycles
Constant amplitude	3	1.0	First five cycles
Constant amplitude	4	1.0	First five cycles
Incremental step	8	0.8	First two repetitions
Incremental step	9	0.8	First two repetitions
Validation	S690#1	-	First five repetitions
Validation	S690#2	-	First five repetitions

Lastly, the recorded experimental data were post-processed and prepared for the subsequent analyses, including estimation of the V–C model parameters and validation of numerical results. To begin, any external disturbances visible in the experimental data (e.g., gripper tightening) were removed from the signals. Next, each signal was filtered using a Savitzky–Golay filter [41] with a window length of five data points and a first-order interpolation function. The filtered signals were set to start at zero by subtracting any offset stress or strain from the respective signals. Finally, the cleaned and filtered force and strain signals were converted to true stress and true strain using Equations (5) and (6). The nominal cross-sectional area of $A_0 = 80 \times 5$ mm^2 was used to calculate the nominal stress applied during the validation experiments. To improve the efficiency of the V–C model parameter optimization procedure and the numerical analysis, the true stress–true strain histories were shortened. The considered portion of each signal is summarized in Table 3.

The post-processed true stress–true strain signals from the material parameter estimation experiments were subsequently used to estimate the elastic modulus E and the offset yield strength $\sigma_{y,0.2\ \%}$ for each material zone and to estimate the V–C model parameters. The experimentally determined elastic modulus E_{exp} is defined as the average elastic modulus estimated from the initial tensile load of each material parameter estimation load protocol. Once the elastic modulus was known, the offset yield stress $\sigma_{y,0.2\ \%}$ was determined according to ASTM standards [42].

The V–C model parameters were determined using the Python library RESSPyLab [29]. The true stress–true strain history of each material model parameter estimation test (see Table 3) was used as input in the optimization tool. To limit the number of V–C model parameters, the number of backstresses was limited to two, as recommended in [15]. For comparison, two different sets of V–C model parameters were determined for each material zone. The first set included the elastic modulus E_{opt} in the vector θ of variables to be determined. The second set of V–C model parameters used the experimentally determined elastic modulus E_{exp} as a constant, removing the elastic modulus from the vector θ. Garcia [11] first proposed to exclude the elastic modulus from the optimized parameters and to define the elastic modulus based on the elastic material response of the material. This approach has also been adopted by Ono et al. [14].

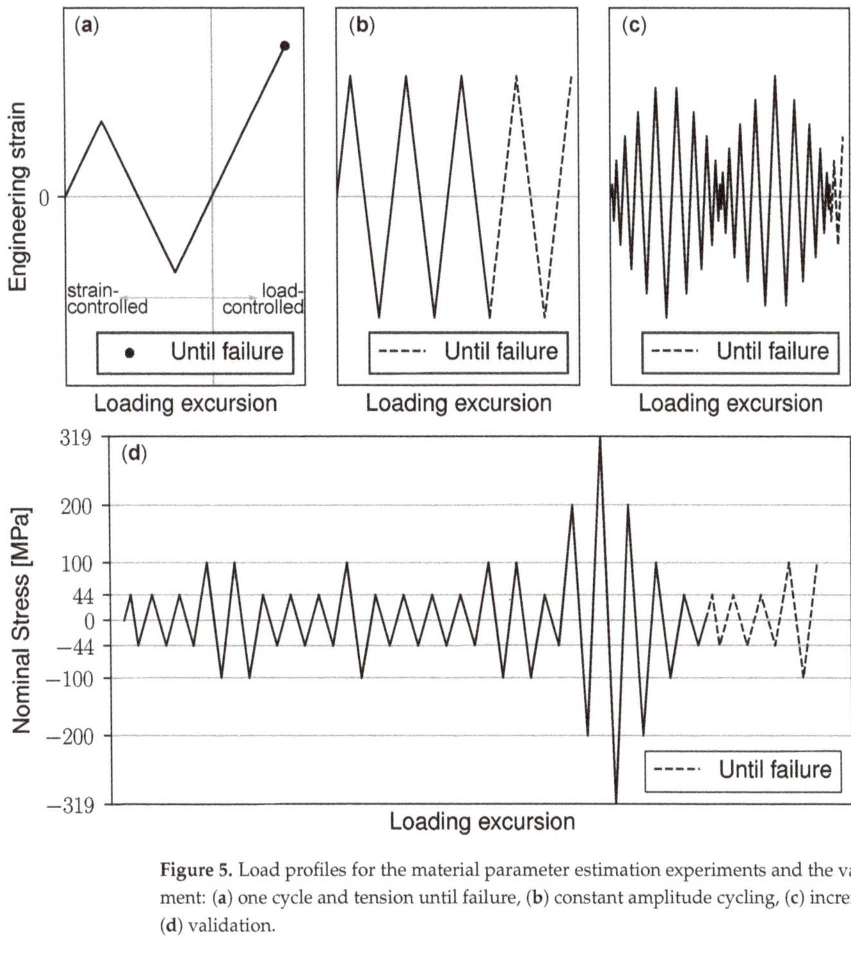

Figure 5. Load profiles for the material parameter estimation experiments and the validation experiment: (**a**) one cycle and tension until failure, (**b**) constant amplitude cycling, (**c**) incremental step test, (**d**) validation.

3.3. Finite Element Analysis

The V–C material model parameters were validated based on two criteria: estimation of the strain range per load cycle and estimation of the damage calculated using the Palmgren–Miner rule. Both criteria require replicating the axial strain measured during the validation experiments. The aim of the FEA is, therefore, to reproduce the validation experiments and to numerically estimate the axial strain recorded in each strain gauge.

The numerical study was conducted through ANSYS Workbench 2021 R2 [43]. The analysis was performed using a static structural analysis in ANSYS Mechanical and the geometry was created in SpaceClaim. The geometry was a replication of the test section of the validation specimen. Due to the nonlinear stress field caused by the validation specimen shape, two geometries, representative of the test sections of both validation specimens, were analyzed. These geometries considered the location of each strain gauge with respect to the stress concentration. The weld of the validation test specimen was not explicitly modeled. As the deformation in the test section was limited to the transition region between elastic and plastic deformation, it was assumed that the deformation in each material zone is mostly linear. On that account, the deformation in each material zone could be analyzed separately. Figure 6 presents the modeled geometry for specimen S690#1. The final geometry was a two-dimensional model, representative of a quarter of a test section, due to symmetry planes and plane strain assumption. The modeled test section was longer than the actual test section to avoid any effect of the boundary conditions on the strain results.

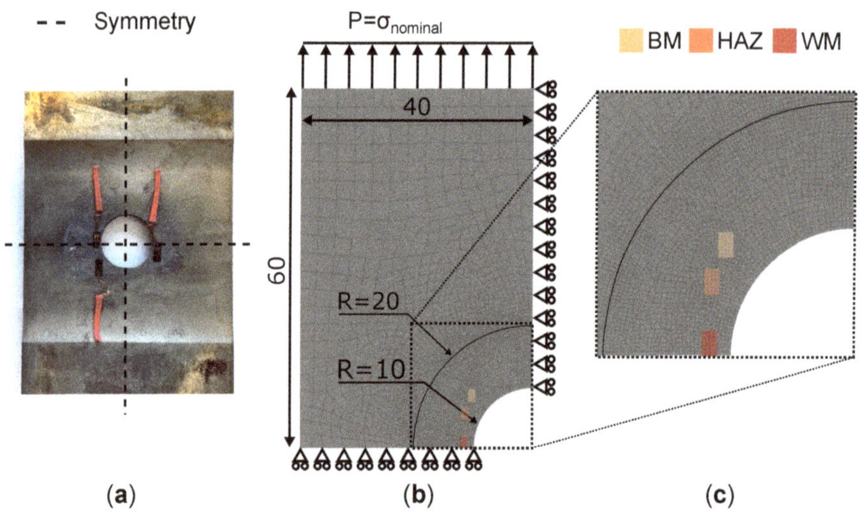

Figure 6. Overview of the different activities of the validation experiment. (**a**) Photo of validation test specimen S690#1. (**b**) Overview of the finite element analysis model used to replicate the validation results. This schematic shows the dimensions of the geometry, the boundary conditions, and the final mesh. The geometry is reduced to a two-dimensional geometry using a plane strain assumption. (**c**) Mesh refinement around the circular notch. The positions of the strain gauges are highlighted in color.

Each geometry was meshed using higher-order two-dimensional PLANE183 elements. The global mesh size was 3 mm. The mesh within an auxiliary arc was reduced to 0.3 mm. This size guarantees a minimum of four elements along the shortest edge of each gauge area. A mesh convergence study showed that a coarser mesh would already show accurate results. The finer mesh was chosen to have additional nodal strain results in each strain gauge area and to have a uniform mesh. Finally, the mesh within each gauge area was mapped to create uniformly distributed quadrilateral elements. The smallest edges of any overlapping areas between two strain gauge areas were meshed with at least one element in the shortest direction to ensure that these features are properly modeled.

The boundary conditions for the geometry are shown in Figure 6b. Two symmetry boundary conditions were applied along the symmetry planes. To replicate the load as accurately as possible, the controlled nominal stress signal from each validation experiment was applied in the form of pressure boundary conditions. In total, 2100 load steps were applied. Four sub-steps were used in between each load step.

The material model was defined using the "isotropic elasticity" module, the " nonlinear isotropic hardening Voce law" module, and the "Chaboche kinematic hardening" module. The "nonlinear isotropic hardening Voce law" implemented in ANSYS differs from the isotropic hardening law considered in this study (see Equation (1)) by a linear term Q_0.

$$\sigma_y = \sigma_{y0} + Q_0\varepsilon^p_{eq} + Q_\infty\left(1 - e^{-b\varepsilon^p_{eq}}\right). \tag{10}$$

This expression can be reduced to Equation (1) by setting the linear coefficient Q_0 to zero. Table 4 summarizes each set of V–C model parameters used in the FEA. The Poisson's ratio was set to $\nu = 0.3$ for each set of parameters.

The axial strain was then obtained by the analyses, and was defined as the average nodal Green–Lagrange strain over each strain gauge area. The average nodal strain was calculated by summing the nodal strain values and dividing them by the number of nodes. The nodal strain was not weighted because the element size within the relevant areas was constant.

Table 4. Summary of the V–C model parameters. The elastic modulus E_{opt} is a result of the V–C model parameter optimization. The elastic modulus E_{exp} is determined based on the material response. The Poisson's ratio $\nu = 0.3$ is used in each parameter set. The remaining parameters are the yield stress σ_y, the isotropic differential stress at saturation Q_∞, the isotropic saturation rate parameter b, the kinematic stress parameter for the 1st and the 2nd backstress C_1 and C_2, and the corresponding kinematic saturation rate parameter γ_1 and γ_2.

Zone	$E_{definition}$	E [MPa]	σ_y [MPa]	Q_∞ [MPa]	b	C_1 [MPa]	γ_1	C_2 [MPa]	γ_2
WM	E_{opt}	176,515	596	0.348	0.397	12,889	105.405	8685	105.405
WM	E_{exp}	204,000	571	0.340	0.394	10,187	100.365	16,296	100.365
HAZ	E_{opt}	178,448	608	0.498	0.539	11,883	96.204	8725	96.204
HAZ	E_{exp}	208,000	586	0.479	0.518	8315	105.299	18,649	105.299
BM	E_{opt}	196,844	694	0.827	0.855	22,748	656.594	25,851	656.594
BM	E_{exp}	209,000	728	0.804	0.829	20,749	459.169	20,749	459.169

4. Results

4.1. V–C Model Parameter Estimation

Figure 7 presents the true stress–true strain results from the material model parameter estimation experiments for the load protocols "one cycle and tension until failure". Noticeable differences in strain amplitude can be observed in the different material zones. For instance, the largest tensile strain was measured in the WM and the largest compressive strain was measured in the HAZ (see Figure 7b). The deformation in the BM was mainly elastic. In addition, although the strain applied over the entire test section (controlled by the extensometer) was fully reversed, the strain in each material zone (measured by a strain gauge) was not symmetric with respect to the origin. The extensometer covered the entire test section of the dog-bone specimens. As a result, the entire test section experienced fully-reversed loading, while the local deformation in each material zone depended on the local mechanical properties, on the size, and on the shape of the material zone. These local differences in deformation do not affect the model parameter optimization using the accumulated true strain approach.

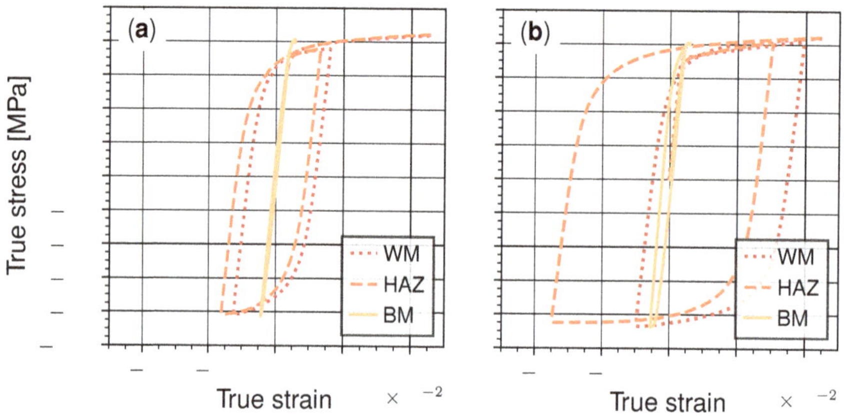

Figure 7. True stress–true strain hysteresis loops recorded during the V–C model parameter estimation experiments: (**a**) Specimen 2, $\varepsilon_a = 1.0\%$, and (**b**) specimen 1, $\varepsilon_a = 2.0\%$. The strain was only recorded up to $\varepsilon = 5\%$.

Figure 8 presents selected experimental and numerical true stress–true strain results from the material model parameter estimation experiments. The individual plots highlight the influence of the elastic modulus on the elastic material response. The set of V–C model parameters with an optimized elastic modulus E_{opt} consistently underestimated the elastic modulus of the considered material zone. Instead, the parameter set with a defined elastic modulus E_{exp} captured the elastic material response more accurately. Nevertheless, the yield stress σ_y was underestimated regardless of the definition of the elastic modulus. Each plot shows early yielding in the material response modeled using the V–C model. The values of the elastic modulus and the yield stress for each material zone are summarized in Table 4, together with the V–C model parameters.

4.2. V–C Model Validation

The V–C model was validated using two different criteria: the relative error in the strain range per load cycle and the normalized number of load block repetitions until failure. These results were calculated for the first five repetitions of the validation load block (see Figure 5d). During the experiments, the HAZ strain gauge in specimen S690#2 was accidentally located in the BM. As a result, specimen S690#1 provided strain histories for the BM, the HAZ, and the WM, while specimen S690#2 provided two strain histories for the BM and one strain history for the WM.

Figure 9 presents a visual comparison of the strain recorded during the validation experiments and the corresponding numerical strain result. Firstly, this figure highlights the differences in strain amplitude between different material zones. The highest strain was measured in the WM since the WM strain gauge was located close to the location of the largest stress concentration. Moreover, the WM has the lowest hardness and is prone to plastic deformation. Secondly, this figure shows that the strain estimated using the V–C model parameter set with the optimized elastic modulus E_{opt} led to an overestimation of the strain amplitude. This observation agrees with previous results. Thirdly, a difference in mean strain between the experimental and numerical strain results was observed in the WM.

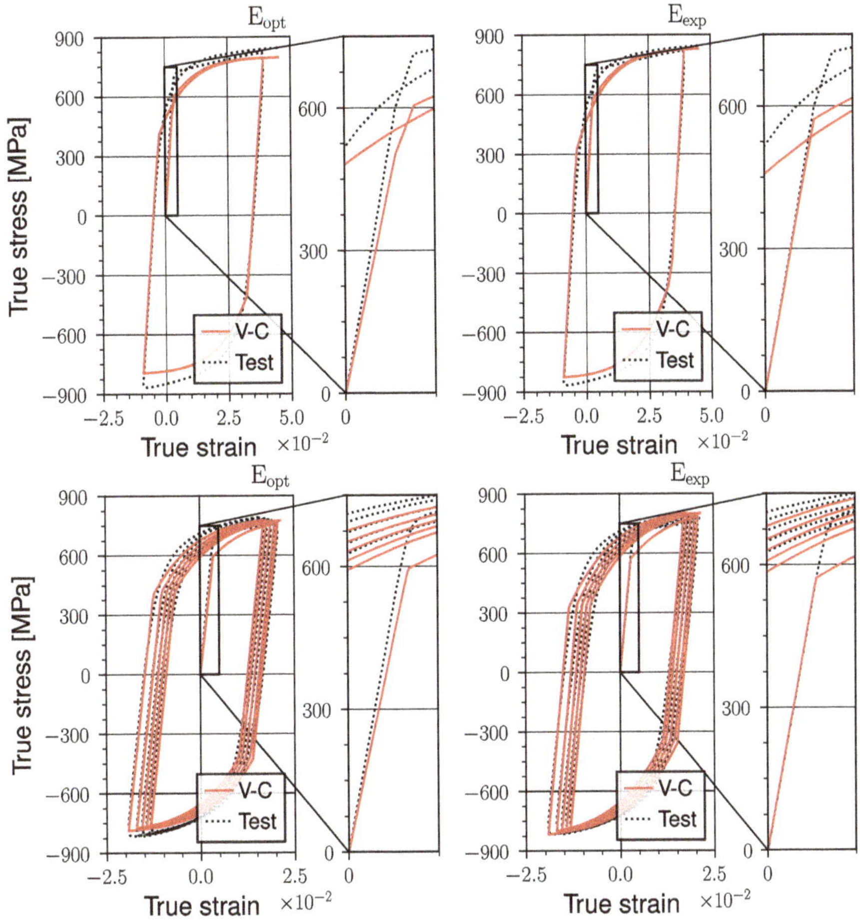

Figure 8. Selected material model parameter estimation results and the corresponding V–C model for the WM. The plots highlight the differences in the elastic material response based on the choice of elastic modulus (E_{opt} or E_{exp}).

It should be noted that, although the applied load was fully reversed (in the validation experiments), the deformation in each material zone was not due to their different mechanical properties. The resulting asymmetric loading allowed ratcheting strain to develop. In the numerical results, ratcheting was likely caused by dissimilar isotropic and kinematic hardening rates at different locations in the test specimen. Since the stress field created by the shape of the validation specimen was nonlinear, only some areas in the test specimen experienced plastic deformation, leading to cyclic hardening or softening. This local change in mechanical properties likely led to the presence of a mean stress. As a result, a ratcheting strain could develop. Nonetheless, the ratcheting strain, present in the experimental and numerical strain results, was small compared to the strain amplitude and did not affect the model validation.

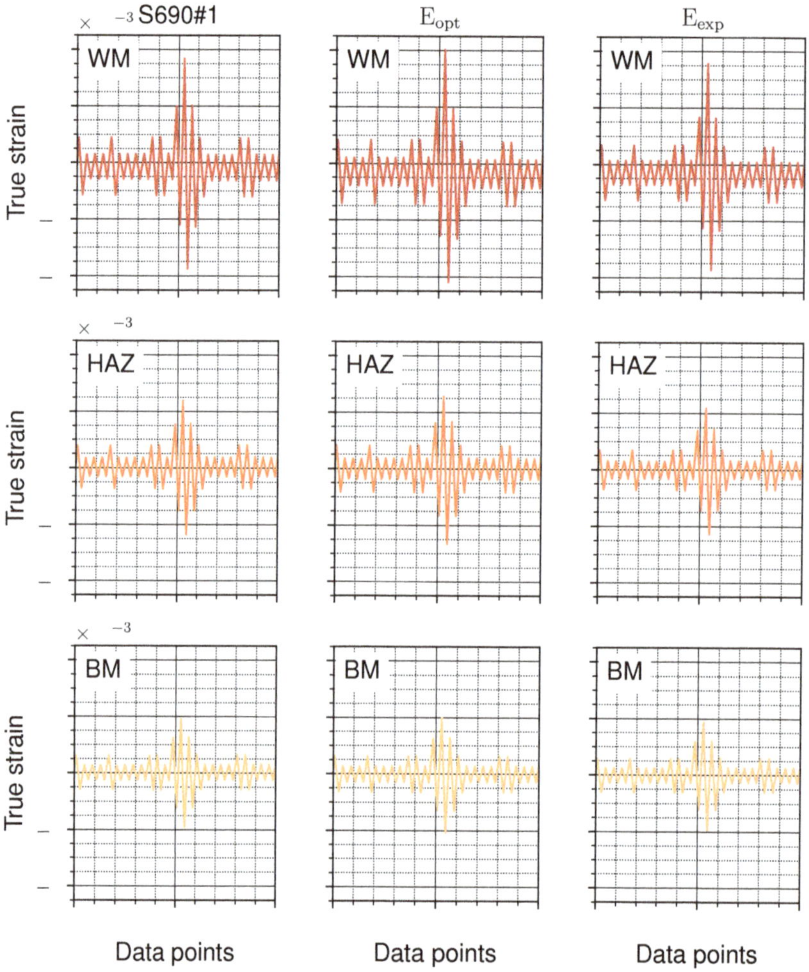

Figure 9. Visual comparison between the measured and simulated (V–C) strain signal of the validation experiment. The strain results correspond to specimen S690#1. Differences in the mean strain are caused by the development of ratcheting strain during the physical and numerical experiments.

The average relative error was calculated for the strain range of each load cycle of a certain load level in the validation load protocol. Here, load cycle refers to the fully-reversed load cycles in the validation load block. The strain range of each load cycle was defined as $\Delta\varepsilon = \varepsilon_{\text{max}} - \varepsilon_{\text{min}}$. The relative error per load cycle was defined as

$$e_{\text{rel}}[\%] = 100 \times \frac{\Delta\varepsilon_{\text{FEA}} - \Delta\varepsilon_{\text{experiment}}}{\Delta\varepsilon_{\text{FEA}}}. \tag{11}$$

Finally, the relative error was averaged over the number of cycles at a certain load level. Since only the first five repetitions of the validation load protocol were considered, this amounted to 60 load cycles at the stress amplitude $\sigma_a = 44$ MPa, 30 load cycles at $\sigma_a = 100$ MPa, 10 load cycles at $\sigma_a = 200$ MPa, and 5 load cycles at $\sigma_a = 319$ MPa.

Figure 10 presents the average relative error per load level. The set of V–C model parameters with an optimized elastic modulus E_{opt} consistently overestimated the strain range for each applied stress amplitude. The relative error was constant across the different applied stress amplitudes σ_a in the BM and the HAZ. The relative error of the WM decreased noticeably at $\sigma_a = 200$ MPa. The set of V–C model parameters with an experimentally determined elastic modulus E_{exp} underestimated the strain range for each applied load amplitude, but it also shows smaller relative errors compared to the optimized set. The largest relative error was observed in the WM at $\sigma_a = 200$ MPa. The relative error of the HAZ increased with increasing applied stress amplitude.

When the deformation is elastic, the relative error in each material zone is constant, as it only depends on the chosen elastic modulus. At an applied stress amplitude of $\sigma_a = 200$ MPa, the WM experienced almost three times higher stresses due to the stress concentration effect of the hole in the specimen. At these stress levels, the WM is about to yield. Since the yield stress used by the V–C model is lower than the experimentally determined offset yield stress of the WM, the numerical result starts to deviate from the experimental result. The relative error reduces at $\sigma_a = 319$ MPa, as both the experimentally measured strain and the numerically simulated strain include plasticity. Considering the relative error within the same material zones under different stress amplitude, the differences in the relative error have high variability for the experimentally determined elastic modulus E_{exp}. The V–C model with an optimized elastic modulus E_{opt} sacrifices some accuracy in the elastic material response for better accuracy when the material response is predominately plastic [15].

Finally, the standard deviation σ_{SD} of the relative error, presented by the error bar, shows the variability in experiments, strain gauge location, and measurement-accuracy-related limitations. The cycling frequency and the sampling frequency used during the validation experiments led to a slight delay between the recorded force and the recorded strain signal. This delay was especially noticeable at the highest load level due to the high loading speed. The resulting uncertainty about the timing of the strain measurement around the peak load led to a larger standard deviation.

To analyze the influence of the difference in strain range on the fatigue life estimation, the number of repetitions until failure of the first five repetitions of the validation load protocol was determined for the experimental and numerical strain histories. The number of repetitions until failure was calculated using the Palmgren–Miner damage $D_{\text{load history}}$,

$$B_{f,\,\text{load history}} = \frac{1}{D_{\text{load history}}}. \tag{12}$$

To calculate the damage $D_{\text{load history}}$, the closed load cycles within the load history were extracted using the rainflow-counting algorithm. Rainflow-counting was performed on the stress signal since the same stress was applied during the experiments and in the FEA. For each closed load cycle, the number of cycles until failure N_f was estimated using the mean stress-corrected strain–life equation,

$$\sigma_{\max}\varepsilon_a^{\text{tot}} = \frac{\left(\sigma_f'\right)^2}{\text{E}}\left(N_f\right)^{2b} + \sigma_f'\varepsilon_f'\left(N_f\right)^{b+c}. \tag{13}$$

The coefficients σ_f', ε_f', b, and c were estimated based on the hardness results. The coefficients were $\sigma_f' = 1317$ MPa and $\varepsilon_f' = 0.39$ for the BM, $\sigma_f' = 1317$ MPa and $\varepsilon_f' = 0.40$ for the HAZ, and $\sigma_f' = 1249$ MPa and $\varepsilon_f' = 0.43$ for the WM. The remaining coefficients b and c were set to -0.09, and -0.56, respectively, for each material zone. The elastic modulus E was defined as E_{exp}. The values $\sigma_{\max}$ and $\varepsilon_a^{\text{tot}}$ were extracted from the load histories using rainflow-counting. The number of cycles until failure N_f was calculated for each closed cycle within the load history by solving the strain–life equation numerically. Once the resulting number of repetitions until failure of each experimental and numerical load history $B_{f,\,\text{load history}}$ were known, the numerical results were normalized with the experimental results, $B_{f,\text{FEA}}/B_{f,\text{experiment}}$.

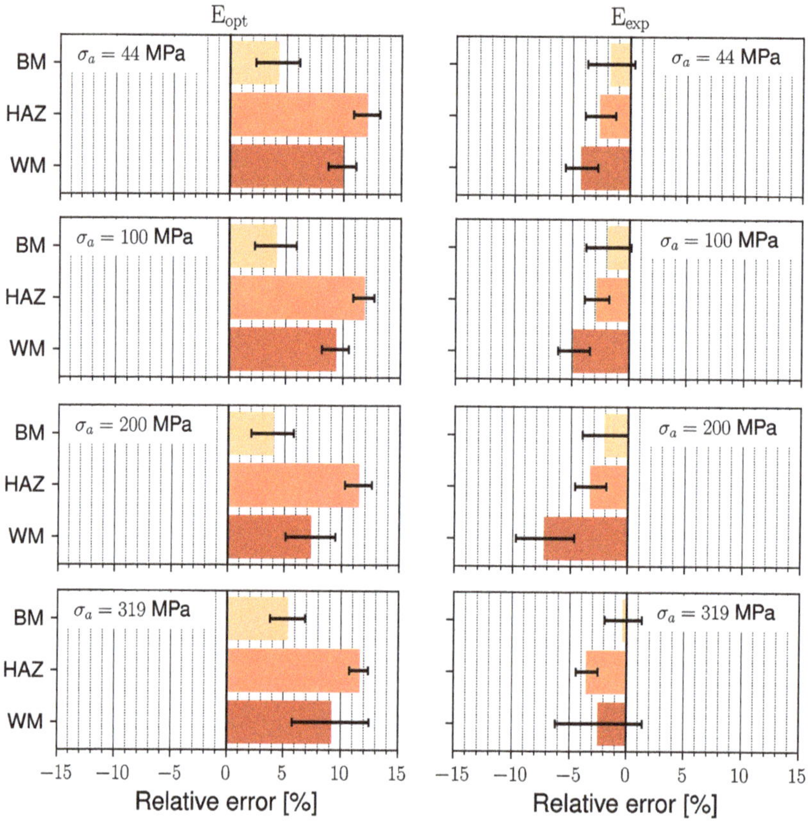

Figure 10. The average relative error in strain range estimation. The standard deviation of the relative error is presented using error bars. A negative relative error indicates that the V–C model underestimates the strain range, while a positive relative error indicates that the V–C model overestimates the strain range. The number of stress–strain signals considered for each material are $n = 3$ for the BM, $n = 1$ for the HAZ, and $n = 2$ for the WM.

Figure 11 presents the average normalized repetitions until failure. The sets of V–C model parameters with an optimized elastic modulus E_{opt} led to conservative damage estimates. Previous results showed that this set of V–C model parameters overestimates the strain range (see Figure 10). The overestimated error per load cycle was accumulated in the damage calculation, which led to conservative fatigue life estimates. The sets of V–C model parameters with an experimentally determined elastic modulus E_{exp} led to non-conservative results, as this set of model parameters tended to underestimate the strain range. Nonetheless, these non-conservative damage estimates were closer to the expected experimental values.

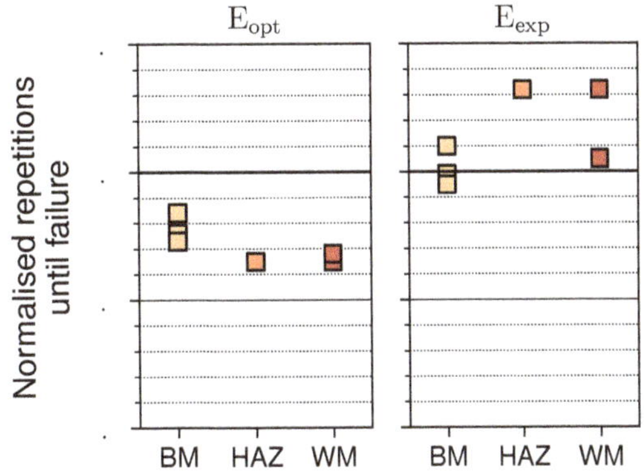

Figure 11. Normalized repetitions until failure for each material zone calculated using the Palmgren–Miner damage rule and the mean stress corrected strain–life equation. A normalized result of less than 1.0 indicates a conservative result, while a result above one 1.0 indicates a non-conservative result.

5. Discussion

As summarized in the literature review presented in Section 1, multiple studies have been conducted on the estimation of the V–C model parameters. These studies have focused on determining the best V–C model parameters to model large deformation. Although the V–C model is extensively used to model a wide range of cyclic loads, Agius et al. [19] and other authors pointed out that there is no consensus on how to define the V–C model parameters. The parameter identification is subjected to uncertainty, and the process can be time-consuming. In addition to that, modeling the deformation of welded joints presents a unique challenge due to differences in the mechanical properties of each material zone. Some material zones may experience plastic deformation, while others are still deforming elastically. In applications where the total deformation of the weld leads to local strains in the transition region between elastic and plastic strain, it is crucial that the chosen material model estimates both the elastic and the plastic material response accurately.

The current study focuses on the V–C model parameter identification in the transition region between the elastic and the plastic strain. Indeed, welded joints in high-strength marine structures are expected to experience strain limited to this transition region. The fit of the V–C model parameters is assessed by comparing the numerical stress–strain results to experimental stress–strain results. Special attention is given to the definition of the elastic modulus because this parameter defines how elastic deformation is modeled. To the best of the authors' knowledge, only Agius et al. [19] have conducted a similar study. In that work, the authors systematically determined the V–C model parameters for aluminum alloy AA 7075-T6 by using a multi-objective genetic algorithm.

The results demonstrate that the accumulated true strain approach, as implemented in the Python library RESSPyLab, was used successfully to estimate the V–C model parameters for each material zone of a butt-welded joint made from S690 steel. The input stress and strain histories were collected from a dog-bone specimen with a weld in the test section. This test specimen design reduced the required number of cyclic tests by a factor of three compared to the original methodology in [15], as one test specimen was sufficient to collect input data for three different material zones. These stress–strain histories were also used to estimate the elastic modulus of each material zone and the offset yield strength of the material zones with the lowest hardness.

The V–C model parameter estimation process seems to systematically underestimate the yield strength of each material zone to achieve a better accuracy in modeling the cyclic material response [15]. Moreover, the definition of the elastic modulus within each set of V–C model parameters has a noticeable effect on the capability to represent the strain response. When the elastic modulus is defined by the optimization algorithm, it is lower than the expected experimental value. A lower elastic modulus leads to an overestimated elastic strain. Since the strain range considered in this study was focused on the transition region where plastic strain is limited, a lower elastic modulus also leads to an overestimated total strain. In addition, any error in strain estimation is accumulated during the calculation of the fatigue life, since the Palmgren–Miner damage rule was used. As a result, an overestimated total strain leads to a conservative fatigue life estimate. In this study, fatigue life estimation was improved by defining the elastic modulus based on tensile tests. This reduced the relative error in the strain range estimation by up to 9%, resulting in better, yet non-conservative, fatigue life results. Based on this observation, it is recommended to exclude the elastic modulus from the parameter estimation, albeit the resulting fatigue life estimation is non-conservative.

Although the present study successfully estimated the V–C model parameters and validated the V–C model using a relevant load profile, some limitations should be addressed. Firstly, the strain-controlled load protocols used in the V–C model parameter estimation experiment did not lead to fully-reversed cyclic loading. The dissimilar end levels of the applied strain led to the development of ratcheting strains in each material zone. The ratcheting strains did not affect the results of this study because only the short-term material response was considered in the V–C model parameter estimation and the model validation. Nonetheless, the development of ratcheting strains can be avoided by performing the material model parameter estimation experiments separately for each material zone. These test specimens should also be used if fatigue crack initiation is expected in the material zone with the highest hardness. Secondly, the validation experiments were performed using only two test specimens and only one load history. Additional load protocols with higher load levels and additional validation load histories should be considered. Lastly, it is essential to define a material model that can account for any change in the stress–strain behavior after an overload, e.g., when a ship navigates through a storm. Thus, future research should investigate the ability of the V–C model to estimate the long-term strain response in the presence of overloads. Current understanding suggests that it is not possible to model memory effects [17]. However, these studies should be revisited for load histories with load levels limited to the transition region between elastic and plastic strain.

6. Conclusions

This paper applied the V–C material model to welded high-strength steel joints subjected to fatigue-relevant loads in large ship structures. The V–C model parameters were determined using the NTR-based accumulated true strain approach developed by de Castro e Sousa et al. [15]. The stress–strain histories, required to determine the V–C model parameters for each material zone, were sampled from a tensile test coupon with a weld in the test section. The material model was validated by comparing the results from load-controlled variable amplitude cyclic loading experiments to numerical results from an FEA.

Based on the results, it was found that the NTR-based accumulated true strain approach could be used to determine the V–C model parameters for different material zones in a welded joint. The elastic modulus used in the V–C model should be defined based on uniaxial tensile tests instead of being optimized as a V–C model parameter. The optimization algorithm proposed by de Castro e Sousa et al. [15] tends to underestimate the elastic modulus and the yield strength, resulting in an overestimated total strain. As the strain in welded joints in marine structures is often limited to the transition region between elastic and plastic strain, the proper definition of the elastic material response is fundamental to improving strain–life estimation.

Author Contributions: Conceptualization and methodology, A.P., P.G. and H.R.; formal analysis, investigation, A.P.; writing—original draft preparation, P.G. and A.P.; writing—review and editing, supervision, P.G. and H.R.; supervision, A.N.; funding acquisition, P.G. and H.R. All authors have read and agreed to the published version of the manuscript.

Funding: This research was funded by the Academy of Finland grant number 321244 and by Meyer Turku Oy (specimen delivery).

Data Availability Statement: Fundamental data are reported in the manuscript. Raw data are available upon reasonable request to authors.

Acknowledgments: The authors would like to acknowledge Matti Rautianen (Meyer Turku Oy) and Abinab Niraula (Aalto University) for their insightful discussions. We are also grateful to Pauli Lehto and Kim Widell (Aalto University) for their invaluable experimental support.

Conflicts of Interest: The authors declare no conflict of interest.

References

1. Mancini, F.; Remes, H; Romanoff, J.; Gallo, P. Influence of weld rigidity on the non-linear structural response of beams with a curved distortion. *Eng. Struct.* **2021** *246*, 113044. [CrossRef]
2. Tailored Products. Available online: https://www.meyerwerft.de/en/technologies/tailored_products/index.jsp (accessed on 9 March 2022).
3. Gallo, P.; Remes, H; Romanoff, J. Influence of crack tip plasticity on the slope of fatigue curves for laser stake-welded T-joints loaded under tension and bending. *Int. J. Fatigue* **2017** *99*, 125–136. [CrossRef]
4. Gallo, P.; Guglielmo, M.; Romanoff, J; Remes, H. Influence of crack tip plasticity on fatigue behaviour of laser stake-welded T-joints made of thin plates. *Int. J. Mech. Sci.* **2018**, *136*, 112–123. [CrossRef]
5. Liinalampi, S.; Remes, H.; Lehto, P.; Lillemäe, I.; Romanoff, J.; Porter, D. Fatigue strength analysis of laser-hybrid welds in thin plate considering weld geometry in microscale. *Int. J. Fatigue* **2016**, *87*, 143–152. [CrossRef]
6. Remes, H.; Korhonen, E.; Lehto, P.; Romanoff, J.; Niemelä, A.; Hiltunen, P.; Kontkanen, T. Influence of surface integrity on the faitgue strength of high strength steels. *J. Constr. Steel Res.* **2013**, *89*, 21–29. [CrossRef]
7. Hobbacher, A.F. *Recommendations for Fatigue Design of Welded Joints and Components*; doc. XIII-2151r4-07/XV-1254r4-07; International Institute of Welding: Paris, France, 2008.
8. Remes, H.; Gallo, P.; Jelovica, J.; Romanoff, J.; Lehto, P. Fatigue strength modeling of high-performing welded joints. *Int. J. Fatigue* **2020**, *135*, 105555. [CrossRef]
9. Devaney, R.J.; O'Donoghue, P.E.; Leen, S.B. Effect of welding on microstructure and mechanical response of X100Q bainitic steel through nanoindentation, tensile, cyclic plasticity and fatigue characterization. *Mater. Sci. Eng. A* **2021**, *804*, 140728. [CrossRef]
10. Hemmesi, K.; Holey, H.; Elmoghazy, A.; Böhm, R.; Farajian, M.; Schulze, V. Modeling and experimental validation of material deformation at different zones of welded structural-steel under multiaxial loading. *Mater. Sci. Eng. A* **2021**, *824*, 140826. [CrossRef]
11. Garcia, M.A.R. Multiaxial Fatigue Analysis of High Strength Steel Welded Joints Using Generalizied Local Approaches. Ph.D. Thesis, École Polytechnique Fédérale de Lausanne, Lausanne, Switzerland, 4 September 2020.
12. Song, W.; Liu, X.; Xu, J.; Fan, Y.; Shi, D.; Khosravani, M.R.; Berto, F. Multiaxial low cycle fatigue of notched 10CrNi3MoV steel and its undermatched welds. *Int. J. Fatigue* **2021**, *150*, 106309. [CrossRef]
13. Chukkan, J.R.; Wu, G.; Fitzpatrick, M.E.; Eren, E.; Zhang, X.; Kelleher, J. Residual stress redistribution during elastic shake down in welded plates. *MATEC Web Conf.* **2018**, *165*, 21004. [CrossRef]
14. Ono, Y.; Yıldırım, H.C.; Kinoshita, K.; Nussbaumer, A. Damage-based assessment of the fatigue rack initiation site in high strength steel welded joints treated by HFMI. *Metals* **2022**, *12*, 145. [CrossRef]
15. de Castro e Sousa, A.; Suzuki, Y.; Lignos, D.G. Consistency in solving the inverse problem of the Voce-Chaboche constitutive model for plastic straining. *J. Eng. Mech.* **2020**, *146*, 04020097. [CrossRef]
16. Hartloper, A.R.; de Castro e Sousa, A.; Lignos, D.G. Constitutive modeling of structural steels: nonlinear isotropic/kinematic hardening material model and its calibration. *J. Struct. Eng.* **2021**, *147*, 04021031. [CrossRef]
17. Court-Patience, D.; Garnich, M. FEA strategy for realistic simulation of buckling-restraint braces. *J. Struct. Eng.* **2021**, *147*, 04021186. [CrossRef]
18. Koo, G.-H.; Ahn, S.-W.; Hwang, J.-K.; Kim, J.-S. Shaking table tests to validate inelastic seismic analysis method applicable to nuclear metal components. *Appl. Sci.* **2021**, *11*, 9264. [CrossRef]
19. Agius, D.; Kajtaz, M.; Kourousis, K.I.; Wallbrink, C.; Wang, C.H.; Hu, W.; Silva, J. Sensiticity and optimization of the Chaboche plasticity model parameters in strain-life fatigue predictions. *Mater. Des.* **2017**, *118*, 107–121. [CrossRef]
20. Chaboche, J.L.; Dang Van, K.; Cordier, G. Modelization of the strain memory effect on the cyclic hardening of 316 stainless steel. In Proceedings of the Transactions of the International Conference on Structural Mechanics in Reactor Technology, Berlin, Germany, 13–21 August 1979.
21. Bari, S.; Hassan, T. Anatomy of coupled constitutive models for ratcheting simulation. *Int. J. Fatigue* **2000**, *16*, 381–409. [CrossRef]
22. Nip, K.H.; Gardner, L.; Davies, C.M.; Elghazouli, A.Y. Extremely low cycle fatigue tests on structural carbon steel and stainless steel. *J. Constr. Steel Res.* **2010**, *66*, 96–110. [CrossRef]
23. Lee, C.; Do, V.; Chang, K. Analysis of uniaxial ratcheting behavior and cyclic mean stress relaxation of a duplex stainless steel. *Int. J. Plasticity* **2014**, *62*, 17–33. [CrossRef]
24. Chaboche, J.L. Time-independent constitutive theories for cyclic plasticity. *Int. J. Fatigue* **1986**, *2*, 149–188. [CrossRef]

25. Smith, C.; Kanvinde, A.; Deierlein, G. Calibration of Continuum Cylic Constitutive Models for Structural Steel Using Particle Swarm Optimization. *J. Eng. Mech.* **2017**, *143*, 04017012. [CrossRef]
26. Rezaiee-Pajand, M.; Sinaie, S. On the calibration of the Chaboche hardening model and a modified hardening rule for uniaxial ratcheting prediction. *Int. J. Solids Struct.* **2009**, *46*, 3009–3017. [CrossRef]
27. Pal, S.; Wije Wathugala, G.; Kundu, S. Calibration of a constitutive model using genetic algorithms. *Comput. Geotech.* **1996**, *19*, 325–348. [CrossRef]
28. Nath, A.; Barai, S.V.; Ray, K.K. Studies on the experimental and simulated cyclic-plasticity response of structural mild steels. *J. Constr. Steel Res.* **2021**, *182*, 106652. [CrossRef]
29. RESSPyLab 1.1.6. Available online: https://pypi.org/project/RESSPyLab/ (accessed on 15 Febuary 2022).
30. Voce, E. The relationship between stress and strain for homogeneous deformations. *J. Inst. Met.* **1948**, *74*, 537–562.
31. Armstrong, P.J.; Frederick, C.O. A mathematical representation of the multiaxial Bauschinger effect. *Mater. High Temp.* **2007**, *24*, 1–26.
32. *ISO 1099:2006(E)*. Metallic Materials—Vickers Hardness Test—Part 1: Test Method; International Organization for Standardization: Geneva, Switzerland, 2018.
33. Boyer, H.E.; Gall, T.L. *Metals Handbook*, Desk edition; ASM Materials: Park Ohio, OH, USA, 1985; p. 1500.
34. Roessle, M.L.; Fatemi, A. Strain-controlled fatigue properties of steels and some simple approximations. *Int. J. Fatigue* **2000**, *22*, 495–511. [CrossRef]
35. *E 606M - 04*. Standard Practice for Strain-Controlled Fatigue Testing; ASTM International: West Conshohocken, PA, USA, 2005.
36. Petry, A. Nonlinear Material Modelling for Fatigue Life Prediction of Welded Joints in High Strength Marine Structures. Master's Thesis, Master of Science in Technology, Aalto University, Espoo, Finland, 1 February 2022.
37. DNV. *Fatigue Assessment of Ship Structures*; DNv-cG-0129; DNV: Høvik, Norway, 2021.
38. Hales, R.; Holdsworth, S.R.; O'Donnell, M.P.; Perrin, I.J.; Skelton, R.P. A Code of Practice for the determination of cyclic stress-strain data. *Mater. High Temp.* **2002**, *19*, 165–185. [CrossRef]
39. *ISO 12106:2003(E)*. Metallic Materials—Fatigue Testing—Axial-Strain-Controlled Method; International Organization for Standardization: Geneva, Switzerland, 2003.
40. *ISO 1099:2006(E)*. Metallic Materials—Fatigue Testing—Axial Force-Controlled Method; International Organization for Standardization: Geneva, Switzerland, 2006.
41. Savitzky, A.; Golay, M.J. Smoothing and differentiation of data by simplified least squares procedures. *Anal. Chem.* **2002**, *36*, 1627–1639. [CrossRef]
42. *E 8M—04*. Standard Test Methods for Tension Testing of Metallic Materials [Metric]; ASTM International: West Conshohocken, PA, USA, 2003.
43. *ANSYS Student 2021 R2*; ANSYS, Inc.: Canonsburg, PA, USA, 2021. https://www.ansys.com/academic/students/ansys-student (accessed on 17 March 2022).

Journal of
Marine Science and Engineering

MDPI

Article

Mechanical Reliability Analysis of Flexible Power Cables for Marine Energy

Pär Johannesson [1], Xiao Lang [2], Erland Johnson [2,3] and Jonas W. Ringsberg [2,*]

1. Division of Materials and Production, Department of Chemistry and Applied Mechanics, RISE Research Institutes of Sweden, 412 79 Gothenburg, Sweden; par.johannesson@ri.se
2. Division of Marine Technology, Department of Mechanics and Maritime Sciences, Chalmers University of Technology, 412 96 Gothenburg, Sweden; xiao.lang@chalmers.se (X.L.); erland.johnson@ri.se (E.J.)
3. Division of Materials and Production, Department of Chemistry and Applied Mechanics, RISE Research Institutes of Sweden, 501 15 Borås, Sweden

* Correspondence: jonas.ringsberg@chalmers.se; Tel.: +46-(0)31-7721489

Abstract: Marine power cables connected to moving devices at sea may experience millions of load cycles per year, and thus they need to be flexible due to the movements of the cable and designed for mechanical loads. In this study, the focus is on the mechanical life of flexible low- and medium voltage power cables connecting devices to hubs. The reliability design method Variational Mode and Effect Analysis (VMEA) is applied, based on identifying and quantifying different types of uncertainty sources, including scatter, model and statistical uncertainties. It implements a load–strength approach that combines numerical simulations to assess the loads on the cable and experimental tests to assess the strength of the cable. The VMEA method is demonstrated for an evaluation of bending fatigue, and is found to be a useful tool to evaluate uncertainties in fatigue life for WEC (Wave Energy Converter) system cables during the design phase. The results give a firm foundation for the evaluation of safety against fatigue and are also helpful for identifying weak spots in the reliability assessment, thereby motivating actions in the improvement process. Uncertainties in terms of scatter, statistical uncertainty and model uncertainty are evaluated with respect to the WaveEL 3.0, a WEC designed by the company Waves4Power, and deployed in Runde, Norway. A major contribution to the overall uncertainty is found to originate from the fatigue life model, both in terms of scatter and model uncertainty.

Keywords: experimental test; fatigue life; numerical simulation; power cable; reliability; uncertainty; VMEA (Variation Mode and Effect Analysis); wave energy

Citation: Johannesson, P.; Lang, X.; Johnson, E.; Ringsberg, J.W. Mechanical Reliability Analysis of Flexible Power Cables for Marine Energy. *J. Mar. Sci. Eng.* **2022**, *10*, 716. https://doi.org/10.3390/jmse10060716

Academic Editor: Fuping Gao

Received: 9 May 2022
Accepted: 22 May 2022
Published: 24 May 2022

1. Introduction

The increased focus on sustainable energy production is a driving force for marine area energy utilization, including offshore wind, wave energy, and tidal energy. The major use of marine power cables has shifted from supplying power to isolated offshore facilities, towards the connection of offshore array systems with a broader field for utilization [1]. There are concerns related to the reliability and long-term serviceability of marine cables [2–5]. Examples of application areas include connecting grids internationally [6] and interconnecting wind turbine generators in offshore wind farms. In the current investigation the focus is on low-voltage power cables for wave energy converters (WECs), which can extract energy from waves. To increase their economic viability, WECs are designed to be grouped into arrays to maximize energy production [7], and power cables transfer the energy from the WECs to a central hub, from which it is then transmitted ashore.

The freely hanging dynamic cables used for WECs are flexible in bending and torsion, with low stiffness properties to cope with the dynamic loads, while the axial stiffness must be designed to handle large axial tension loads. The structural integrity of the flexible power cable is influenced by fatigue damage caused by fluctuating loads due to motions of the

WEC and the hub, and also due to direct loads along the cable caused by the environmental waves, currents, and wind. In view of cost-effectiveness, reliable fatigue life predictions for the cable are essential. The typical flexible power cable consists of several layers combined in a cylindrical or a helical configuration, which leads to complex mechanical behavior [8]. The analyses must include the whole wave energy system, including all mechanical couplings, to consider all these mechanical loads.

Previous research has developed a numerical approach to analyze mechanical loads and fatigue life [9]. It has also presented valuable data on cable curvature under real environmental conditions in the petroleum industry [10]. However, relevant field experiences in marine renewable energy are lacking. There is a need to develop a methodology to enhance the reliability and fatigue life assessment process of flexible power cables for WECs.

Failure Mode and Effect Analysis (FMEA) is often used in industry to achieve reliable mechanical systems and components design [11]. FMEA focuses on identifying and eliminating known or potential failures. However, it is a qualitative method, and it does not measure the resulting reliability. The failure modes are most often triggered by unwanted variation [12], thus a general design philosophy has been developed, including all different sources of unavoidable variation. This reliability and robust design methodology, called Variation Mode and Effect Analysis (VMEA), was first presented in [13,14] and was further developed in [15–17]. A more general presentation of the methodology is found in [18–22]. The VMEA method concept takes into account quantitative measures of failure causes. The method is based on statistics, reliability, and robust design, which can guide engineers to identify critical areas of unwanted variation. The technique has been successfully implemented for fatigue design and maintenance in the automotive and aeronautic industries [16,17], as well as in the marine energy field [22–24]. A main uncertainty in failure analysis of WECs is the shortage of knowledge and experiment data [9], which makes assessment design difficult. An effective reliability method must thus be able to take "lack of experience" into account. The VMEA method can combine different types of uncertainties and is applied in this study addressing the fatigue design of flexible power cables.

The flexible power cable used in this study is a low-voltage power cable mounted on the full-scale floating point-absorber WEC prototype WaveEL 3.0, developed by the company Waves4Power, see [25] for details. The WaveEL 3.0 has been deployed in Runde, Norway; see Figure 1 (right). An illustration of the WaveEL 3.0 system installation is shown in Figure 1 (left). The mooring system consists of three mooring legs. Each mooring leg has two segments connected by a submerged floater. Gravity anchors are used to fix the WEC to the seabed. The cable is freely hanging and transfers power from the WEC to a hub. The WEC and the hub are both equipped with bend restrictors to protect the cable from large bending movements of loads that may cause deformation and clashing [26]. The design service life of the WEC system is over 25 years.

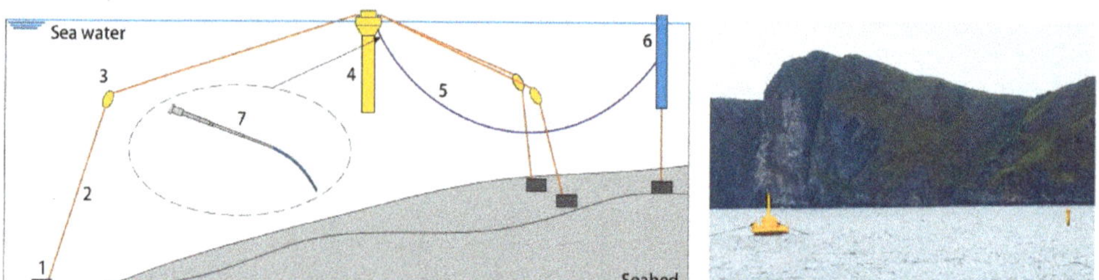

Figure 1. (**Left**) Illustration of the WaveEL 3.0 installation in Runde: (1) gravity anchor, (2) segment of a mooring line, (3) floater, (4) WEC, (5) cable, (6) hub, (7) bending restrictor, and (**right**) the full-scale WaveEL 3.0 system (photograph taken by the fourth author).

The life and ageing of cables mainly depend on chemical, electrical, and mechanical properties, and degradation. Cable manufacturers typically perform life testing and evaluation giving attention to chemical and electrical degradation [26]. The focus of this study is the fatigue life of dynamic cables subjected to repeated load cycles. The work aims to adapt and apply the VMEA methodology to quantify all uncertainty sources for life predictions of cables used in WEC array systems. Experimental tests and numerical simulations are used for quantification, and the method is demonstrated with an evaluation of bending fatigue life.

The next section introduces methodologies for numerical simulation of cable motion, fatigue life, and VMEA. The VMEA analysis results of the fatigue damage evaluation are presented in the third section, including the fatigue life model of the cable, the analysis of its uncertainties, the numerical simulation model, and the sensitivity of cable properties. A discussion and the conclusions close the paper.

2. Methodology

In this study of wave energy application, the focus is on cable fatigue life due to mechanical loads. The dynamic cable was subjected to repeated load cycles with varying bending, tension, and rotation loads, resulting in mechanical stress histories. The fatigue calculation followed the typical life evaluation process for wave energy converters cable [27]. It involved the input of marine loads, numerical simulation of WECs system, and the fatigue life model; see Figure 2. The dominant failure mode was altered according to the different platforms, cables and locations. Initial modelling of the dynamic cable and the surrounding system indicated that the bending load dominated this study. Thus, the most critical failure mode for the mechanical life of the dynamic cable was assumed to be caused by the bending loads on the cable.

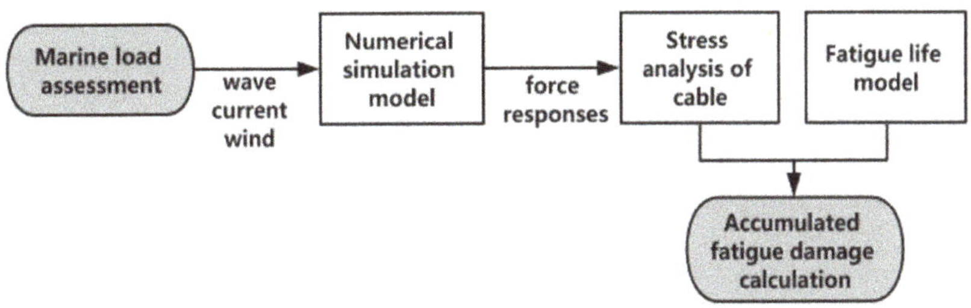

Figure 2. The typical life evaluation process for a wave energy converter cable.

The study has an emphasis on models used in numerical simulations, fatigue analysis and statistical assessment. The models used in the numerical simulations of the WEC system, and for fatigue damage calculation of the cable, have been validated in previous research work by the authors. Section 2.1 presents a brief overview of these studies, followed by a description of each model in Figure 2 in Sections 2.2–2.4.

2.1. Full-Scale Measurements on WaveEL 3.0 and Power Cable Fatigue Tests

The design of a numerical simulation model of a WEC system such as WaveEL 3.0 requires an understanding of the physics that can be represented by commercial software, in order to accurately capture the real physics ultimately affecting the motion responses of the WEC and its components, and the force responses in e.g., the power cable and the mooring lines. The simulation procedure (including the numerical simulation models) in this study was validated against WEC model tests (on a WEC system similar to WaveEL 3.0) carried out in model scale in a laboratory ocean basin [25,27]. There are, however, scaling effects that need to be considered between model and full scale. The representation and modelling of environmental loads (e.g., wave, wind, and ocean current) at full scale are also more challenging compared to model test conditions.

Ringsberg et al. [28] presented results from a full-scale measurement campaign on Waves4Power's WaveEL 3.0 WEC, deployed at a test site off the coast of Runde, Norway. The measurement campaign lasted from June to November 2017. The WEC system was instrumented with several sensors and equipment that recorded several factors, such as the WEC's motions and position, axial forces in the mooring lines, the power performance of the WEC, and the responses of the power cable from the WEC to the power-collecting hub; see Figure 1. The authors designed a simulation model of this WEC system, based on the simulation procedure validated in [25,27] and presented a comparison between the full-scale measurements and the numerical simulations regarding WEC motions and mooring-line forces. It was found that the measured and simulated WEC motion responses were in good agreement, as were the measured and simulated axial forces in the mooring lines. The numerical simulation results of the mooring-line forces were mostly 10% higher than the measurements, which was within an acceptable range of error due to uncertainties in system instrumentation, environmental conditions and tidal influence that changed the pre-tension force of the mooring lines. Therefore, the predictability of the numerical simulation model of WaveEL 3.0 was found to be good.

As part of the measurement campaign, the motions in a short segment of the power cable were recorded. Unfortunately, the resolution of the recorded data was not sufficient to compare with the results from the numerical simulation model. Nevertheless, it was possible to see that it was subjected to repeated motions resulting in load cycles with varying bending, tension, and rotation loads, resulting in mechanical stress histories that fatigue the power cable. According to [26], for wave energy applications with a power cable between the buoy and the hub, the cyclic variation of axial and bending stresses is the essential cause of the power cable's fatigue damage, where the bending stress often dominates the cyclic total stress.

A test rig generating a rotating-bending cyclic load was developed to simulate a large number of load cycles for the cable in a short time; see the test setup illustrated in Figure 3. The cable was subjected to a rotational cycle at the lower end A attached to a rail, while the upper end B of the cable was fixed. When the rail rotated around a vertical axis through point B, end A was rotated one turn back around the horizontal axis by utilizing a mechanical drive to avoid torque along the cable. This design allowed the required measurements of conductivity to be performed during testing. The cable was bent 90°. The rig permitted the radius R to be mechanically adjusted in the interval from 400 to 800 mm, defined as the distance from the connection point to the intersection point as Figure 3 shows. The test rig was set to operate at a frequency of about 2 Hz because of the internal heating limitation. The interruption criterion was interpreted as either survival after a large number of cycles which was considered as run-out, or when a failure criterion was achieved. The run-out level was decided to be 8,600,000 cycles, while an increase in electrical resistance of 15% was stated as the failure criterion. Electrical resistance was calculated by the voltage drop at a fixed electrical current, at this voltage drop many of the copper threads should have failed due to fatigue.

A numerical simulation model of the test rig was presented in [29]. It presented a methodology for modelling umbilical power cables, which was verified against a benchmark study in the literature. In the same study, the power cable used in the current study was used in the numerical model of the test rig setup. A comparison of results from the real fatigue tests verified the numerical models of the test rig and the power cable. To summarize, the numerical simulation model of the WEC system used in the study was validated in model scale in [25,27], verified against a full-scale installation in [28], and the numerical model of the dynamic power cable was verified in [29].

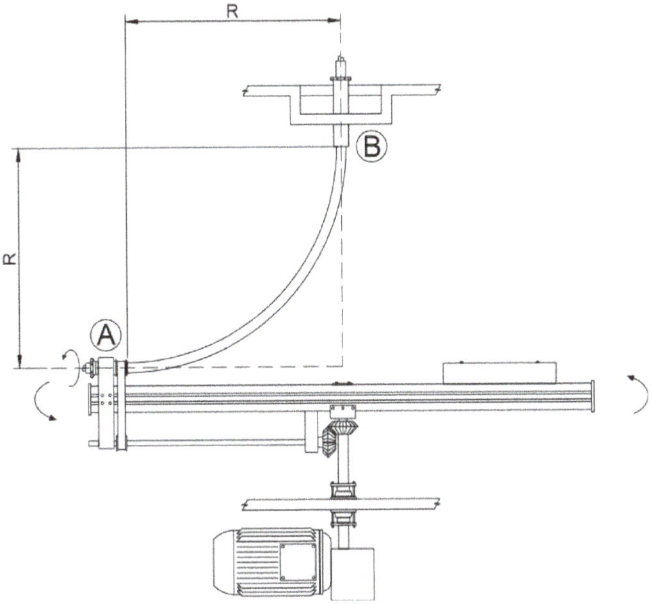

Figure 3. The test rig setup with cable mounted. The cable is fixed at the top, and the rotating end is controlled via a mechanical drive visible at the bottom of the picture (A and B are end points, see the text for clarification).

2.2. Numerical Simulation of Cable Motion

The numerical model of WaveEL 3.0 was used in the simulations, with the commercial software package DNV GL SESAM [30,31]. Due to the coupling between the components of the WEC system, the time-domain coupled simulation procedure was adopted to solve the motions of all components in the WEC system simultaneously, considering the interactions between the WEC buoy, the mooring lines, and the power cable. The coupled simulation model was developed and adapted to the Runde test-site conditions and installation [25,28].

2.3. Fatigue Life Model

A fatigue model evaluates fatigue tests and performs damage calculation for simulated loads. In this study, a Basquin-like relation was considered in conjunction with the linear Miner rule for damage accumulation; cycle amplitudes were resolved using the rainflow method. This provides a conservative damage calculation compared to using, for example, fatigue limit or the Haibach approach. In the fatigue life model, the number of cycles to failure N is related to the stress amplitude S following the Basquin equation:

$$N = N_0 \cdot \left(\frac{S}{S_0}\right)^{-m}, \tag{1}$$

where S_0 represents the fatigue strength (in terms of stress amplitude) at N_0 cycles, here selected as $N_0 = 1 \times 10^6$.

2.4. Variation Mode and Effect Analysis

VMEA is a probabilistic method that studies the variation and uncertainty around a nominal design, and an adaption to marine energy applications is found in [22]. Based on all variation and uncertainty sources, the methodology determines a statistically based safety factor that, together with an optional additional safety factor based on engineering risk judgments, gives an overall safety factor against eventual failure. The statistical safety factor is constructed through a confidence interval, which is determined from an overall standard deviation of the defined target function. The standard deviation is generally obtained from the square root of overall variance, which can be calculated using Gauss's approximation formula. It gives the variance of the target function f as the sum of variance contributions from different influencing variables x_i, each described by its variance together with its influence of the target, employing its sensitivity coefficient c_i; see Equation (2).

$$\mathrm{Var}[f(x_1, x_2, \ldots, x_n)] \approx \sum_{i=1}^{n} c_i^2 \mathrm{Var}[x_i] + \mathrm{Cov} \tag{2}$$

where the last term is the covariances between the influencing variables. The covariances need to be included when relevant, however, in most cases they can be neglected or be avoided by a re-parametrization of the model. This probabilistic approach represents a first-order, second-moment reliability method.

The VMEA method progresses through three different design phases; see Figure 4. The first and crudest approximation is called the "basic" VMEA, and is used in the early design phase when little is known about variations. The standard deviations and sensitivity coefficients are simply replaced by scores, on a scale from 1 to 10, based on engineering judgments about uncertainty and sensitivity, respectively. The refinement process in the next design phase is named "enhanced" VMEA. The uncertainties are quantified by judging standard deviations via standard rules, while fundamental physical knowledge is used to assess sensitivities. Further refinement in the later detailed design phase, called "probabilistic" VMEA, is developed by obtaining more information about the most critical uncertainty sources. More detailed empirical results are used to obtain the standard deviations, and sensitivity coefficients are taken from numerical sensitivity studies or differentiation of physical, mathematical models. This analysis gives an estimate of the resulting total uncertainty, and a corresponding statistical safety factor can be derived. Further, the VMEA work process is grouped into four activities: "Define-Analyse-Evaluate-Improve", and can be described by seven steps, as illustrated in Figure 5.

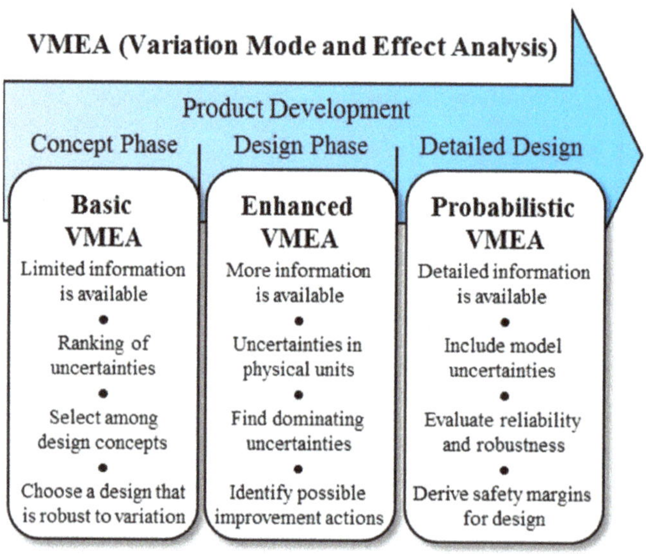

Figure 4. VMEA in different design stages.

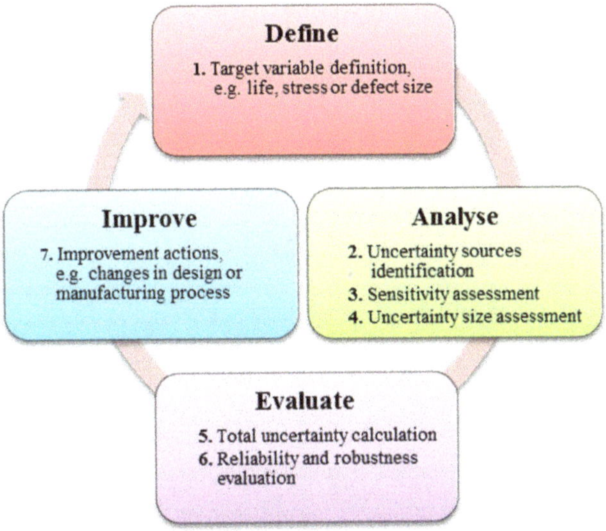

Figure 5. VMEA in the design and improvement cycle.

3. Results

First, the target variable and sources of uncertainties were identified within the basic VMEA. The results from the basic VMEA were then used as a starting point to establish a full probabilistic VMEA. The probabilistic VMEA was based on probabilistic descriptions of equivalent strength and equivalent load on the cable. These were first derived, based on the same methods as described in [17–22,24], but adapted to the current application on dynamic cables. Experimental tests of cable bending life tests were used to assess the equivalent strength, and the equivalent load was assessed using numerical cable simulation and subsequent fatigue damage calculations. The fatigue strength was evaluated from rotating bending fatigue tests, and sensitivities to parameters were investigated by numerical simulation, which was used in the probabilistic VMEA. Finally, a reliability evaluation was performed. All results are presented below.

3.1. Target Variable and Sources of Uncertainties

A basic VMEA was established, following the procedure in Figure 5, with the involvement of a cross-functional team of experts with different views and competencies within cable manufacturing, mechanics, numerical simulations, laboratory testing, and statistics. The main objective of this activity was to define the target variable and to identify sources of uncertainties.

The target variable in this study was chosen as the fatigue life of the flexible power cable. Following the life evaluation process shown in Figure 2, the uncertain sources within the cable fatigue life calculation were divided into five categories: marine load, cable motion, cable properties, fatigue life model, and laboratory testing; see the Ishikawa diagram [32] in Figure 6. The waves and the ocean currents were the two main marine loads. The WEC system and the cable were sensitive to the induced motions. The boundary conditions representing the cable connections to the buoy and the hub were modelled but exhibited uncertainties, as did the influence of marine growth. There are a few cable property parameters in the simulation model whose values affect the fatigue life. The main parameters considered are the outer diameter of the cable, the mass per meter, and stiffnesses in axial, bending and torsional directions.

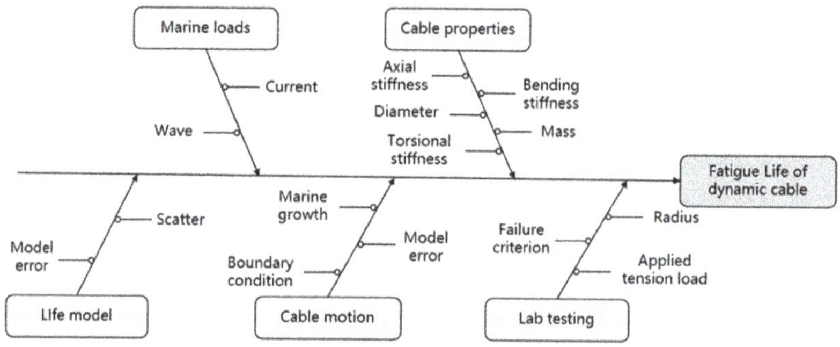

Figure 6. Ishikawa diagram [32] for the given uncertainty sources for the VMEA application of flexible power cable reliability and fatigue assessment process design.

A random variation in fatigue life and other uncertainties exist due to estimation of the life model and possible model errors in the fatigue life model. Several potential uncertainty sources were present in the laboratory testing due to the boundary conditions in the cable's attachment in the test rig, the evaluation of the failure criterium in terms of conductivity drop, the cable bending radius, and the applied axial load that should be zero. The influence of stress from a possible unwanted axial load in the test rig should be negligible compared to the contribution of the bending load.

3.2. Equivalent Load and Strength Variables

The load and strength variables were defined as stress amplitudes measured in unit MPa based on fatigue equivalent load and strength amplitudes corresponding to N_{eq} number of cycles, here chosen to $N_{eq} = 1 \times 10^6$ cycles. The choice of equivalent number of cycles is arbitrary, but was chosen to acquire meaningful interpretations of the equivalent load and strength variables. The equivalent load, L_{eq}, is defined as the fatigue equivalent stress amplitude at N_{eq} cycles corresponding to the target life, T_{life}, which is chosen to $T_{life} = 25$ years in this study. The equivalent strength, S_{eq}, is defined as the stress amplitude in the Wöhler curve at N_{eq} cycles to failure. Note that the equivalent load and strength variables are connected through the same equivalent number of cycles, $N_{eq} = 1 \times 10^6$, together with the damage exponent of the Wöhler curve, chosen to $m = 6.236$, which is the value from [8], also used in [26]. It can be observed that the parameters for the bending stiffness, EI, and the cable diameter, d, are common for equivalent load and strength definitions. The target variable Y is defined as the difference between the logarithmic strength and load, namely $Y = \ln(S_{eq}) - \ln(L_{eq})$.

The strength of the cable was investigated experimentally by life testing at different bending radii on the cable. The results were modelled by a life–strength relation following the Basquin Equation (1), where the number of cycles to failure N were related to the bending curvature k:

$$N = N_0 \cdot \left(\frac{k}{k_0}\right)^{-m}, \tag{3}$$

where k_0 represents the fatigue strength (in terms of the curvature of cable) at $N_0 = 1 \times 10^6$ number of cycles.

For the VMEA, strength and load measures were compared, requiring a common dimension, here chosen as stress. Pure bending was considered, employing beam theory to relate the curvature to maximum stress S according to:

$$S = \frac{32EI}{\pi d^3} \cdot k \tag{4}$$

with d being the diameter and EI the bending stiffness of the cable.

Assuming a fixed damage exponent m, using Equation (3), each observed life can be recalculated into a damage-equivalent curvature, $k_{eq,i}$, at N_{eq} cycles, and the logarithmic mean estimated from the fatigue tests as:

$$\ln k_{eq} = \frac{1}{n}\sum_{i=1}^{n}\ln\left(k_{eq,i}\right), \quad k_{eq,i} = k_i \cdot \left(\frac{N_i}{N_{eq}}\right)^{\frac{1}{m}} \tag{5}$$

with N_i being the observed life at the curvature k_i. Defining the equivalent stress S_{eq} as the stress amplitude in the Wöhler curve that corresponds to N_{eq} cycles to failure, as Equation (4) gives:

$$S_{eq} = \frac{32EI}{\pi d^3} \cdot k_{eq} \tag{6}$$

The estimated Wöhler curve for the fatigue bending tests are presented in Figure 7. The equivalent curvature was estimated to $k_{eq} = 1.64\ \mathrm{m}^{-1}$, using Equation (5) with exponent $m = 6.236$, which is consistent with the test data illustrated in Figure 7. The nominal value of the equivalent strength was then estimated to $S_{eq,nom} = 1.41$ MPa, using Equation (6). Additional tests would be beneficial to obtain additional valuable knowledge. However, the limited amount of data is reflected in the uncertainty size assessment as demonstrated below.

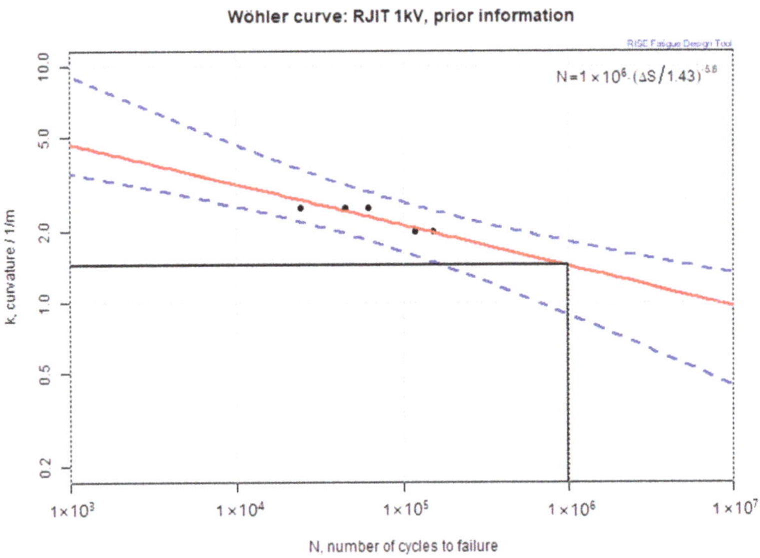

Figure 7. Estimated Wöhler curve of the cable, using prior knowledge on the exponent m, where the circles are experimental data, the solid line is the estimated Wöhler curve, and the dashed lines the 95% confidence intervals.

In parallel, a corresponding equivalent load is needed. In the simulations, the fatigue damage was calculated based on the wave climate at Runde, defined by the wave scatter diagram. The calculations were performed as follows. The wave scatter diagram is defined by the matrix $F = (f_{ij})$, where the indices represent the significant wave height, $H_{s,i}$, and the wave period, $T_{p,j}$, and these are discretized in steps. The matrix position f_{ij} specifies the number of hours that each sea state occurs during one year. For each sea state $(H_{s,i}, T_{p,j})$, the cable was simulated for one hour using the finite element method, and the accumulated fatigue damage D_{ij} calculated using the Palmgren-Miner rule, the rainflow counting method, and the S-N-curve, Equation (1), according to:

$$D_{ij} = \sum_k \frac{n_k}{N(L_k)} = \sum_k \frac{n_k}{N_0}\left(\frac{L_k}{S_0}\right)^m = \frac{1}{N_0 S_0^m}\sum_k n_k L_k^m = \frac{1}{N_0 S_0^m} d_{ij} \tag{7}$$

where n_k is the number of stress cycles with amplitude L_k, and $d_{i,j}$ is the so-called pseudo damage for one hour. The total damage for one year can now be calculated as:

$$D_1 = \sum_{i,j} f_{ij} \cdot D_{ij} = \frac{1}{N_0 S_0^m}\sum_{i,j} f_{ij} \cdot d_{ij} = \frac{d_1}{N_0 S_0^m} \tag{8}$$

Note that d_1, the total pseudo damage for one year, only depends on the simulated load and parameter m, but not on the fatigue strength S_0. The most critical position along the cable with the maximum damage value is chosen. The fatigue damage corresponding to the target life T_{target} becomes:

$$D_{target} = D_1 \cdot T_{target} \tag{9}$$

We now introduce the equivalent load L_{eq} as the stress amplitude that, for N_{eq} cycles, gives the same fatigue damage during the target life as D_{target}, and by using the Basquin equation in Equation (1), the equivalent damage is calculated as:

$$D_{eq} = \frac{N_{eq}}{N_0 \cdot \left(\frac{L_{eq}}{S_0}\right)^{-m}} = \frac{N_{eq} L_{eq}^m}{N_0 S_0^m} \tag{10}$$

Consequently, by solving $D_{target} = D_{eq}$, the equivalent load is calculated according to:

$$L_{eq} = \left(\frac{d_1 \cdot T_{target}}{N_{eq}}\right)^{\frac{1}{m}} = \left(T_{target}\right)^{\frac{1}{m}} \cdot L_{eq,1} \tag{11}$$

with $L_{eq,1}$ being the equivalent load for one year. Using results from numerical simulations of the motions of the cable, the nominal value of the equivalent load was estimated at $L_{eq,nom} = 0.47$ MPa, using Equation (11).

Based on the equivalent strength, S_{eq}, and the equivalent load, L_{eq}, the target function Y is defined according to:

$$Y = \ln(S_{eq}) - \ln(L_{eq}) \tag{12}$$

where negative values of Y indicate failure, while the safe region corresponds to positive values. The equivalent load, L_{eq}, the equivalent strength, S_{eq}, and the target function, Y, are stochastic variables, and their variations are due to uncertainties in their underlying dependent variables. These are be investigated in the next subsection.

The logarithmic scale was chosen since experience has shown that it reduces non-linearities of the target function, thus reducing errors due to linearization of the target function. Further, the logarithmic scale makes the variance representation of uncertainty less dependent on the scale. This is taken advantage of in the VMEA analysis in the next subsection.

3.3. *Evaluation of Sensitivities and Uncertainty Sizes*

The probabilistic VMEA was established based on the uncertainty sources identified in the basic VMEA, and the developed models for equivalent strength and load. The variation of the target function $Y(X_1, X_2, \ldots X_p)$ depends on several uncertainty sources, $X_1, X_2, \ldots X_p$, where p is the number of variables. If the uncertainty sources are considered as independent random variables, the Gauss approximation formula for the target function gives its variance as a function of the variances of each uncertainty source according to:

$$\tau^2 = \mathrm{Var}\left[Y(X_1, X_2, \ldots X_p)\right] = \sum_{i=1}^{p} \left(\frac{\partial Y}{\partial X_i}\right)^2 \cdot \mathrm{Var}[X_i] = \sum_{i=1}^{p} c_i^2 \cdot s_i^2 \tag{13}$$

where s_i is the standard deviation of uncertainty source X_i. The sensitivity coefficients c_i are the partial derivatives of Y with respect to X_i. The logarithmic scale of the target function in Equation (12) implies that the uncertainties represent relative standard deviations that can be judged in terms of uncertainty in percentage.

The sensitivities and sizes of the uncertainty sources judged to be of most importance in the basic VMEA were evaluated. The assessment of each uncertainty source is detailed below, and the results are summarized in a so-called VMEA table; see Table A1 in Appendix A, where the total uncertainty is also shown. The uncertainty sources were grouped in the same way as for the basic VMEA, as shown in Figure 6.

Many uncertainties must be assessed by engineering judgements. Engineers are sometimes unfamiliar with statistical properties like standard deviations, but may find it easier to make judgements about the possible uncertainty interval of a certain property. From such a judgement of an uncertainty interval of ±x%, the standard deviation of the uncertainty source can be calculated assuming a uniform distribution, namely:

$$s = \frac{\mathrm{x}\%}{\sqrt{3}}. \tag{14}$$

The uniform distribution interpretation of an uncertainty interval has been used throughout this study, unless indicated otherwise.

3.3.1. Marine Loads

Wave climate in Runde: Based on simulations in [26], sensitivity coefficients due to marine load parameters were evaluated. Significant wave height varied between values $H_{S,1} = 1.5$ m and $H_{S,2} = 7.5$ m, and the sensitivity depended only on the load and was evaluated to $c_{H_s} = -1.31$. Experts judged the relative uncertainty in significant wave height to be around ±10% and consequently, assuming a uniform distribution, the uncertainty became $10\%/\sqrt{3} = 5.8\%$. Similarly, the influence of ocean current was evaluated, and the sensitivity coefficient became 0.67, and the relative uncertainty 5.8%.

3.3.2. Cable Motion

Cable boundary conditions and model error: In the simulations in [26], the boundary condition of the buoy and the hub was judged to be well modelled, and the total system was verified in [28,33]. The model error due to cable boundary conditions was judged to be at most ±5%, which translates to an uncertainty of 2.9%, assuming a uniform distribution. The model error due to the cable motion simulation model was judged to be ±10%, which translates to an uncertainty of 5.8%. Since these were judged to be total uncertainties in the target function, the corresponding sensitivity coefficients equal unity.

Marine growth: In the simulations, see [26] for data, different models for marine growth were assumed. The resulting uncertainty due to marine growth was estimated to 10.5%. This value was achieved by considering two extreme cases and assuming a uniform distribution between these. Since the uncertainty was evaluated with respect to the target function, the corresponding sensitivity coefficient equals unity.

3.3.3. Cable Properties

The applied and installed cable was a 1.2 kV low-voltage cable, consisting of three conductors laid together with Kevlar ropes. The outer sheathing comprised two layers of polyurethane (PUR) material.

Cable diameter: The cable diameter was 39 mm. The variation due to manufacturing was judged by production engineers as max variation intervals ±x mm. This interval was interpreted as a uniform distribution giving a standard deviation $x/\sqrt{3}$, and finally, the relative uncertainty, see Table A1 in the Appendix A for estimated values. It should be noted that for different sizes of cables, the uncertainty may vary.

The diameter of a cable impacts the load and the strength. Since the same basic cable model was used for the cable simulation (for load evaluation) and the conversion between bending radius in test and bending stress (the strength evaluation), the corresponding sensitivities due to the diameter was evaluated zero.

Cable stiffnesses: The cable stiffness parameters EI, EA and GKv were evaluated based on tests; see [28,29,34], and the corresponding uncertainties are shown in Table A1 in Appendix A for estimated values.

The parametric sensitivity analysis within a total of 27 cases with different cable bending stiffness ($EI = 2,\ 4,\ 6\ \mathrm{Nm}^2$), axial stiffness ($EA = 2,\ 4,\ 6$ MN) and torsion stiffness ($GKv = 1.5,\ 3.0,\ 4.5\ \mathrm{Nm}^2/\mathrm{rad}$) were simulated; see [26,29,34] for simulation results and stress analyses. The bending stress component was extracted from these simulations, and sensitivity coefficients for cable stiffness were assessed. A multivariate regression model was fitted to all the simulation results in the evaluation, representing a linear response surface. The sensitivities to EA and GKv were close to zero, as expected since only the bending stress was used in the evaluation. Sensitivity depends on load and strength; the equivalent strength was evaluated from the bending radius according to Equation (9). From this relation, one notes that the equivalent nominal stress depends linearly on EI but has no dependence on EA and GKv. Consequently, the strength sensitivities in log-scale become one for EI and zero for EA and GKv. The total sensitivities thus become approximately zero.

Mass of cable: Based on simulation in [26], the sensitivity coefficient due to mass was evaluated. The parameter varied from low, to medium, to high as 2 kg/m, 2.75 kg/m, and 6 kg/m.

The load sensitivity coefficient was similar to the above, defined in log-scale and where the equivalent load was calculated using Wöhler exponent $b = 6.238$. The load sensitivity coefficient was calculated to 1.52. The equivalent nominal strength does not depend on the mass of the cable, $c_M = 0 - 1.52 = -1.52$. The maximum variation of the mass was judged to ±2%, which is interpreted as uniform distribution and giving the uncertainty $2\%/\sqrt{3} = 1.2\%$.

3.3.4. Life Model

Uncertainties due to fatigue model: The identified uncertainty sources include fatigue scatter, estimation uncertainty and fatigue model error. These influence only the strength aspect and not the load aspect of the target function.

Uncertainty due to fatigue scatter was evaluated from the fatigue tests. The standard deviation of the logarithmic equivalent curvature, $\ln(k_{eq,i})$, was evaluated to $s = 5.9\%$. The uncertainty in VMEA required adjustment by a t-correction factor, taking the limited number of tests into account, $t_{corr} = 1.4$, corresponding to 4 DOF (number of tests under consideration, 5, minus the number of estimated model parameters, 1). The resulting uncertainty due to fatigue scatter is thus $t_{corr} \cdot s = 8.3\%$. The estimation uncertainty was evaluated by the standard deviation of Equation (5), giving $t_{corr} \cdot s/\sqrt{5} = 3.7\%$. The sensitivity coefficient equals unity since the uncertainties were evaluated in terms of $\ln(S_{eq})$.

The possible model error in fatigue life, here considered as an uncertainty, was judged by experience as a factor of 2 in life, corresponding to an uncertainty of 40%, assuming a uniform distribution. From Equation (1), one observes that the stress amplitude depends on the life (number of cycles) raised to the power of $1/m$. The sensitivity coefficient with respect to logarithmic fatigue scatter and logarithmic fatigue model error thus becomes $1/m = 0.16$.

3.3.5. Laboratory Testing

Bending radius uncertainty: The description and results from the rotating-bending fatigue cable tests are presented in [29,34]. The bending radius in the test influenced the strength but not the load. Equation (4) shows that the equivalent strength is inversely proportional to the bending radius, $R = 1/k$. The logarithmic sensitivity thus becomes -1. The uncertainty in the bending radius originated from cable mounting in the test rig and dynamic effects evaluated by simulations of the cable rig. The total uncertainty contribution was evaluated at 5%. Other possible systematic errors in the test setup are assumed to be negligible and were thus neglected. The random uncertainties due to lab testing are assumed to have been captured by the estimated scatter in fatigue.

3.4. *Probabilistic VMEA Table*

The results from the probabilistic VMEA table are summarized and presented in the Appendix A, Table A1. The total uncertainty of the target variable Y was estimated to $\tau = 19.4\%$, i.e., the uncertainty in the predicted life of the dynamic cable was estimated to 19.4% in terms of relative standard deviation. The total uncertainty is used below to derive safety factors for design.

Further, it can be observed that the five most considerable resulting uncertainties are due to wave climate, cable model error, marine growth, fatigue scatter, and fatigue model error. Since these uncertainty sources dominate the total uncertainty, they should be the candidates for further studies, e.g., investigating if they can be more accurately assessed or if by some measure they can be reduced.

3.5. *Reliability Evaluation*

The target function, also called limit state function, should exceed zero with a proper safety margin. The evaluated total uncertainty of the target function, together with nominal values of equivalent strength and load, can be used to calculate a reliability index and to derive safety factors.

First, we present safety factors derived through the Cornell reliability index, and then we introduce the concept of an extra safety factor. More details are found in [22]. Together with the total uncertainty of Y from the VMEA analysis, τ, the Cornell reliability index is defined as:

$$\beta = \frac{\ln(S_{eq,nom}) - \ln(L_{eq,nom})}{\tau} \tag{15}$$

where the nominator is the nominal difference between the logarithm of the nominal values of strength and load, respectively, and the denominator, τ, is the total uncertainty from the VMEA. For our case, the reliability index becomes:

$$\beta = \frac{0.34 - (-0.76)}{19.4\%} = 5.7 \tag{16}$$

The reliability index is denoted as the safety index or distance from failure mode, since it can be interpreted as the number of standard deviations from the failure mode [12,35].

A requirement of 95% reliability corresponds to a reliability index $\beta > \beta_{req} = 1.64$, assuming a normal distribution. This reliability level is fulfilled in the current case. Further, it can also be converted into a required safety factor:

$$SF_{95\%} = \exp(\beta_{req} \cdot \tau) = \exp(1.64 \cdot 0.194) = 1.4 \tag{17}$$

However, for an engineering design, a 95% reliability is often not sufficient. This can be addressed by increasing the required reliability index. However, knowledge about rare events, representing the tail of the statistical distribution, is often relatively weak. Instead, the reliability model can address this problem in the following way. The required distance between logarithmic load and strength is formulated as:

$$\ln(S_{eq,nom}) - \ln(L_{eq,nom}) > \beta_{req} \cdot \tau + \delta_E \quad (18)$$

where β_{req} is related to the statistical safety and δ_E defines an extra safety distance. From this expression, a safety factor SF can be calculated according to:

$$SF = \exp(\beta_{req} \cdot \tau) \cdot \exp(\delta_E) = SF_{95\%} \cdot SF_E \quad (19)$$

The safety factor is thus subdivided into two parts. The first part, $SF_{95\%}$, representing 95% reliability, is related to the combined uncertainty of all possible uncertainty sources, and is found through the probabilistic VMEA procedure. The second extra safety factor, SF_E, is related to unknown and extreme events and estimated by engineering experience combined with judgment about the severity of the risk, i.e., the likelihood of rare detrimental events and the consequence of failure. Guidance and further details can be found in [22]. The concept of reliability evaluation using the limit state function is illustrated in Figure 8.

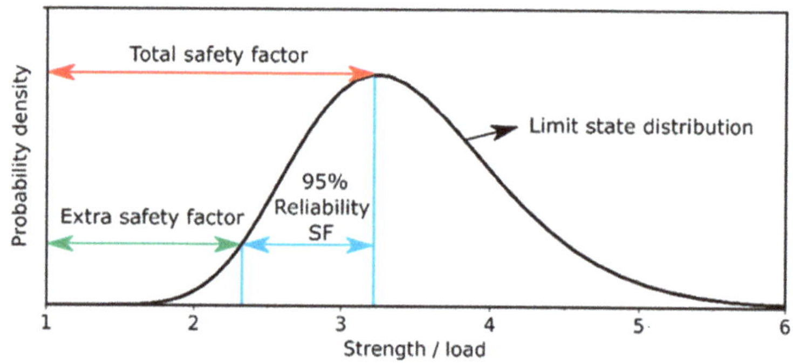

Figure 8. Load–strength limit state function.

4. Discussion of Results

The VMEA methodology has been introduced and adapted to marine cables in WEC systems. The methodology has been described in terms of a seven-step procedure, and its capacity to identify and quantify all uncertainties demonstrated. With these uncertainties as a basis, the safety against fatigue within a load–strength reliability framework has been evaluated. Uncertainties originate from different sources including scatter, model uncertainties and statistical uncertainties. A unique feature of the VMEA methodology is that it incorporates all these different sources in a unified framework. The methodology can be applied at different stages in the design process, and it has here been adapted to two different design stages, an early stage of basic VMEA with the emphasis on defining the target function and identifying the sources of uncertainty, and a later design stage of probabilistic VMEA, which has been the main part of the work.

The basic VMEA was initially performed with a group of experts with complementary perspectives, knowledge, and experiences. This group identified the sources of uncertainty relating to bending fatigue failure, and made judgements about their magnitudes and sensitivities. Based on these estimations, the respective contributions of each source to the uncertainty in the load–strength fatigue model were calculated. The uncertainty sources were grouped into five categories. The fatigue life model's uncertainty was found to contribute most to the overall uncertainty in the predicted fatigue life. These uncertainty evaluations gave important information for preparation of the more detailed probabilistic VMEA. Since the fatigue model's uncertainty was found to contribute more than other factors, in the next stage action was taken to reduce uncertainties in the fatigue model.

In the probabilistic VMEA phase, the uncertainty sources are assessed using quantitative information about uncertainties and sensitivities. These may be obtained through a combination of experimental tests, numerical simulations or mathematical relations.

For the experimental investigations, a test rig was developed for rotating cable fatigue tests, and tests were performed to evaluate cable properties including the bending fatigue strength.

A numerical simulation model of the whole WEC system was developed based on the DNV GL software package. The simulations were essential for many reasons. First, they were used to calculate the cable motion and to identify the most critical fatigue region on the cable. Simulations were also important to obtain a quantitative correction of the dynamic effects in the test rig that were not possible to measure during the live tests. By putting the marine field load conditions into the WEC system simulations, the fatigue load on the cable could be evaluated during different environmental conditions. Finally, parametric sensitivity analyses were performed with the numerical model to evaluate sensitivities for most of the uncertainty sources, including load- and strength-related sources.

For the uncertainty sources directly related to the fatigue load-strength and the fatigue life models, mathematical expressions were used directly based on simpler beam models or empirical relations, and these enabled simple calculations of the corresponding sensitivity factors.

These three different approaches were thus integrated with the VMEA methodology and the uncertainties in terms of scatter, statistical uncertainty and model uncertainty were evaluated for the different uncertainty sources. It was also demonstrated how the resulting total uncertainty can be used to calculate reliability indices, and to evaluate safety factors against fatigue failure. It should be noted that the limited amount of data (e.g., cable experimental tests and numerical simulations) is reflected by the assessed uncertainties.

A major contribution to the overall uncertainty comes from the fatigue life model, both in terms of scatter and model uncertainty. This was also the major uncertainty source identified with the basic VMEA. The second and third largest uncertainty sources switched order compared to the basic VMEA, and so did also the two least significant uncertainty sources.

The identification of the dominating uncertainty sources is useful for several reasons. The more significant uncertainty sources from a basic VMEA show where attention should be focused in the following more detailed probabilistic VMEA. It also shows where to apply focus if the uncertainty needs to be reduced.

In addition to the identification of the dominating uncertainty sources, the VMEA results also gave a basis for calculation of safety factors and reliability indices. It has been demonstrated how to derive safety factors based on the probabilistic VMEA. For a 95% reliability requirement, the load unit's safety factor became 1.4.

The connection between the safety factor and all uncertainty sources enabled the identification of the dominating uncertainties. From these, guidelines can be drawn to reduce the overall uncertainty and thereby decrease the required safety factor.

The specific cable in this study was a 1.2 kV low-voltage cable with a small diameter of 39 mm. It should be noted that other cables, including those with larger radius for higher voltage ratings, may have very different properties and different uncertainty levels. Therefore, both the relevant uncertainty sources and their sizes need to be re-evaluated for each specific case; the current study should be useful as an assessment template. Obviously, the results of this study represent not only a specific cable, but also a specific geographic location (Runde in Norway) and the specific platforms (WEC and hub of the WaveEL 3.0 prototype) used in the simulation. Thus, new numerical simulations need to be performed when investigating different set-ups. Future work should include applying the methodology to different case studies to consider the effect of changing locations, platform designs and cable designs.

5. Conclusions

A model framework has been developed for analyzing the mechanical life of flexible marine cables. It is based on the VMEA reliability design method, together with a load–strength approach that combines numerical simulations to assess load on the cable with experimental tests to assess cable strength. Experiments and calculations are also included for calculations of uncertainties and sensitivities. The main key findings and highlights from the investigation are the following:

The framework enables identification and calculations of all types of uncertainty sources, such as scatter, model uncertainties and statistical uncertainties. This is demonstrated for an evaluation of bending fatigue. The method was found to be a useful tool for evaluating uncertainties in fatigue life during the design phase. The uncertainty results gave a firm foundation for the evaluation of safety against fatigue, helpful for identifying weak spots in reliability assessment to motivate actions in the improvement process.

Specifically, the model allows the dominating uncertainty sources to be obtained, which is useful information when a reduction of overall uncertainty or an increase of safety factor is needed. The framework will be of interest for future studies of other types of wave energy systems at different locations and under different environmental conditions. However, a limitation is that for applications where only minor experience is available, it could be difficult to identify all the important uncertainty sources.

Author Contributions: The work presented in this research article engaged several researchers who had different responsibilities and contributions to the entire project, and this research article in particular. P.J. was the main contributor to the establishment of the VMEA reliability methodology, and performed the statistical analyses and the VMEA uncertainty assessment. He also contributed to the conceptualization, formal analyses, and contributed to writing main parts of the paper, reviewing and editing. X.L. carried out a significant number of numerical simulations using the SESAM software, contributed to the formal analyses, investigations, and visualization of the results. He was also responsible for the preparation of the article and writing the original draft. E.J. was deeply involved in the conceptualization and establishment of the methodology. He contributed to the statistical and formal analyses, and assisted X.L. in writing parts of the paper, reviewing and editing the final draft of the paper. J.W.R. was involved in funding acquisition and project administration. He was responsible for the supervision of X.L., and he contributed actively to the conceptualization and establishment of the methodology. He contributed to all analyses and investigations presented in the paper, including writing parts of the paper, reviewing and editing. All authors have read and agreed to the published version of the manuscript.

Funding: The research was partially funded by strategic internal funding from Chalmers University of Technology and from RISE Research Institutes of Sweden, from Chalmers University of Technology Foundation for the strategic research project "Hydro- and aerodynamics", and by the Swedish Energy Agency projects "R&D of dynamic low voltage cables between the buoy and floating hub in a marine energy system" under contract No. 41240-1, and "Simulation model for operation and maintenance strategy of floating wave energy converters–analysis of fatigue, wear, and influence of biofouling for effective and profitable energy harvesting" under contract No. 36357-2.

Institutional Review Board Statement: Not applicable.

Informed Consent Statement: Not applicable.

Data Availability Statement: Not applicable.

Acknowledgments: The authors want to acknowledge the support from the company NKT cables (Sweden) with the delivery of cables for testing and for cooperation throughout the work. The authors also acknowledge Shun-Han Yang for her contributions to the study and the projects.

Conflicts of Interest: The authors declare no conflict of interest. The funders had no role in the design of the study, in the collection, analyses, or interpretation of data, in the writing of the manuscript, or in the decision to publish the results.

Appendix A

Table A1. Probabilistic VMEA for dynamic cable.

Input				Result	
				Variation Contribution	
Uncertainty Components	**Sensitivity**	**Uncertainty**	**Resulting Uncertainty**	**Variance/10^4**	**Proportion**
Marine Loads					
Wave climate at site (Runde)	−1.31	5.8%	7.6%	57	15%
Ocean currents at site (Runde)	−0.67	5.8%	3.9%	15	4%
Total Marine Loads			**8.5%**	**72**	**19%**
Cable Motion					
Cable boundary conditions	−1.00	2.9%	2.9%	8	2%
Cable model error	−1.00	5.8%	5.8%	33	9%
Marine growth	−1.00	10.5%	10.5%	110	29%
Total Cable Motion			**12.3%**	**152**	**40%**
Cable Properties					
Diameter, within batch variation	0.00	0.3%	0.0%	0	0%
Diameter, batch variation	0.00	0.7%	0.0%	0	0%
Diameter, non-spherical	0.00	2.2%	0.0%	0	0%
Axial stiffness, *EA*	0.01	8.1%	0.1%	0	0%
Bending stiffness, *EI*	−0.03	12.2%	0.4%	0	0%
Torsional stiffness, GK_v	0.01	4.4%	0.0%	0	0%
Mass [kg/m]	−1.52	1.2%	1.8%	3	1%
Total Cable Properties			**1.8%**	**3**	**1%**
Life Model					
Fatigue life scatter	1	8.4%	8.4%	70	18%
Fatigue life estimation	1	3.7%	3.7%	14	4%
Fatigue life model error	0.16	40.0%	6.4%	41	11%
Total Life Model			**12.9%**	**125**	**33%**
Laboratory testing					
Bending radius of cable [mm]	−1.00	5.0%	5.0%	25	7%
Total Laboratory Testing			**5.0%**	**25**	**7%**
Total Uncertainty			**19.4%**	**377**	**100%**

References

1. Worzyk, T. Application of submarine power cable. In *Submarine Power Cables: Design, Installation, Repair, Environmental Aspects*, 1st ed.; Springer: Berlin, Germany, 2009; pp. 1–7. ISBN 978-3-642-01269-3.
2. DNV GL. Offshore Wind Industry Joins Forces to Reduce Costs of Cable Failures. 2018. Available online: https://www.dnvgl.com/news/offshore-wind-industry-joins-forces-toreduce-costs-of-cable-failures-117811 (accessed on 18 May 2022).
3. Offshore Wind Programme Board. Offshore Wind Programme Board. Export Cable Reliability Description of Concerns. Technical Report. July 2017. Available online: https://ore.catapult.org.uk/app/uploads/2018/02/Export-Cable-Reliability-Step-1-v7-UPDATE-Jul-17.pdf (accessed on 18 May 2022).
4. 4COffshore. Joint Industry Project Looks to Reduce Cable Failures. 2018. Available online: https://www.4coffshore.com/news/joint-industry-project-looks-to-reduce-cable-failuresnid7457.html (accessed on 18 May 2022).
5. Marazzato, H.; Barber, K.; Jansen, M.; Graeme, B. Cable Condition Monitoring to Improve Reliability. 2004. Available online: https://www.nexans.co.nz/NewZealand/2012/22_1.04.2004%20-%20Cable%20Condition%20Monitoring.pdf (accessed on 18 May 2022).

6. Hammons, T.J.; Woodford, D.; Loughtan, J.; Chamia, M.; Donahoe, J.; Povh, D.; Bisewski, B.; Long, W. Role of HVDC transmission in future energy development. *IEEE Power Engng. Rev.* **2000**, *20*, 10–25. [CrossRef]
7. Folley, M.; Whittaker, T.J.T. The effect of sub-optimal control and the spectral wave climate on the performance of wave energy converter arrays. *Appl. Ocean Res.* **2009**, *31*, 260–266. [CrossRef]
8. Nasution, F.P.; Sævik, S.; Gjøsteen, J.K.Ø.; Berg, S. Experimental and finite element analysis of fatigue performance of copper power conductors. *Int. J. Fatigue* **2013**, *47*, 244–258. [CrossRef]
9. Thies, P.R.; Johanning, L.; Smith, G.H. Assessing mechanical loading regimes and fatigue life of marine power cables in marine energy application. *Proc. Inst. Mech. Eng. Part O J. Risk Reliab.* **2011**, *226*, 18–32. [CrossRef]
10. Trarieux, F.; Lyons, G.J.; Patel, M.H. Investigation with bandwidth measure for fatigue assessment of the Foinaven dynamic umbilical including VIV. *Engng. Struct.* **2006**, *28*, 1671–1690. [CrossRef]
11. Stamatis, D.H. *Failure Mode and Effect Analysis: FMEA from Theory to Execution*, 2nd ed.; American Society for Quality, Quality Press: Milwaukee, WI, USA, 2003; pp. 21–81. ISBN 0-87389-598-3.
12. Davis, T.P. Science, engineering and statistics. *Appl. Stoch. Models Bus Ind.* **2006**, *22*, 401–430. [CrossRef]
13. Chakhunashvili, A.; Johansson, P.; Bergman, B. Variation mode and effect analysis. In Proceedings of the Annual Reliability and Maintainability Symposium, Los Angeles, CA, USA, 26–29 January 2004.
14. Johansson, P.; Chakhunashvili, A.; Barone, S.; Bergman, B. Variation mode and effect analysis: A practical tool for quality improvement. *Qual. Reliab. Engng. Int.* **2006**, *22*, 865–876. [CrossRef]
15. Chakhunashvili, A.; Barone, S.; Johansson, P.; Bergman, B. Robust product development using variation mode and effect analysis. In *Robust Design Methodology for Reliability: Exploring the Effects of Variation and Uncertainty*, 1st ed.; Bergman, B., de Mare, J., Loren, S., Svensson, T., Eds.; John Wiley & Sons Ltd.: West Sussex, UK, 2009; pp. 57–70. ISBN 978-0-470-71394-5.
16. Johansson, P.; Svensson, T.; Samuelsson, L.; Bergman, B.; de Mare, J. Variation mode and effect analysis: An application to fatigue life prediction. *Qual. Reliab. Engng. Int.* **2009**, *25*, 167–179. [CrossRef]
17. Svensson, T.; de Mare, J.; Johansson, P. Predictive safety index for variable amplitude fatigue life. In *Robust Design Methodology for Reliability: Exploring the Effects of Variation and Uncertainty*, 1st ed.; Bergman, B., de Mare, J., Loren, S., Svensson, T., Eds.; John Wiley & Sons Ltd.: West Sussex, UK, 2009; pp. 85–96. ISBN 978-0-470-71394-5.
18. Bergman, B.; de Mare, J.; Loren, S.; Svensson, T. (Eds.) *Robust Design Methodology for Reliability: Exploring the Effects of Variation and Uncertainty*, 1st ed.; John Wiley & Sons Ltd.: West Sussex, UK, 2009; ISBN 978-0-470-71394-5.
19. Johannesson, P, P.; Speckert, M. (Eds.) *Guide to Load Analysis for Durability in Vehicle Engineering*, 1st ed.; John Wiley & Sons Ltd.: West Sussex, UK, 2013; ISBN 978-1-118-64831-5.
20. Svensson, T.; Johannesson, P. Reliability fatigue design, by rigid rules, by magic or by enlightened engineering. *Procedia Eng.* **2013**, *66*, 12–25. [CrossRef]
21. Johannesson, P.; Bergman, B.; Svensson, T.; Arvidsson, M.; Lönnqvist, Å.; Barone, S.; de Maré, J. A robustness approach to reliability. *Qual. Reliab. Engng. Int.* **2013**, *29*, 17–32. [CrossRef]
22. Johannesson, P. (Ed.) Reliability Guidance for Marine Energy Converters. Report RiaSoR (Reliability in a Sea of Risk). December 2016. Available online: http://riasor.com/wp-content/uploads/2016/12/ReliabilityGuidanceMECs_v1.0_20161216.pdf (accessed on 18 May 2022).
23. Jia, C.; Ng, C.; McKeever, P.; Johannesson, P.; Svensson, T.; Buck, E.; Shanks, A. Improving reliability in a sea of risk. In Proceedings of the 12th European Wave and Tidal Energy Conference, Cork, Ireland, 27 August–1 September 2017.
24. Johannesson, P.; Svensson, T.; Gaviglio, H. Reliability evaluation using variation mode and effect analysis: Application to CorPower's mooring pre-tension cylinder. In Proceedings of the 13th European Wave and Tidal Energy Conference, Napoli, Italy, 1–6 September 2019.
25. Yang, S.-H.; Ringsberg, J.W.; Johnson, E.; Hu, Z.Q.; Duan, F.; Bergdahl, L. Experimental and numerical investigation of a taut-moored wave energy converter—A validation of simulated buoy motions. *Proc. Inst. Mech. Eng. Part M J. Engng. Marit. Env.* **2018**, *232*, 97–115. [CrossRef]
26. Yang, S.-H.; Ringsberg, J.W.; Johnson, E. Parametric study of the dynamic motions and mechanical characteristics of power cables for wave energy converters. *J. Mar. Sci. Technol.* **2018**, *23*, 10–29. [CrossRef]
27. Yang, S.-H.; Ringsberg, J.W.; Johnson, E.; Hu, Z.Q. Experimental and numerical investigation of a taut-moored wave energy converter: A validation of simulated mooring line forces. *Ships Offshore Struct.* **2021**, *15* (Suppl. 1), S55–S69. [CrossRef]
28. Ringsberg, J.W.; Yang, S.-H.; Lang, X.; Johnson, E.; Kamf, J. Mooring forces in a floating point-absorbing WEC system—A comparison between full-scale measurements and numerical simulations. *Ships Offshore Struct.* **2021**, *15* (Suppl. 1), S70–S81. [CrossRef]
29. Kuznecovs, A.; Ringsberg, J.W.; Yang, S.-H.; Johnson, E. A methodology for design and fatigue analysis of power cables for wave energy converters. *Int. J. Fatigue* **2019**, *122*, 61–71. [CrossRef]
30. DNV GL. *Subsea Power Cables in Shallow Water, DNVGL-RP-0360*; Det Norske Veritas (DNV): Høvik, Norway, 2016.
31. DNV GL. *Sesam SIMA V3.4-00*; Det Norske Veritas (DNV): Høvik, Norway, 2017.
32. Ishikawa, K.; Loftus, J. *Introduction to Quality Control*; 3A Corporation: Tokyo, Japan, 1990.
33. Lang, X.; Yang, S.-H.; Ringsberg, J.W.; Johnson, E.; Guedes Soares, C.; Rahm, M. Comparison between full-scale measurements and numerical simulations of mooring forces in a floating point-absorber WEC system. In Proceedings of the 3rd International Conference on Renewable Energies Offshore, Lisbon, Portugal, 8–10 October 2018; pp. 865–876.

34. Hindrum, K.; Hüffmeier, J. *R&D of Dynamic Low Voltage Cables between the Buoy and Floating Hub in a Marine Energy System*; Report; Swedish Energy Agency: Eskilstuna, Sweden, 2018.
35. O'Connor, P.; Kleyner, A. *Practical Reliability Engineering*, 5th ed.; John Wiley & Sons: Hoboken, NJ, USA, 2012.

MDPI
St. Alban-Anlage 66
4052 Basel
Switzerland
Tel. +41 61 683 77 34
Fax +41 61 302 89 18
www.mdpi.com

Journal of Marine Science and Engineering Editorial Office
E-mail: jmse@mdpi.com
www.mdpi.com/journal/jmse

www.ingramcontent.com/pod-product-compliance
Lightning Source LLC
Chambersburg PA
CBHW040836250926
51850CB00007BA/1333
9783036575964